A publication of the British Geomorphological Research Group

River Channel Changes

Edited by

K. J. Gregory
Professor of Geography
University of Southampton

A Wiley–Interscience Publication

JOHN WILEY & SONS

Chichester · New York · Brisbane · Toronto

Copyright © 1977, by John Wiley & Sons, Ltd.

Library of Congress Cataloging in Publication Data:

Main entry under title:

River channel changes.

Based on papers presented at a symposium organised by the
British Geomorphological Research Group during the annual
conference of the Institute of British Geographers at Lanchester
Polytechnic in Jan. 1976.
'A publication of the British Geomorphological Research
Group.'
'A Wiley–Interscience publication.'
Includes index.
1. River channels—Congresses. I. Gregory,

Kenneth John.
GB561.R58   551.4′83   77-4342
ISBN 0 471 99524 X

THE UNIVERSITIES PRESS (BELFAST) LTD.
BELFAST, N. IRELAND
PRINTED BY PHOTOLITHOGRAPHY AND
BOUND AT THE PITMAN PRESS, BATH

Contributing Authors

J. R. L. ALLEN — *Professor of Geology, Sedimentology Research Laboratory, Department of Geology, University of Reading, Whiteknights, Reading, RG6 2AB*

B. J. BRINDLE — *Formerly Department of Geography, University College of Wales, Aberystwyth, Llandinam Building, Penglais, Aberystwyth, Dyfed, SY23 3DB*

B. E. DAVIES — *Lecturer in Geography, University College of Wales, Aberystwyth, Llandinam Building, Penglais, Aberystwyth, Dyfed, SY23 3DB*

G. H. DURY — *Professor of Geography and Geology, Department of Geography, University of Wisconsin, Science Hall, Madison, Wisconsin 53706*

R. FERGUSON — *Lecturer in Earth and Environmental Science, University of Stirling, Stirling FK9 4LA, Scotland*

W. FROEHLICH — *Institute of Geography, Polish Academy of Sciences, Kraków*

K. J. GREGORY — *Professor of Geography, Department of Geography, University of Southampton, SO9 5NH*

A. M. HARVEY — *Lecturer in Geography, Department of Geography, University of Liverpool, Roxby Building, P.O. Box 147, Liverpool, L69 3BX*

E. J. HICKIN — *Associate Professor of Geography, Department of Geography, Simon Fraser University, Burnaby, British Columbia, Canada V5A156*

D. HITCHCOCK — *Lecturer in Geography, Department of Geography, Polytechnic of Central London, 32–38 Wells Street, London, W1P 3FG*

J. M. HOOKE — *Lecturer in Geography, Manchester Polytechnic, Aytoun Street, Manchester 1*

D. J. HUGHES — *Formerly Department of Geography, University of Liverpool (42 Westbourne Avenue, Leigh, Lancs., WN7 1TL)*

L. KASZOWSKI — *Institute of Geography, Jagiellonian University, Kraków*

M. J. KIRKBY   *Professor of Physical Geography, Department of Geography, University of Leeds, LS2 9JT*

A. D. KNIGHTON   *Lecturer in Geography, Department of Geography, University of Sheffield, Sheffield, S10 2TN*

I. A. LACZAY   *Research Associate, Research Institute for Water Resources Development, Budapest* (H-1428, Budapest, P.Box 44, Hungary)

J. LEWIN   *Lecturer in Geography, Department of Geography, University College of Wales, Aberystwyth, Llandinam Building, Penglais, Aberystwyth, Dyfed, SY23 3DB*

E. MYCIELSKA-DOWGIAŁŁO   *Institute of Geography, University of Warsaw, Warsaw, Krakowski Przedmiesie 30, Poland*

C. C. PARK   *Lecturer in Geography, Department of Geography, St. David's University College, Lampeter, Cardiganshire, Dyfed, SA48 7ED*

G. E. PETTS   *Lecturer in Geography, Dorset Institute of Higher Education, Wallisdown Road, Poole, Dorset BH12 5BB*

G. PICKUP   *Lecturer in Geography, University of Papua New Guinea, Department of Geography, P.O. Box 4820, Papua New Guinea*

K. S. RICHARDS   *Lecturer in Geography, Department of Geography, The University, Hull, Yorkshire, HU6 7RX*

S. J. RILEY   *Lecturer in School of Earth Sciences, School of Earth Sciences, Macquarie University, North Ryde, New South Wales, 2113, Australia*

L. STARKEL   *Professor of Geography, Institute of Geography, Polish Academy of Sciences, Kraków*

J. B. THORNES   *Lecturer in Geography, Department of Geography, London School of Economics, Houghton Street, London, WC2A 2AE*

P. J. WOLFENDEN   *Formerly Department of Geography, University College of Wales, Aberystwyth, Llandinam Building, Penglais, Aberystwyth, Dyfed, SY23 3DB*

R. WOOD   *Severn-Trent Water Authority, Directorate of Resource Planning, Abelson House, Coventry Road, Sheldon, Birmingham*

Contents

Preface xiii

INTRODUCTION

1 The context of river channel changes   1
K. J. Gregory
The geomorphological context.. 2

MECHANICS AND SEDIMENTOLOGY

2 Changeable rivers: some aspects of their mechanics and sedimentation   15
J. R. L. Allen
The river system.. 16
Uniform steady flow in a rectilinear channel. 17
Uniform steady flow in a uniformly curved channel 22
Sediment transport and bed forms under uniform steady conditions 31
Conclusion 40

3 Simulation modelling of river channel erosion   47
G. Pickup
Model development 47
Simulation of eroding channel development.. 52
The limits of channel erosion 57
Conclusions 58

4 Peak flows, low flows, and aspects of geomorphic dominance   61
G. H. Dury
The hurricane Agnes floods 62
Application to the magnitude-frequency scale 65
Application to former rates of fluvial denudation 70
Problems of geomorphic dominance in mass-movement 71
Summary and conclusions.. 72

5 Channel pattern changes during the Last Glaciation and Holocene, in the northern part of the Sandomierz basin and the middle part of the Vistula Valley, Poland   75
E. Mycielska-Dowgiałło
The end of the last Glaciation.. 83
Late glacial and Holocene 84

CHANNEL GEOMETRY CHANGES

6 Hydraulic geometry and channel change   91
J. B. Thornes

7 Short-term changes in hydraulic geometry   101
A. D. Knighton
Sources of short-term variation 102
Changes in channel state 112
Conclusion 117

8 Man-induced changes in stream channel capacity   121
C. C. Park
Types of channel change induced by man 122
Feasibility study of channel morphometric relationships 125
Examples of man-induced changes in stream channel capacity .. 128
Conclusion 141

9 Channel response to flow regulation: The case of the River Derwent, Derbyshire   145
G. E. Petts
Causes and consequences of an altered flow regime 145
The River Derwent, Derbyshire 147
Channel response to flow regulation 151
Conclusion 160

RIVER CHANNEL PATTERN

10 Channel pattern changes   167
J. Lewin
Defining geometry 167
Available evidence and available timespans 172
Pattern changes 175
Some conclusions .. .. 179 ,

11 Channel pattern changes of Hungarian rivers: The example of the Hernád River   185
I. A. Laczay
The Hernád River 186
Patterns of meander development 186
Conclusions 192

12 Rates of erosion on meander arcs 193
D. J. Hughes
   Field observations 193
   Erosion rates 195
   Interpretation 201

13 Channel pattern changes in divided reaches: An example in the coarse bed material of the Forest of Bowland 207
D. Hitchcock
   Measurement of change 209
   Analysis of change 213
   Discussion 217

14 Confined meanders 221
J. Lewin and B. J. Brindle
   Definition and extent 222
   Degrees of confinement 224
   Patterns of confinement 227
   Processes in confinement 228
   Conclusions 232

15 Meander migration: equilibrium and change 235
R. I. Ferguson
   Steady state and dynamic equilibrium 237
   Statistical characterization of irregular meanders 240
   A model for meander spectra 244
   Research needs 247

16 The analysis of river-planform responses to changes in discharge 249
E. J. Hickin
   The field study area 250
   Methods of meander planform analysis 251
   Results of the Beatton river planform analysis 255
   Interpretations and conclusions.. 259

17 The distribution and nature of changes in river channel patterns: The example of Devon 265
J. M. Hooke
   Sources of evidence 266
   Analysis 267
   Conclusions 279

18 Underfit streams: Retrospect, perspect, and prospect   281
  G. H. Dury
     The twenty-five year retrospect 281
     The state of the art 287
     Immediate future prospects 289

NETWORK CHANGE AND THEORY

19 Network adjustments and progress towards theory   297
  K. J. Gregory

20 Event frequency in sediment production and channel change   301
  A. M. Harvey
     Observations and methodology.. 304
     Event frequency and sediment production 306
     Event frequency and channel change 308
     Discussion 310

**21 Channel changes in ephemeral streams: Observations, problems
   and models**   317
  J. B. Thornes
     A case observation: southeast Spain 318
     Controls of runoff, sediment movement, and morphology.. .. 323
     Dynamic models for ephemeral channel changes.. 328
     Conclusion 333

**22 Some downstream trends in the hydraulic, geometric, and
   sedimentary characteristics of an inland distributary system** 337
  S. J. Riley
     The Namoi–Gwydir distributary network 338
     Similarity patterns among stream sites 348
     Discussion 351

**23 Interactions between channel change and historic mining sedi-
   ments**   353
  J. Lewin, B. E. Davies and P. J. Wolfenden
     River Ystwyth study sites 355
     Discussion 363
     Conclusions 366

**24 Urbanization, water redistribution, and their effect on channel
   processes**   369
  K. S. Richards and R. Wood

Hydrological effects of water transfer 370
Channel form and process.. 380
Conclusion 385

25 Channel and network metamorphosis in northern New South Wales   389
K. J. Gregory
Progress towards a general approach 389
Northern New South Wales: New England 395
Conclusion 406

26 Studies of present-day and past river activity in the Polish Carpathians   411
W. Froehlich, L. Kaszowski, and L. Starkel
Typology and moulding of river channels 413
The course of fluvial transport.. 417
Reconstruction of fluvial processes in the past 422

27 Maximum sediment efficiency as a criterion for alluvial channels   429
M. J. Kirkby
The general hypothesis 429
Bed forms 430
Analysis of a channel reach in graded time-spans 433
Hydraulic geometry 436
Conclusion 441

Index   443

Preface

The subject for this volume emerged as the theme for a one-day symposium organized by the British Geomorphological Research Group during the annual conference of the Institute of British Geographers at Lanchester Polytechnic in January 1976. Many of the contributors to that symposium have offered papers for this book and the scope has been broadened by including contributions from Australia, Europe, and North America. The objective of the volume is to demonstrate the various approaches available for the study of river channel change, to illustrate the results that have been obtained against the progress of the last decade, and hopefully to point the way forward for future work. As many of the contributions illustrate a particular approach, it seemed desirable to ask experts to write general reviews to introduce the major sections of the book and I am grateful to Professor John R. L. Allen, Dr John B. Thornes, and Dr John Lewin for kindly providing such reviews in Chapters 1, 6, and 10. The first section (Chapters 2 to 5) is designed to illustrate some of the ways in which the analysis of sedimentary sequences and deposits offers potential for the geomorphologist near the interface of his subject with sedimentology. Studies of river channel geometry have been the basis for many of the early approaches to change and are introduced in the second section in Chapters 6 to 9. River channel pattern has been the focus for many studies of change and the breadth of approaches is illustrated by a third section composed of Chapters 10 to 18. A fourth group of papers (Chapters 19 to 27) is devoted to studies of change within networks and to theoretical deductions and modelling of change. The reader will readily appreciate that these four sections are not independent and discrete, but the groupings of the past often help to identify the profitable research areas for the future. The subject of river channel changes is of interest to a number of disciplines but it is central to geomorphology which provides the underlying theme for this volume.

It is a great pleasure to acknowledge the helpful comments and advice from five referees, the secretarial assistance provided by Miss J. A. Coull, the editorial help given by my wife, and the work of all the contributors. I hope that they enjoy the final result!

New Year's Day, 1977   KENNETH J. GREGORY

INTRODUCTION

K. J. GREGORY

Professor of Geography
University of Southampton

1

The Context of River Channel Changes

'In the space of one hundred and seventy-six years the Lower Mississippi has shortened itself two hundred and forty two miles. That is an average of a trifle over one mile and a third per year. Therefore, any calm person, who is not blind or idiotic, can see in the Old Oölitic Silurian Period, just a million years ago next November, the lower Mississippi River was upwards of one million three hundred thousand miles long, and stuck out over the Gulf of Mexico like a fishing rod. And by the same token any person can see that seven hundred and forty two years from now the Lower Mississippi will be only a mile and three quarters long'.—Mark Twain, *Life on the Mississippi*, 1884.

The dangers of the injudicious extrapolation of rates of change are clearly underlined by this statement by Samuel L. Clemens. Although a considerable amount of work has been done on the Mississippi since Clemens wrote in 1884, a number of recent reports have culminated in a study of the middle Mississippi to show the physical effects of river construction works on river morphology and behaviour (Simons, Schumm, and Stevens, 1974). These workers showed how the surface area of the river has been reduced, how the lateral migration of the river has been restricted, and how the river stages and river depths have been altered. The deliberate alteration of river channels is now known to have increased in the twentieth century and to have had effects beyond those areas where such deliberate changes are engineered. The character and distribution of such effects upon river channels merits further study.

Direct channel manipulation has been shown by Mrowka (1974) to include dam and reservoir construction, channelization, bank manipulation and levee construction, and irrigation diversions. He has indicated the distribution of dams and aqueducts or canals, and of local flood, shore protection works, and levees in 1969 in the United States. Less obvious effects on river channels have followed from changes of drainage basin characteristics because alterations in land use such as afforestation, clearing of woodland, agricultural land development, and urbanization occasion changes in runoff and sediment production that in turn can affect the river channel and induce change. It is likely that the amount of river channel change, both directly and indirectly induced, has previously been underestimated. Because changes of river channels are significant in relation to flooding, rural and urban land use, as well as in relation to interpretations of past river development and estimations of future river behaviour, it appears timely to present a volume which collects together the results of recent geomorphological studies of channel change. River channel change is taken to include any changes of river channel geometry within the context of the cross section, the pattern or network in a drainage basin. The emphasis tends towards studies of man-induced changes.

The Geomorphological Context

Despite the clear descriptions of river channel change by writers like Clemens (1884), it is surprising that the *suspected changes* inferred before 1900 were not investigated in subsequent research. After 1900 the Davisian interpretation prevailed and dominated much geomorphological thinking about rivers. This interpretation involved *assumed changes* which occurred during the cyclic progression of landscape evolution. During this period a number of papers, such as the one by Sonderreger (1935), did indicate the significance of modifying the physiographic balance. For the Rio Grande, Texas, he showed how storage operations initiated in 1915 had promoted river regulation so that lower peak discharges downstream of Elephant Butte dam induced changes in channel size. A further phase of geomorphological enquiry began after 1945 with increased recognition of the significance of hydrology, and this prompted a number of works which involved the *measurement of change*. Valuable studies indicated the extent to which change occurred and although much early research was founded upon the detailed investigation of small areas, this catalysed the growth of fluvial geomorphology including a very significant milestone provided by the book written by Leopold, Wolman, and Miller (1964). Subsequent studies of rivers were placed in a drainage basin hydrological context depending upon an understanding of measurement methods (Gregory and Walling, 1973) but in 1973 there had been relatively few studies of changes of river channels.

A fourth phase of progress has appeared as studies have been directed towards *predicted change*. One of the greatest outstanding problems for the geomorphologist is to distinguish between evolutionary sequences of chance and adjustments inspired by direct and indirect human activities. Increased interest in river channel changes is shown by the appearance of general papers which have indicated the consequences of sudden changes of sediment load (Beckinsale, 1972) and of the modification of the hydrologic regime (Mrowka, 1974). In basins draining to the Atlantic coast of the U.S.A. changes in sediment loads have been related to the construction of reservoirs and to changes in land use (Meade and Trimble, 1974). It was suggested that although sediment loads decreased immediately downstream from reservoirs, the persistence of large loads further downstream indicates that channels are now being degraded. Meade and Trimble (1974) argue that this situation obtains because main rivers are still receiving sediment from large tributaries that accumulated sediment during accelerated erosion following deforestation and farming. Collections of papers on river morphology (Schumm, 1972), and on environmental geomorphology and landscape conservation (Coates, 1973) demonstrate that some research was completed during the past century which is pertinent to river channel change. Progress has also been made by the results of specific symposia on fluvial geomorphology (Morisawa, 1973), by progress made towards environmental geomorphology (Coates, 1970), and by exploration of the relations between geomorphology and engineering (Coates, 1976). Perhaps the most important developments in studies proceeding towards predicted channel change have been the elaboration of palaeohydrology (Schumm, 1965), of river metamorphosis (Schumm, 1969, 1971) and of relations between meandering valleys and discharge as summarized by Dury in this volume (Chapter 18). Employing regime theory, Blench (1972) has used applied morphometrics to deduce the consequences of river diversion, of river channel straightening, and of reservoir construction. The net result of a decade of investigation of the nature and causes of channel change has provided the foundation for future studies which can be the basis for improved models as attempted for specific areas by Wolman (1967) and by Strahler (1956). Such models must be constructed with due regard for the degrees of freedom which a channel system possesses. Hey (1974) has argued that there are five degrees of freedom, namely velocity, slope, flow depth, plan shape, and width, and these are the basic variables which can demonstrate change.

A number of methods are available for the identification of channel changes and these are evident in the ensuing chapters. *Empirical measurements* can provide valuable data on sequences of change at specific locations and are illustrated in Chapters 12, 13, 14, 20, and 26. Because such measurements are restricted to relatively short periods of observation which may not be representative of longer-term periods, it is also necessary to employ *historical methods* which employ information from early topographic surveys and maps, photographic or remote sensing surveys, and from documentary records. Such techniques are employed in Chapters 11, 14, 15, and 17. Because historical methods relate to information from a number of specific time periods, for the purposes of analysis it is often necessary to assume a uniform change over the period. Such an assumption is not completely warranted so that it is necessary to assess the contribution of large events as explored in Chapter 4, and the short term changes as illustrated for hydraulic geometry in Chapter 7. Historical methods often need to be complemented by other techniques which can substantiate the conclusions obtained and ensure that the problems of map convention, projection, and state of preservation do not invalidate the conclusions. To this end a number of *dating techniques* have been adopted to facilitate the study of river channel changes. In addition to use of ^{14}C and pollen analysis of materials preserved in former channels and in flood-plain sequences, botanical methods have included tree-ring dating and the analysis of tree root exposure, and sedimentological investigations embrace analysis of sequences of sediments and of those produced by a single event. Thus Helley and Lamarche (1973) were able to study the deposits produced by the floods of 1964 in northern California as an aid to the comparison and identification of ancient flood deposits. Progress has also been made employing human artifacts so that motor vehicle licence plates enabled Costa (1975) to date the rate of overbank deposition in the Piedmont Province, Maryland. In Death Valley, California, Beaumont and Oberlander (1973) were able to use tin cans to date a recent diversion across an alluvial fan. Such dating techniques are exemplified by sediment characteristics employed in Chapter 5, by dendrochronology in Chapter 9, by lichenometry in Chapter 25, and by the presence in floodplain sediments of material derived from mining in Chapter 23.

A fourth group of methods used to infer channel change relies upon the idea of *space–time substitution* which is sometimes referred to as the ergodic hypothesis. This depends upon establishing a relationship for unmodified areas so that this relationship can then be extended and compared with actual field measured values. The difference between the actual measured value and the expected one can be the basis for the estimation of the nature and amount of change. The technique has been employed in a number of previous studies (e.g. Hammer, 1972; Gregory and Park, 1976) and in this volume is illustrated in Chapters 8, 9, and 25. The optimum use of the technique depends upon a detailed elaboration of the downstream variation in a river system so that we have a basis for the evaluation of change. Such detailed analysis is illustrated for semi-arid systems in Chapter 22. The improved understanding provided by empirical measurements can be the foundation for improved models of change and such *theoretical approaches* may be accomplished in at least two ways. Simulation is illustrated by a model developed in Chapter 3 and mathematical modelling is elaborated for the adjustment of channel geometry in Chapter 27. Although no studies based upon laboratory hardware models are included, this approach has been well represented in work by engineers (Shen, 1971; International Association for Hydraulic Research, 1973). Ultimately contemporary process response systems should be defined with such clarity that parameters can be varied to indicate the likely sequences of possible change. This has been attempted by Schumm in river metamorphosis and specifically Rango (1970) derived relationships between measures of sediment character, channel dimensions, and precipitation/runoff and the utilized these relationships to indicate the consequences of precipitation modification for channel dimensions.

A significant number of results of channel change has been obtained for a variety of world areas and these results are summarized, with the techniques employed and the causes of change, in Table 1.1. These examples, although selective, emphasize the distinction between evolutionary change and the adjustment to a new hydrologic regime (see Chapter 10). A modified hydrologic regime can be induced by land-use change including deforestation, agricultural drainage or building activity and urbanization. These all have feedback effects on river channels. Conservation land-use measures could lead to regulated flows with lower discharges but there are few studies of such effects on river channels except where stock ponds have led to reduced flows. More direct effects (Table 1.1) arise through regulation of the channel discharge which may give reduced peak discharges. Although the lack of available sediment immediately downstream of dams has induced scour which has had to be considered in dam design, it is now appreciated that the effects of dams and reservoirs on discharge produces effects which may persist for a considerable distance downstream. River channel regulation is also significant, and this is manifested not only through the protection and reinforcement of river banks and regulation of the river course but also through the effects of gravel extraction, dredging, and the diversion of water (Table 1.1). The consequences of such changes may be found in changes of geometry and/or changes of pattern or profile (Table 1.1). Whereas many studies have one or more consequences for the river channel (Table 1.2), few investigations have explored the interrelationship of the several possible effects.

Several outstanding problems remain although they have beome less prominent than they were a decade ago because of the research which has been completed. *Complex response* has been discussed by Schumm (1973) to characterize the fact that two basins can respond in different ways or at different rates to the same stimulus for change. In the same paper Schumm (1973) elaborated the significance of thresholds and it is evident that change will only occur once particular threshold values have been exceeded. This underlines the need for research to focus upon the significance of hydrological events of a particular magnitude because such events may be instrumental in triggering a sequence of change as instanced in a study by Mosley (1975). Applied research may therefore be directed towards the identification of thresholds below which change can be constrained. The *relationship between river channel, drainage network, and drainage basin* requires further

TABLE 1.1 SOME TECHNIQUES FOR THE STUDY OF, REASONS FOR, AND CONSEQUENCES OF, RIVER CHANNEL CHANGE

Summarizing results elaborated in more detail in Table 1.2

	Notation in Table 1.2
TECHNIQUES FOR STUDY	
Field survey and empirical measurements including	
surveyed profiles	FS
Historical methods	HM
including maps of different dates and	
editions	HM MAP
air photographs	HM PHOTO
documentary records	HM DOC
Dating techniques	DT
including Botanical techniques	DT BOT
Sedimentological evidence	DT SED
Space–time substitution or ergodic hypothesis	ST
REASONS FOR CHANNEL CHANGE	
Changes of Basin Land Use giving increased peak discharges	
Removal of Forest	Def
including Fire	Deff
Logging	Def L
Land use changes	Lus
including Potato cultivation	Lus Pot
Effect of man on river banks	
and flood plains	Lus FP
Land drainage	Lus Drain
Urbanization	Urb
Direct Regulation of Channel Discharge	
Water storage by reservoirs giving reduced peak	
flows	Dam*
and higher base flows	Damb
seepage from floodwater structures	Seep
River Channel Regulation	
Regulation of part of river channel	Reg
for flood protection and navigation	Reg Flood navig
improvements	
artificial straightening to prevent flood	
damage	Reg st
gravel extraction from river channel	Reg gr
dredging	Reg dr
diversion of water to millraces	Reg mill
artificial capture	Reg cap
NATURE OF RIVER CHANNEL CHANGE	
Changes of Channel Geometry	Geom
including channel capacity or size	
(increase +, decrease −)	cap
channel width	W
channel scour	sc
intense bank erosion	be
sediment accumulation on stream	
bed, aggradation	se
channel shape changed	shape

TABLE 1.1 (*continued*)

Changes of Pattern and Profile

including development of pattern	patt
new channel formed or distributaries	chann, distrib
cutoff	cut
sinuosity (decreased sinuosity)	F (F⁻)
old channel reactivated	old chann
channel incision	in
floodplain destruction	fpd
floodplain aggradation	fpa
overbank deposition (increased)	od (+)
slope (decreased)	S(−)
erosion (upstream) (downstream)	e (u) (d)
deposition (upstream) (downstream)	d (u) (d)
terrace formed	t
rapids	r

investigation because it provides a context for the results of research from different areas, representing several scales and aspects of the river channel. The numerous studies of gully development are not considered here but have been the subject of an important recent work by Cooke and Reeves (1976). Further progress in the analysis and investigation of river channel changes will profit from *improved measuring techniques.* Whereas the last two decades have allowed a surge in the use of hydrological equipment for the documentation of water and sediment discharge, there is scope for further progress in the continuous monitoring of river channel behaviour. Finally *improving modelling* of river channel behaviour is basic to our improved interpretation of past changes and the prediction of future ones because such improved models should provide the path for the refinement of the methods and objectives of river metamorphosis. Studies to date have been generated particularly by situations in North America and in Europe, and in future work there is a need to obtain results representative of diverse world areas.

River channel changes merit investigation for a number of reasons amongst which the human significance is exemplified by Mark Twain...

'When the river is rising fast, some scoundrel whose plantation is back in the country, and therefore of inferior value, has only to watch his chance, cut a little gutter across the narrow neck of land some dark night, and turn the water into it, and in a wonderfully short time a miracle has happened: to wit, the whole Mississippi has taken possession of that little ditch and placed the countryman's plantation on its bank (quadrupling its value), and that other party's formerly valuable plantation finds itself away out yonder on a big island . . . . .'

K. J. Gregory

TABLE 1.2 SOME STUDIES OF RIVER CHANNEL CHANGE

CHANGE	AREA STUDIED	METHOD USED	CHANGES		SOURCE
	Beatton River, B.C. Canada	DT BOT	patt		Hickin and Nanson (1975)
	Central Indiana	HM MAP Doc	patt geom		Edgar and Melhorn (1974)
	Erap River, Papua New Guinea	FS	chann		Knight (1975)
	Danube, Sarkoz section	HM MAP	F		Somogyi (1974)
	Cimarron River, Kansas	HM MAP PHOTO	W^+		Schumm and Lichty (1963)
	Central Wales	HM MAP PHOTO	patt		Lewin and Hughes (1976)
	Vistula, Plock-Torun	HM MAP	be fpa		Koc (1972)
	Arroyo de Frijoles, New Mexico	FS	se		Leopold, Emmett, and Myrick (1966)
	Black Bottom, Ohio River, S. Illinois	DT SED	fpa		Alexander and Prior (1971)
	River Bollin Cheshire	FS	cut		Mosley (1975)
	Gila River, California	HM MAP PHOTO FS	fpd fpa		Burkam (1972)
Induced changes: Change of Basin Land Use:				Cause of Change	
	Monroe Canyon, S. California	FS	eu dd	Deff	Orme and Bailey (1971)
	South Island, New Zealand	HM PHOTO	cap^+ dd	Def	O'Loughlin (1970)
	British Columbia	FS	Geom s^+ s^-	Def L	Slaymaker and Gilbert (1972)
	Piedmont Province, Maryland	FS DT SED	od sc	Lus	Costa (1975)
	Dunajec River, Beskid Sadecki Mts, Poland	HM PHOTO	distrib	Def Lus Pot	Klimek and Trafas (1972)
	Wisoka River, W. Carpathians, Poland	FS HM MAP	be	Lus FP	Klimek (1974)
	River Bollin. Cheshire	HM MAP	cut F^-	Lus Drain Urb	Mosley (1975)

INDUCED CHANGES	AREA STUDIED	METHOD USED	CHANGES	CAUSE OF CHANGE	SOURCE
Urbanization:					
	Denver area	HM MAP	fpa in	Urb	Graf (1975)
	Watts, Branch, Maryland	FS	cap^-	Urb	Emmett (1974)
	Philadelphia, Pennsylvania	ST	cap^+	Urb	Hammer (1972)
	Piedmont, U.S.A.	ST	W^+	Urb	Wolman (1967)
	Maryland	FS	cap^- then cap^+	Urb	Leopold (1973)

TABLE 1.2 (*continued*)

Regulation of Channel Discharge:					
	North Platte River	FS HM MAP	W⁻ F⁺	Dam	Schumm (1969)
	Republican River, Nebraska	FS	W⁻ cap⁻	Dam	Northrup (1965)
	Western Oklahoma	FS	shape	Seep	Bergman and Sullivan (1963)
	Tone, Somerset	ST	cap⁻	Dam	Gregory and Park (1974)
	Willow Creek, Montana		Geom⁻	Dam	Frickel (1972)
	Colorado, New Mexico	FS	cap⁻	Damb	Sonderreger (1935)
	U.S.S.R.	FS	du ed t	Dam	Makkavayev (1972)
	Colorado, Grand Canyon	FS	ed⁺	Dam	Dolan, Howard, and Gallenson (1974)
River Channel Regulation					
	Furnace Creek Wash, California	FS	eu s⁻ in	Reg cap	Troxel (1974)
	Ouse, U.K.	ST	W⁻ cap⁻	Reg mill	Dury (1973)
	Mississippi	FS HM	be patt cap⁺	Reg Flood Reg Navig	Kesel *et al.* (1974) Simons, Schumm, and Stevens (1974)
	Near Ames, Iowa	HM	F⁺	Reg st	Noble and Palmquist (1968)
	Tucson, Arizona	FS	cap⁺	Reg gr	Bull and Scott (1974)
	Tujunga Wash, California	FS	old chann	Reg gr	Bull and Scott (1974)
	Blackwater, Missouri	FS	cap⁺ e	Reg st Reg dr	Emerson (1971) –
	Warta, Poland	HM MAP	patt	Reg	Piasecka (1974)
	Willow County Ditch, Iowa	HM	in	Reg st	Daniels (1960)

References

– Alexander, C. S. and Prior, J. C. 1971. Holocene sedimentation rates in overbank deposits in the Black Bottom of the Lower Ohio River, Southern Illinois. *Amer. J. Sci.*, **270,** pp. 361–372.

Beaumont, P. and Oberlander, T. M. 1973. Litter as a geomorphological aid—Death Valley, California. *Geography*, **58,** pp. 136–141.

Beckinsale, R. P. 1972. The effect upon river channels of sudden changes in load. *Acta Geographica Debrecina*, **X,** pp. 181–186.

Bergman, D. L. and Sullivan, C. W. 1963. Channel changes on Sandstone Creek near Cheyenne, Oklahoma. *U.S. Geol. Surv. Prof. Paper 475-C*, pp. 145–148.

Blench, T. 1972. Morphometric Changes. In Oglesby, R. T., Carlson, C. A., and McCann, J. A. (eds.), *River ecology and man.* Academic Press, New York, pp. 287–308.

Bull, W. B. and Scott, K. M. 1974. Impact of mining gravel from urban stream beds in the South Western United States. *Geology*, **2,** pp. 171–174.

Burkham, D. E. 1972. Channel changes of the Gila River in Safford Valley, Arizona, 1846–1870. *U.S. Geol. Surv. Prof. Paper 655 G*, pp. G1–G24.

Clemens, S. L. (Mark Twain). 1884. *Life on the Mississippi,* J. R. Osgood, New York.

Coates, D. R. 1970. *Environmental Geomorphology.* State University of New York, Binghamton.

Coates, D. R. (Ed.) 1973. *Environmental Geomorphology and Landscape Conservation,* Vol. III, Dowden, Hutchinson and Ross, Pennsylvania.

Coates, D. R. (Ed.) 1976. *Geomorphology and Engineering.* State University of New York, Binghampton.

Cooke, R. U. and Reeves, R. W. 1976. *Arroyos and environmental change in the American South-West.* Clarendon Press, Oxford.

Costa, J. E. 1975. Effects of agriculture on erosion and sedimentation in Piedmont province, Maryland. *Geol. Soc. Amer. Bull.,* **86,** pp. 1281–1286.

Daniels, R. B. 1960. Entrenchment of the Willow County ditch, Harrison Co. Iowa. *Amer. J. Sci.,* **258,** pp. 161–176.

Dolan, R., Howard, A., and Gallenson, A. 1974. Man's impact on the Colorado river in the Grand Canyon. *American Scientist,* **62,** pp. 392–401.

Dury, G. H. 1973. Magnitude frequency analysis and channel morphometry. In Morisawa, M. E. (Ed.) *Fluvial Geomorphology.* State University of New York, Binghamton, pp. 91–121.

Edgar, D. E. and Melhorn, W. N. 1974. Drainage basin response: documented historical change and theoretical considerations. *Purdue University Water Resources Research Center Technical Report No. 3,* Studies in Fluvial Geomorphology.

Emerson, J. W. 1971. Channelization: A case study. *Science,* **173,** pp. 325–326.

Emmett, W. W. 1974. Channel changes. *Geology,* **2,** pp. 271–272.

Frickel, D. G. 1972. Hydrology and effects of conservation structures, Willow Creek Basin, Valley County, Montana 1954–68. *U.S. Geol. Surv. Water Supply Paper 1532-G.*

Graf, W. L. 1975. The impact of suburbanization on fluvial geomorphology. *Water Resour. Res.,* **11,** pp. 690–692.

Gregory, K. J. and Park, C. C. 1974. Adjustment of river channel capacity downstream from a reservoir. *Water Resour. Res.,* **10,** pp. 870–873.

Gregory, K. J. and Park, C. C. 1976. Stream channel morphology in northwest Yorkshire. *Revue de Géomorph. dyn.,* **25,** pp. 63–72.

Gregory, K. J. and Walling, D. E. 1973. *Drainage basin form and process.* Edward Arnold, London.

Hammer, T. R. 1972. Stream channel enlargement due to urbanization. *Water Resour. Res.,* **8,** pp. 1530–40.

Helley, E. J. and Lamarche, V. C. 1973. Historic flood information for northern California streams from geological and botanical evidence. *U.S. Geol. Prof. Paper 485-E.*

Hey, R. D. 1974. Prediction and effect of flooding in alluvial systems. *Geol. Soc. Misc. Paper,* **3,** pp. 42–56.

Hickin, E. J. and Nanson, G. C. 1975. The character of channel migration on the Beatton River, northeast British Columbia, Canada. *Geol. Soc. Amer. Bull.,* **86,** pp. 487–494.

International Association for Hydraulic Research, 1973. *International Symposium on River Mechanics,* Bangkok, Thailand (3 volumes).

Kesel, R. H., Dunne, K. C., McDonald, R. C., Allison, K. R., and Spicer, B. E. 1974. Lateral erosion and overbank deposition in the Mississippi river in Louisiana caused by 1973 flooding. *Geology,* **2,** pp. 461–464.

Klimek, K. 1974. The retreat of alluvial river banks in the Wisłoka valley (South Poland). *Geographica Polonica,* **28,** pp. 59–75.

Klimek, K. and Trafas, K. 1972. Young Holocene changes in the course of the Durajec river in the Beskid Sadecki Mts. (Western Carpathians). *Studia Geomorphologica Carpatho-Balcanica,* **VI,** pp. 85–92.

Knight, M. J. 1975. Recent crevassing of the Erap river, New Guinea. *Australian Geog. Studies*, **13**, pp. 77–82.

Koc, L. 1972. Nineteenth and twentieth centuries changes in Vistula channel between Ptock and Torun (Abstr.). *Przeglad Geograficzny*, **44**, pp. 703–719.

Leopold, L. B. 1973. River channel change with time: an example. *Geol. Soc. Amer. Bull.*, **84**, pp. 1845–1860.

Leopold, L. B., Emmett, W. W., and Myrick, R. M. 1966. Channel and hillslope processes in a semi-arid area, New Mexico. *U.S. Geol. Surv. Prof. Paper 352-G*, pp. 193–253.

Leopold, L. B., Wolman, M. G., and Miller, J. P. 1964. *Fluvial processes in Geomorphology*. W. Freeman, San Francisco.

Lewin, J. and Hughes, D. 1976. Assessing channel change on Welsh rivers. *Cambria*, **3**, pp. 1–10.

Makkavayev, N. I. 1972. The impact of large water-engineering projects in geomorphic processes in stream valleys. *Soviet Geogr.*, **13**, pp. 387–393.

Meade, R. H. and Trimble, S. W. 1974. Changes in sediment loads in rivers of the Atlantic drainage of the United States since 1900. *Effects of Man on the Interface of the Hydrological Cycle with the Physical Environment* 1AHS-A1SH Publ. No. 113, pp. 99–104.

Morisawa, M. E. (Ed.) 1973. *Fluvial Geomorphology*. State University of New York, Binghamton.

Mosley, M. P. 1975. Channel changes on the river Bollin, Cheshire 1872–1973. *East Midland Geographer*, **6**, pp. 185–199.

Mosley, M. P. 1975. Meander cutoffs on the River Bollin, Cheshire, in July 1973. *Revue de Géomorph. Dyn.*, **24**, pp. 21–31.

Mrowka, J. P. 1974. Man's impact on stream regimen and quality. In Manners, I. R. and Mikeswell, M. W. (Eds.) *Perspectives on Environment*, Ass. of Amer. Geogrs Publication No. 13, pp. 79–104.

Noble, C. A. and Palmquist, R. C. 1968. Meander growth in artificially straightened streams. *Proc. Iowa Acad. Sci.*, **75**, pp. 234–242.

Northrup, W. L. 1965. Republican river channel deterioration. *U.S. Dept. Agric. Misc. Pub.*, **970**, pp. 409–424.

O'Loughlin, C. L. 1970. Streambed investigations in a small mountain catchment *N.Z. Jl. Geol. Geophys.*, **12**, pp. 684–706.

Orme, A. R. and Bailey, R. G. 1971. Vegetation conversion and channel geometry in Monroe Canyon, California. *Yearbook of Pacific Coast Geographers*, **33**, pp. 65–82.

Piasecka, J. E. 1974. Hydrographic changes in the Warta Valley during the last 200 years. *Czasopismo Geograficzne*, **45**, pp. 229–238.

Rango, A. 1970. Possible effects of precipitation modification on stream channel geometry and sediment yield. *Water Resour. Res.*, **6**, pp. 1765–1770.

Schumm, S. A. 1965. Quaternary Palaeohydrology. In Wright, H. E. and Frey D. G. (Eds.) *The Quaternary of the United States*, Princeton Univ. Press, Princeton, pp. 783–794.

Schumm, S. A. 1969. River metamorphosis. *Proc. Amer. Soc. civ. Engrs. J. Hyd. Div.*, **95**, pp. 251–273.

Schumm, S. A. 1971. Channel adjustment and river metamorphosis. In Shen, H. W. (Ed.) *River Mechanics*, Vol. I, pp. 5–1 to 5–22, Water Resources Publications, Fort Collins.

Schumm, S. A. (Ed.) 1972. *River morphology*. Dowden, Hutchinson and Ross, Pennsylvania.

Schumm, S. A. 1973. Geomorphic thresholds and complex response of drainage systems. In Morisawa, M. E. (Ed.) *Fluvial Geomorphology*. State University of New York, Binghamton, pp. 299–310.

Schumm, S. A. and Lichty, R. W. 1963. Channel widening and flood plain construction along Cimarron river, in south western Kansas. *U.S. Geol. Surv. Prof. Paper 352D*.

Shen, H. W. (Ed.) 1971. *River Mechanics*. Water Resources Publications, Fort Collins.

Simons, D. B., Schumm, S. A., and Stevens, M. A. 1974. *Geomorphology of the Middle Mississippi river*. Colorado State University Engineering Research Center, Fort Collins, Final Report.

Slaymaker, H. O. and Gilbert, R. E. 1972. Geomorphic processes and land use changes in the west coast mountains of British Columbia: a case study. *Processus périglaciaires etudes sur le terrain*. Symposium Internationale de Geomorphologie, Liège, pp. 269–279.

Somogyi, S. 1974. Channel and flood plain development in the Sárkòz section of the Danube as shown by mapping between 1782 and 1950. (Abstract) *Foldrajzi Ertesito*, **23**, pp. 27–36.

Sonderreger, A. L. 1935. Modifying the physiographic balance by conservation measures. *Trans. Amer. Soc. civ. Engrs. Paper No. 1897*, pp. 284–304.

Strahler, A. N. 1956. The nature of induced erosion and aggradation. In Thomas, W. L. (Ed.) *Man's role in changing the face of the earth*. Chicago, pp. 621–638.

Troxel, B. W. 1974. Man made diversion of Furnace Creek Wash, Zabriskie Point, Death Valley, California. *California Geology*, **27**, pp. 221–223.

Wolman, M. G. 1967. Two problems involving river channel changes and background observations. *Northwestern Studies in Geography*, **14**, pp. 67–107.

Section I

MECHANICS AND SEDIMENTOLOGY

Graf, W. H. 1971. *Hydraulics of Sediment Transport.* McGraw–Hill, New York.

Grass, A. J. 1971. Structural features of turbulent flow over smooth and rough boundaries. *J. Fluid Mech.,* **50,** pp. 233–255.

Guy, H. P., Simons, D. B., and Richardson, E. V. 1966. Summary of alluvial channel data from flume experiments, 1956–61. *U.S. Geol. Surv. Prof. Paper 462-I.*

Handy, R. L. 1972. Alluvial cutoff dating from subsequent growth of a meander. *Geol. Soc. Amer. Bull.,* **83,** pp. 475–480.

Harms, J. C. and Fahnestock, R. K. 1965. Stratification, bed forms, and flow phenomena (with an example from the Rio Grande). *Spec. Publns Soc. econ. Palaeont. Miner., Tulsa,* **12,** pp. 84–110.

Herbertson, J. G. 1969. A critical review of conventional bed load formulae. *J. Hydrol.,* **8,** pp. 1–26.

Hey, R. D. and Thorne, C. R. 1975. Secondary flow in river channels. *Area,* **7,** pp. 191–195.

Hickin, E. J. 1974. The development of meanders in natural river-channels. *Amer. J. Sci.,* **274,** pp. 414–442.

Hickin, E. J. and Nanson, G. C. 1975. The character of channel migration on the Beatton River, northeast British Columbia, Canada. *Geol. Soc. Amer. Bull.,* **86,** pp. 487–494.

Hill, H. M., Srinivasan, V. S., and Unny, T. E. 1969. Instability of flat bed in alluvial channels. *J. Hydraul. Div. Amer. Soc. civ. Engrs.,* **95,** pp. 1545–1558.

Ikeda, S. 1974. On secondary flow and dynamic equilibrium of transverse bed profile in alluvial curved open channel. *Proc. Japan Soc. civ. Engrs. No. 229,* pp. 55–65.

Ikeda, S. 1975. On secondary flow and bed profile in alluvial curved open channel. *Proc. 16th Congr. Int. Assoc. Hydraulic Res.,* **2,** pp. 105–112.

Jackson, R. G. 1975. Velocity-bed form-texture patterns of meander bends in the lower Wabash River of Illinois and Indiana. *Geol. Soc. Amer. Bull.,* **86,** pp. 1511–1522.

Jackson, R. G. 1976. Large scale ripples of the lower Wabash River. *Sedimentology,* **23,** pp. 593–623.

Karcz, I. 1974. Reflections on the origin of small-scale longitudinal streambed scours. In Morisawa, M. E. (Ed.), *Fluvial Geomorphology.* State University of New York, Binghamton, pp. 149–173.

Kennedy, J. F. 1961. Stationary waves and antidunes in alluvial channels. *W. M. Keck Laboratory for Hydraulics and Water Resources, California Institute of Technology, Report KH-R-2.*

Kennedy, J. F. 1963. The mechanics of dunes and antidunes in erodible-bed channels. *J. Fluid Mech.,* **16,** pp. 521–544.

Kondrat'yev, N. Y. 1969. Hydromorphological principles of computations of free meandering. 1. Signs and indexes of free meandering. *Soviet Hydrology: Selected Papers 1968,* pp. 309–335.

Langbein, W. B. 1964. Geometry of river channels. *J. Hydraul. Div. Amer. Soc. civ. Engrs.,* **90,** pp. 301–311.

Leliavsky, S. 1959. *An Introduction to Fluvial Hydraulics.* Constable, London.

Leopold, L. B. and Maddock, J. T. 1953. The hydraulic geometry of stream channels and some physiographic implications. *U.S. Geol. Surv. Prof. Paper 252,* pp. 1–57.

Leopold, L. B. and Wolman, M. G. 1957. River channel patterns: braided, meandering and straight. *U.S. Geol. Surv. Prof. Paper 282-B,* pp. 39–85.

Leopold, L. B. and Wolman, M. G. 1960. River meanders. *Geol. Soc. Amer. Bull.,* **71,** pp. 769–794.

McGowen, J. H. and Garner, L. E. 1970. Physiographic features and stratification types of coarse-grained point bars: modern and ancient examples. *Sedimentology,* **14,** pp. 77–111.

Mercer, A. G. 1971. Analysis of alluvial bed forms. In Shen, H. W. (Ed.), *River Mechanics*, Vol. I. H. W. Shen, Fort Collins, Colorado, pp. 10.1–10.26.

Meyer-Peter, R. and Müller, R. 1948. Formulas for bed-load transport. *Int. Assoc. Hydraulic Structures Research*, Second Meeting, Stockholm, pp. 39–64.

Myers, R. C. and Elsawy, E. M. 1975. Boundary shear in channel with flood plain. *J. Hydraul. Div. Amer. Soc. civ. Engrs.*, **101,** pp. 933–946.

Nasner, H. 1974. Über das Verhalten von Transportkörpern im Tidegebiet. *Mitt. Franzius-Institut*, **40,** pp. 1–149.

NEDECO (Netherlands Engineering Consultants). 1959. *River Studies and Recommendations on Improvement of Niger and Benue*. North-Holland, Amsterdam.

Neill, C. R. 1969. Bed forms of the Lower Red Deer River. *J. Hydrol.*, **7,** pp. 58–85.

Nilsson, G. and Martvall, S. 1972. The Öre River and its meanders. *University of Uppsala, UNGI Report No. 19.*

Onishi, Y., Jain, S. C., and Kennedy, J. F. 1972. Effects of meandering on sediment discharges and friction factors of alluvial streams. *Iowa Institute of Hydraulic Research, University of Iowa, Report No. 141.*

Paulissen, P. 1973. De morfologie en de kwatarstratigrafie van de Maasvallei in Belgisch Limburg. *Verh. K. Acad. Wet. Lett. Sch. Kunste Belg.*, **35,** pp. 5–266.

Popov, I. V. 1965. Hydromorphological principles of the theory of channel processes and their use in hydrotechnical planning. *Soviet Hydrology: Selected Papers 1964*, pp. 188–195.

Prasad, R. 1970. Numerical methods of calculating flow profiles. *J. Hydraul. Div. Amer. Soc. civ. Engrs.*, **96,** pp. 75–86.

Pratt, C. J. and Smith, K. V. H. 1972. Ripple and dunes phases in a narrowly graded sand. *J. Hydraul. Div. Amer. Soc. civ. Engrs.*, **98,** pp. 859–874.

Pretious, E. S. and Blench, T. 1951. Final Report on Special Observations in Lower Fraser River at Ladner Reach during 1950 Freshet. *National Research Council of Canada*, Vancouver.

Raudkivi, A. J. 1967. *Loose Boundary Hydraulics*. Pergamon, Oxford.

Richards, K. S. 1973. Hydraulic geometry and channel roughness—a non-linear system. *Amer. J. Sci.*, **273,** pp. 877–896.

Richards, K. S. 1976. Complex width-discharge relations in natural rivers sections. *Geol. Soc. Amer. Bull.*, **87,** pp. 199–206.

Rouse, H. 1938. *Fluid Mechanics for Hydraulic Engineers*. McGraw–Hill, New York.

Rozovskii, I. L. 1961. *Flow of Water in Bends of Open Channels*. Israel Program for Scientific Translations, Jerusalem.

Schumm, S. A. 1960. The shape of alluvial channels in relation to sediment type. *U.S. Geol. Surv. Prof. Paper 352-B.*

Schumm, S. A. 1968. River adjustment to altered regimen—Murrumbidgee River and paleochannels, Australia. *U.S. Geol. Surv. Prof. Paper 598.*

Sellin, R. H. J. 1964. A laboratory investigation into the interaction between the flow in the channel of a river and that over its flood plain. *Houille Blanche*, **19,** pp. 793–801.

Sellin, R. H. J. 1969. *Flow in Channels*. Macmillan, London.

Shen, H. W. 1971. Total sediment load. In Shen, H. W. (Ed.), *River Mechanics*, Vol. I. H. W. Shen, Fort Collins, Colorado, pp. 13.1–13.26.

Simons, D. B. and Richardson, E. V. 1962. The effect of bed roughness on depth-discharge relations in alluvial channels. *Wat.-Supply Irrig. Pap. Wash. 1498-E.*

Simons, D. B. and Richardson, E. V. 1966. Resistance to flow in alluvial channels. *U.S. Geol. Surv. Prof. Paper 422-J.*

Simons, D. B., Richardson, E. V., and Nordin, C. F. 1965. Sedimentary structures generated by flow in alluvial channels. *Spec. Publns. Soc. econ. Palaeont. Miner.*, *Tulsa*, **12,** pp. 34–52.

Singh, I. B. and Kumar, S. 1974. Mega- and giant ripples in the Ganges, Yamuna, and Son rivers, Uttar Pradesh, India. *Sediment. Geol.*, **12**, pp. 53–66.

Smith, T. R. 1974. A derivation of the hydraulic geometry of steady-state channels from conservation principles and sediment transport laws. *J. Geol.*, **82**, pp. 98–104.

Southard, J. B. 1971. Representation of bed configurations in depth-velocity-size diagrams. *J. sedim. Petrol.*, **41**, pp. 903–915.

Southard, J. B. and Boguchwal, L. A. 1973. Flume experiments on the transition from ripples to lower flat bed with increasing sand size. *J. sedim. Petrol.*, **43**, pp. 1114–1121.

Stein, R. A. 1965. A Laboratory study of total load and apparent bed load. *J. geophys. Research*, **70**, pp. 1831–1842.

Stückrath, T. 1969. Die Bewegung von Grossrippeln an der Sohle des Rio Paraná. *Mitt. Franzius-Institut.*, **32**, pp. 267–293.

Suga, K. 1967. The stable profiles of the curved open channel beds. *Proc. 12th Congr. Int. Assoc. Hydraulic Research*, **1**, pp. 487–495.

Sundborg, Å. 1956. The River Klarälven: a study of fluvial processes. *Geogr. Annlr.*, **38**, pp. 127–316.

Varshney, D. V. and Garde, R. J. 1975. Shear distribution in bends in rectangular channels. *J. Hydraul. Div. Amer. Soc. civ. Engrs.*, **101**, pp. 1053–1066.

Williams, G. P. 1967. Flume experiments on the transport of a coarse sand. *U.S. Geol. Surv. Prof. Paper 562-B*.

Williams, G. P. 1970. Flume width and water depth effects in sediment-transport experiments. *U.S. Geol. Surv. Prof. Paper 562-H*.

Williamson, M. 1972. *The Analysis of Biological Populations*. Arnold, London.

Wolman, M. G. 1954. A method of sampling coarse bed material. *Trans. Amer. Geophys. Union*, **35**, pp. 951–956.

Yalin, M. S. 1964. Geometrical properties of sand waves. *J. Hydraul. Div. Proc. Amer. Soc. civ. Engrs.*, **90**, pp. 105–119.

Yalin, M. S. 1972. *Mechanics of Sediment Transport*. Pergamon, Oxford.

Yen, B. C. 1965. *Characteristics of Subcritical Flow in a Meandering Channel*. Institute of Hydraulic Research, University of Iowa.

Yen, C. L. 1970. Bed topography effect on flow in a meander. *J. Hydraul. Div. Amer. Soc. civ. Engrs.*, **96**, pp. 57–73.

Yen, C. L. and Overton, D. E. 1973. Shape effects on resistance in floodplain channels. *J. Hydraul. Div. Amer. Soc. civ. Engrs.*, **99**, pp. 219–238.

Zeller, J. 1967. Flussmorphologische Studie zum Mäanderproblem. *Geogr. Helvet.*, **22**, pp. 57–95.

G. PICKUP

Lecturer in Geography
University of Papua New Guinea

3

Simulation Modelling of River Channel Erosion

When a river is affected by a change in discharge or sediment load, it may develop sediment transport discontinuities such as knickpoints. The discontinuities move along the channel, resulting in erosion or deposition which changes channel morphology and gradually restores stability to the system. Because of the time scale involved, the behaviour of a sediment transport discontinuity is difficult to study in the field. It can, however, be modelled and the purpose of this paper is to describe the development and results of a simulation model of river channel erosion.

The first stage of simulation is to construct a model of the system which is being studied. Most mathematical models of channel development have been developed by hydraulic engineers attempting to predict degradation below dams and other hydraulic structures (see, for instance, Tinney, 1962; Komura and Simons, 1967; Hales and coworkers, 1970) although a model of meandering channel processes was recently described by Bridge (1975). Usually these models are based on the continuity equation for sediment transport and are restricted to long profile development. This restriction is unreal because the degradation process usually involves changes in channel cross-section as well as long profile. The model described below therefore allows for changes in channel size and shape as well as slope as erosion proceeds.

Model Development

At its simplest, the development of an eroding channel is the morphologic expression of the behaviour of a sediment transport discontinuity with time. The discontinuity is created by changes in channel morphology at one end of a reach and results in the addition or removal of sediment from a storage reservoir made up of the bed and banks of the channel. Because sediment is removed from the channel, however, channel morphology changes. This change affects the rate of sediment transport and thus feeds back to and modifies the transport discontinuity.

To represent this process, the model should reproduce four sets of relationships. These are:

(i) The relationship between channel slope, shape and hydraulic characteristics and flow conditions;
(ii) The relationship between flow conditions and the rate of sediment transport;
(iii) The relationship between downstream variations in sediment transport and the rate of erosion or deposition;
(iv) The relationship between erosion or deposition and changes in channel shape and slope.

Relationship (i)

In a uniform channel, flow conditions may be calculated from one of the uniform flow formulas such as the Strickler equation:

$$v = kR^{\frac{2}{3}}S_f^{\frac{1}{2}} \tag{1}$$

where v is velocity; k is the roughness coefficient; R is hydraulic radius; and S_f is the friction slope. In an eroding channel, channel characteristics will vary with distance so the flow will be gradually varied. The determination of flow conditions therefore becomes more complex and requires computation of the flow profile (Chow, 1959). Water surface profile computations are usually based on an integration of the one-dimensional gradually varied flow equation which may be stated as:

$$\frac{dy}{dx} = \frac{S_0 - S_f}{1 - \dfrac{\alpha Q^2 W}{gA^3}} \tag{2}$$

where y is water surface elevation; x is a longitudinal coordinate; S_0 is the slope of the bed; S_f is the friction slope; α is a velocity distribution coefficient; Q is discharge; W is channel width; g is the gravity constant, and A is channel area. Closed-form solution of equation (2) is rarely possible so it is necessary to use a numerical procedure such as the method of successive substitutions (Prasad, 1970) or the bisection method (McBean and Perkins, 1975). The method of successive substitutions is the most frequently used technique in water profile calculation by computer, but occasionally it becomes unstable and fails to converge on the correct solution. It is therefore better to use the slower but more reliable bisection method, and this was incorporated into the model.

Relationship (ii)

Water surface profile computations provide data on flow depth, velocity and energy slope at a series of points along the channel. With this information it is possible to estimate the rate of sediment transport using one of the many sediment transport equations available (see, for instance, Hubbel and Matejka, 1959; Vanoni, Brooks, and Kennedy, 1961; Colby, 1964, and the Committee on Sedimentation, 1971, for reviews). Some of these equations are more reliable than others and all apply to a limited range of conditions so the equation must be chosen with care.

In the simulation described below, flow is fully turbulent and the river bed material is mainly fine gravel. The most suitable transport equation for these conditions is that of Meyer-Peter and Müller (1948) which may be stated as:

$$\gamma_w \cdot \left(\frac{Q_s}{Q}\right) \cdot \left(\frac{k_b}{k_r}\right)^{\frac{3}{2}} \cdot S_f = 0.047 \cdot (\gamma_s - \gamma_w) \cdot d_m + 0.25 \cdot \left(\frac{\gamma_w}{g}\right)^{\frac{1}{3}} \cdot G'^{\frac{2}{3}} \qquad (3)$$

where Q_s/Q is a bank friction correction factor; k_b is the Strickler bed roughness coefficient; k_r is the Strickler particle roughness coefficient $= 26/d_{90}^{\frac{1}{6}}$; S_f is the friction slope; γ_s is the specific weight of sediment; γ_w is the specific weight of water; d_m is the mean particle size; G' is bed-material discharge per second per unit width measured under water.

Relationship (iii)

The third relationship, between the rate of erosion or deposition and downstream variations in sediment transportation may be expressed by the continuity equation for sediment transport. This equation is based on the assumption that sediment transport varies with distance along the channel, x, and time t. The sediment input to a reach, dx, during the time increment, dt, is given by:

$$G\,dt$$

while the output from the reach is:

$$\left(G + \frac{\partial G}{\partial x}\,dx\right)dt$$

where G is the rate of sediment transport. The sediment balance of the reach for the time increment, dt, is the difference between sediment input and output, and is given by:

$$G\,dt - \left(G + \frac{\partial G}{\partial x}\,dx\right)dt$$

If the sediment balance is not equal to zero there is a difference between input and output, and material is added to or removed from storage in the

G. Pickup

reach. The equation for the change in storage, S, may therefore be written in the following form:

$$\frac{\partial S}{\partial t}\,dxdt = \left\{ Gdt - \left(G + \frac{\partial G}{\partial x}\,dx \right)dt \right\} \tag{4}$$

which simplifies to:

$$\frac{\partial S}{\partial t} = -\frac{\partial G}{\partial x} \tag{5}$$

which is the continuity equation for sediment transport and gives the rate of erosion or deposition.

Where the channel is wide and uniform, it may be possible to derive an analytical expression for the right hand side of equation (5) (see, for instance, Peters and Bowler, 1968). If the channel changes during the course of erosion or deposition, this becomes very difficult and it is necessary to use a finite difference evaluation of equation (5). The difference scheme used here is as follows. The average rate of change of storage with time in a particular reach, $x_{n-1} - x_n$ is given by:

$$\frac{\Delta S}{\Delta t}(x_{n-1} - x_n, t) = \{G(x_{n-1}, t) - G(x_n, t)\}/\Delta x \tag{6}$$

The rate of change of storage with time at a particular point, x_n, is the average of the rates of change of storage in the two adjacent reaches, $x_{n-1} - x_n$ and $x_n - x_{n+1}$, thus:

$$\frac{\Delta S}{\Delta t}(x_n, t) = \left\{ \frac{\Delta S}{\Delta t}(x_{n-1} - x_n, t) + \frac{\Delta S}{\Delta t}(x_n - x_{n+1}, t) \right\}/2 \tag{7}$$

Substituting equation (6) into (7) and simplifying gives:

$$\frac{\Delta S}{\Delta t}(x_n, t) = \{G(x_{n-1}, t) - G(x_{n+1}, t)\}/2\,\Delta x \tag{8}$$

This equation may be used to calculate the rate of erosion or deposition at a series of discrete points along the channel. The amount of erosion or deposition in a particular time increment is then given by:

$$\Delta S(x_n, t) = \frac{\Delta S}{\Delta t}(x_n, t) \cdot \Delta t \tag{9}$$

Computations of erosion are carried out for a series of time increments so that:

$$S(x_n, t + \Delta t) = S(x_n, t) + \frac{\Delta S}{\Delta t}(x_n, t) \cdot \Delta t \tag{10}$$

Relationship (iv)

The relationship between the rate of erosion or deposition and the changes which occur in channel slope and shape is a key element in unstable channel simulation because it determines the way in which the channel develops over time. Changes in channel morphology resulting from erosion or deposition affect the rate of sediment transport and therefore change the characteristics of the sediment transport discontinuity. This, in turn, affects the rate of erosion or deposition, providing a feedback relationship between changes in process and form.

The effect of erosion or deposition on channel characteristics is likely to vary widely from stream to stream. It depends on such factors as the size, erosion resistance, and angle of repose of the bed and bank sediment, the effects of vegetation, the sediment load of the stream, and the magnitude and frequency of discharge (see, for instance, Schumm, 1961, 1971).

In the case of erosion, channel response tends to fall between two extremes. Firstly there is the case where the shape of the channel remains essentially the same and sediment is scoured only from the bed. This type of channel response tends to be restricted to channels with highly cohesive banks or bank protection works and channel adjustment is basically a change in slope. At the other extreme is the case where most of the erosion

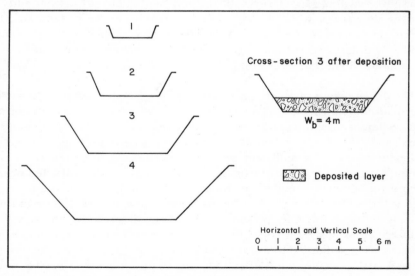

FIGURE 3.1 CHANGES IN CHANNEL CROSS-SECTION WITH EROSION AND DEPOSITION USED IN THE SIMULATION MODEL

The cross-sections on the left show how cross-section changes as the depth of incision increases while the cross-section on the right illustrates the effects of deposition

occurs by channel widening without any real change in slope. In some circumstances widening may be excessive and although net erosion has occurred, the level of the bed actually rises. This sometimes occurs with eroding sand bed streams.

In most cases, channel response lies somewhere between the two extremes and involves changes in both cross-section and slope. This type of channel response has been assumed in the simulation described below. The scheme of channel changes used is based on those observed in Crawfords Creek, a small stream in eastern New South Wales (Pickup, 1975) and is illustrated in Figure 3.1. When erosion begins, a small rectangular channel is cut in the bed of the main channel. As erosion continues, the new channel becomes wider and its banks become less steep until it gradually merges into the original channel. Although erosion may be the dominant process, it does not occur continuously. Instead erosion may be succeeded by short periods of deposition as sediment transport discontinuities move along the system. When this deposition occurs, it is assumed to occur only on the channel bed (Figure 3.1) which appears to be the case in Crawfords Creek itself.

Other Relationships in Channel Erosion

The feedback between channel slope and shape, i.e. the rate of sediment transport and the characteristics of the sediment transport discontinuity, is not the only factor which tends to restore stability to the system. Another important process is bed armouring in which the sediment in the channel is sorted by differential erosion leaving a layer of coarse material which forms an armour coat, thus preventing erosion of the finer fractions which lie beneath it (Livesey, 1965).

The development of an armoured bed is often a major factor in restricting degradation below dams and an algorithm for the computation of changes in bed sediment size distribution as scour proceeds has been presented by Gessler (1970). Whether bed armouring occurs or not depends on whether sediment particles of sufficient size to resist erosion are present. It also depends on whether the sediment supply from upstream is restricted or not. Where the sediment supply is restricted, as with a dam or weir, an armour coat will develop as long as sufficiently large sediment particles are present. If large particles are only present in small quantities, and if the supply of sediment from upstream is not restricted and new sediment is continuously injected as a sediment transport discontinuity moves through the system, there is less opportunity for an armour coat to develop and it may be absent altogether (Pickup, 1975). In these circumstances bed armouring as a factor in channel adjustment may be ignored and it has not been allowed for here.

Simulation of Eroding Channel Development

A model of the erosion process based on relationships (i), (ii), (iii), and (iv) has been translated into the FORTRAN IV computer program,

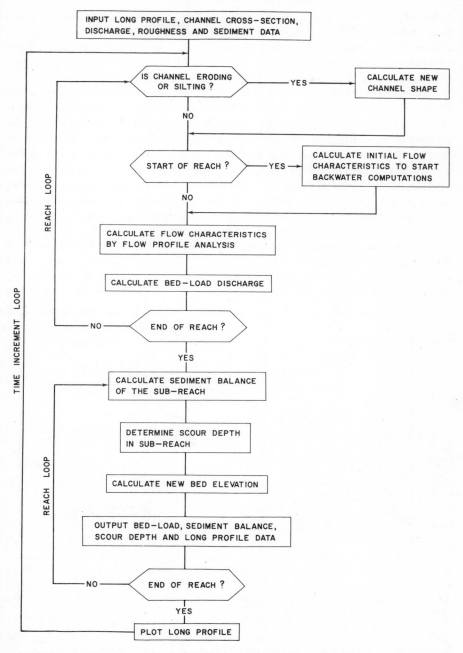

FIGURE 3.2 FLOW CHART ILLUSTRATING THE OPERATION OF THE COMPUTER PROGRAM ERODE

G. Pickup

ERODE. The structure and operation of this program is illustrated by the flow chart in Figure 3.2. To illustrate the behaviour of an eroding channel, the model has been applied to a channel with a uniform slope except for a straight knickpoint at its downstream end which creates a sediment transport discontinuity.

The changes which occur in the long profile of the reach through time are illustrated in Figure 3.3. Apparently the system develops by knickpoint retreat rather than by rotation, for the knickpoint maintains its initial steepness right through the simulation. The knickpoint also maintains its height which indicates two things. Firstly, the erosion just upstream of the knickpoint, which tends to flatten its upper part, is not as intense as the erosion down the knickpoint face which tends to cut it back. Secondly, there is no general build-up of sediment in the reach below the knickpoint face. The reach downstream of the knickpoint is therefore capable of evacuating most of the sediment delivered to it from upstream.

Because the knickpoint maintains its height and steepness, it undergoes parallel retreat and the reach below it therefore tends to have a more or less uniform gradient similar to that of the original long profile. If the new and old long profiles have the same gradient then the depth of incision is the

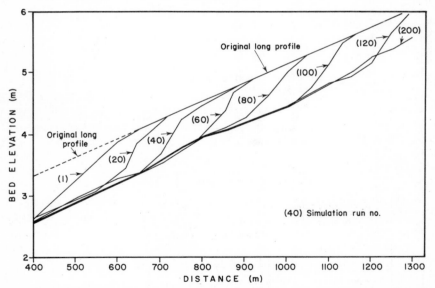

FIGURE 3.3 SIMULATED CHANGES IN THE LONG PROFILE OF AN ERODING CHANNEL WITH TIME

The profile marked (1) is the knickpoint at the beginning of the simulation run. The reach between 0 and 400 m was affected by boundary conditions namely the assumed values of flow depth and energy slope used to start the flow profile computations. It has therefore been excluded

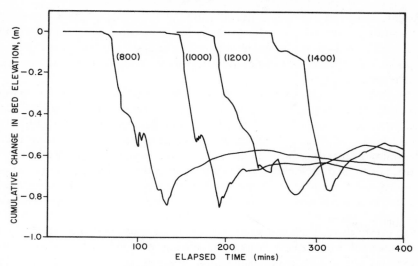

FIGURE 3.4 CUMULATIVE CHANGES IN BED ELEVATION WITH TIME AT FOUR LOCATIONS IN THE CHANNEL BEFORE, DURING AND AFTER THE PASSAGE OF THE KNICKPOINT

The figure in parentheses on each line refers to the distance of each location from the downstream end of the system. Note the lack of variation in knickpoint height, the period of deposition after the knickpoint has passed and renewed erosion which follows deposition

same right along the new channel. Channel cross-section characteristics depend on the depth of incision, so not only does the channel have a uniform bed-slope, it also has a uniform cross-section. The new channel is therefore likely to be in a state of bed-load transport continuity.

The pattern of changes in bed elevation which occurs in the eroding reach, as time passes, is shown in Figure 3.4 for points 800, 1000, 1200, and 1400 m from the downstream end of the system. The pattern, remains markedly uniform with increasing distance, reflecting the way in which the knickpoint height and steepness are maintained throughout the simulation. It consists of five basic elements:

(i) A period of time during which bed elevation did not change because the knickpoint was still downstream;

(ii) An initial, slight decrease in bed elevation as erosion began in response to the drawdown created by increased flow velocity over the knickpoint;

(iii) A rapid decrease in bed elevation down to a minimum level as the knickpoint passed;

(iv) A rapid, then a more gradual increase in bed elevation followed by;

(v) A slow decrease in bed elevation.

There are also some small irregularities within this pattern. These ir-regularities occur because, of necessity, the simulation uses a finite differ-ence approach to what is essentially a continuous process. They may be suppressed by the use of smaller steps in the calculations but at the expense of increased computer time.

The pattern of changes in bed elevation with time shows that, as channel development proceeds, the intense erosion occurring with the passage of a knickpoint is succeeded by a period of rather less intense deposition. This deposition is, in turn, succeeded by a period of erosion indicating that a knickpoint reach does not have just the two sediment transport discon-tinuities described by Pickup (1975) but at least three. The development and maintenance of three discontinuities is not unique to the simulation model but has also been observed in flume experiments (Pickup and Holland, unpublished report).

The presence of three sediment transport discontinuities indicates that the restoration of equilibrium conditions during and after the passage of a knickpoint does not occur smoothly. Instead it involves a series of progressively-damped oscillations in bed elevation in which increasingly minor adjustments are made to channel characteristics which gradually restore transport continuity to the system. These oscillations are built into the system and occur for the reasons given below.

The increase in sediment transport capacity at the knickpoint creates a deficiency of load which is satisfied by erosion. Downstream of the knick-point, channel slope decreases and the cross-section changes so that trans-port capacity is reduced and the supply of load from upstream becomes excessive, resulting in deposition. As the channel develops, both the eroding knickpoint and the zone of deposition below it move upstream and the sediment supply is no longer excessive. Instead, the increased slope which is developed on the downstream side of the zone of deposition results in an increase in transport capacity and erosion is renewed.

Comparison of the results of the simulation with patterns of erosion observed in natural rivers is difficult because of the absence of data. It is, however, possible to compare simulation-derived erosion patterns with those observed in flume experiments.

Knickpoint retreat occurred in the simulation because the characteristics of the channel cross-section were dependent on the depth of incision. The conditions of uniform slope and cross-section required to satisfy transport continuity could therefore only exist if the depth of incision was uniform right along the channel. It was therefore necessary for knickpoint height to be maintained.

This pattern of development is different from that observed in some flume experiments with eroding channels (Lewis, 1945, Brush and Wolman, 1960, and Gessler, 1971). Here knickpoints tend to flatten or rotate as in Figure 3.5 rather than maintain their height. Some of these experiments, however,

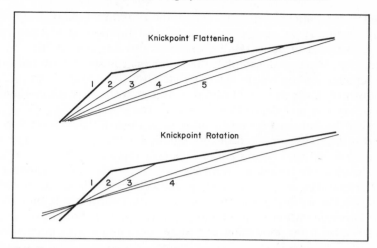

FIGURE 3.5 PATTERNS OF EROSION OBSERVED IN FLUME EXPERIMENTS IN WHICH
CHANNEL WIDTH IS NOT A FUNCTION OF DEPTH OF INCISION

used fixed-width channels in which cross-sectional adjustment could not
occur. Others were carried out in sand or only slightly cohesive sediment in
which bank erosion could occur freely and the channel could achieve a
uniform width without necessarily having a uniform depth of incision right
along the channel. In such a case, it is not necessary for knickpoint height to
be maintained for the restoration of sediment transport continuity. The
difference in the pattern of erosion is therefore a result of the way in which
the channel cross-section responds to erosion.

The Limits of Channel Erosion

In the simulation just described, the erosion process operates through a
process of negative feedback to destroy the sediment transport discontinuity
so that as time increases, $(\partial G/\partial x) \to 0$. Erosion ceases when a state of
sediment transport continuity is reached, i.e. $(\partial G/\partial x) = 0$. This state can exist
under two sets of conditions, these being:

$$G_n = G_{n+1} = G_{n+2} = G_{n+3} \cdots > 0$$
$$G_n = G_{n+1} = G_{n+2} = G_{n+3} \cdots = 0$$

In the run described in the previous section, channel stability was restored as
the system approached the first condition where $G > 0$. Other runs with the
model have produced systems in which the second stability condition was
developed, i.e. where the flow was no longer competent to transport
available bed sediment.

In virtually all of the simulation work which has been presented in the
literature so far, the two sediment transport continuity conditions have been

accepted as the limits of erosion. Indeed, the condition $(\partial G/\partial x) = 0$ has been used by a number of authors for operational forecasting of the depth of scour downstream of dams (Hales, Shindala and Denson, 1970).

Within the limits of the model, the sediment transport continuity condition is all that is required for channel stability. In reality, however, other conditions may have to be met and it is here that the existing mathematical theory of erosion and deposition is inadequate.

Probably the most important of these conditions is that channels must be adjusted to the prevailing hydrologic regime, for there is a growing body of evidence that channel conditions are closely related to flow variability as well as the absolute magnitude of discharge (see, for instance, Kilpatrick and Barnes, 1964; Harvey, 1969; Pickup and Warner, 1976). At present no quantitative theory linking channel shape with hydrologic regime has appeared. However, when such a theory is developed, it should prove to be a useful advance in the modelling and understanding of river behaviour.

Conclusions

A simulation model of fluvial erosion has been developed in which changes in both bed elevation and channel cross-section are allowed for. The model is based on the continuity equation for sediment transport, flow profile analysis, and the Meyer-Peter and Müller (1948) bed-load equation. The model was applied to a hypothetical watercourse in which the shape and size of the channel cross-section varied systematically with the depth of incision. The simulation runs were begun assuming that the watercourse had a uniform slope except for a knickpoint at its downstream end. During the simulation runs the system developed by knickpoint retreat and it developed three sediment transport discontinuities.

The end product of the simulation was a uniform channel in which a state of sediment transport continuity was developed. In real channels, however, sediment continuity is only one of the conditions a stable channel must fulfil. Further developments in simulation await the development of a more complete quantitative theory of channel adjustment.

Acknowledgements

The author is grateful to W. A. Rieger for his helpful discussion and for donating computer time. The initial research was carried out while the author held a Commonwealth Scholarship and additional financial assistance was provided by the Department of Geography, University of Sydney. Later work was supported financially by the University of Papua New Guinea.

References

Bridge, J. S. 1975. Computer simulation of sedimentation in meandering streams. *Sedimentology*, **22,** pp. 3–43.

Brush, L. M. and Wolman, M. G. 1960. Knickpoint behaviour in non-cohesive material: a laboratory study. *Geol. Soc. Amer. Bull.*, **71,** pp. 59–74.

Chow, V. T. 1959. *Open Channel Hydraulics.* McGraw-Hill, New York.

Colby, B. R. 1964. Practical computations of bed material discharge. *J. Hydraul. Div. Amer. Soc. Civ. Engrs.*, **90,** pp. 217–246.

Committee on Sedimentation, A.S.C.E. 1971. Sediment transportation mechanics, H. sediment discharge formulas. *J. Hydraul. Div. Amer. Soc. Civ. Engrs.*, **97,** pp. 523–570.

Gessler, J. 1970. Self-stabilizing tendencies of alluvial channels. *J. Waterways and Harbours Div. Amer. Soc. Civ. Engrs.*, **96,** pp. 235–249.

Gessler, J. 1971. Aggradation and degradation. In Shen, H. W. (ed) *River Mechanics*, H. W. Shen, Fort Collins, Colorado, Chapter 8.

Hales, Z. L., Shindala, A., and Denson, K. H. 1970. Riverbed degradation prediction. *Water Resour. Res.*, **6,** pp. 549–556.

Harvey, A. M. 1969. Channel capacity and the adjustment of streams to hydrologic regime. *J. Hydrol.*, **8,** pp. 82–98.

Hubbel, D. W. and Matejka, D. Q. 1959. Investigations of sediment transportation Middle Loup River at Dunning, Nebraska. *U.S. Geol. Survey Water Supply Paper*, 1476.

Kilpatrick, F. A. and Barnes, H. H. 1964. Channel geometry of Piedmont streams as related to frequency of floods. *U.S. Geol. Surv. Prof. Paper* 422E.

Komura, S. and Simons, D. B. 1967. Riverbed degradation below dams. *J. Hydraul. Div. Amer. Soc. Civ. Engrs.*, **93,** pp. 1–13.

Lewis, W. V. 1945. Knickpoints and the curve of water erosion. *Geol. Mag.*, **82,** pp. 256–266.

Livesey, R. H. 1965. Channel armouring below Fort Randall Dam. Procs. Federal Inter-Agency Sedimentation Conference, *U.S. Department of Agriculture Miscellaneous Publication 970*, pp. 461–470.

McBean, E. A. and Perkins, F. E. 1975. Convergence schemes in water profile computation. *J. Hydraul. Div. Amer. Soc. Civ. Engrs.*, **101,** pp. 1380–1384.

Meyer-Peter, E. and Müller, R. 1948. Formulas for bedload transport. *International Association for Hydraulic Structures Research, 2nd Meeting*, Stockholm, pp. 39–64.

Peters, J. C. and Bowler, R. A. 1968. Discussion of "Riverbed degradation below dams". *J. Hydraul. Div. Amer. Soc. Civ. Engrs.*, **94,** pp. 590–593.

Pickup, G. 1975. Downstream variations in morphology, flow conditions and sediment transport in an eroding channel. *Z. Geomorph. N.F.*, **19,** pp. 443–459.

Pickup, G. and Warner, R. F. 1976. Effects of hydrologic regime on magnitude and frequency of dominant discharge. *J. Hydrol.*, **29,** pp. 51–75.

Prasad, R. 1970. Numerical method of computing flow profiles. *J. Hydraul. Div. Amer. Soc. Civ. Engrs.*, **96,** pp. 75–86.

Schumm, S. A. 1961. Effect of sediment characteristics on erosion and deposition in ephemeral stream channels. *U.S. Geol. Surv. Prof. Paper 352-C.*

Schumm, S. A. 1971. Fluvial geomorphology: Channel adjustment and river metamorphosis. In Shen, H. W. (Ed.) *River Mechanics.* H. W. Shen, Fort Collins, Colorado, Chapter 5.

Tinney, E. R. 1962. The process of channel degradation. *J. Geophys. Res.*, **67,** pp. 1475–1480.

G. Pickup

Vanoni, V. A., Brooks, N. H., and Kennedy, J. F. 1961. Lecture notes on sediment transportation and channel stability. *W. M. Keck Laboratory of Hydraulics and Water Resources, California Institute of Technology Report KH-R-1.*

List of Symbols

d_m	mean size of bed-material;
d_{90}	size below which 90% of bed-material is finer;
g	gravity;
k	Strickler roughness coefficient;
k_b	Strickler bed roughness coefficient;
k_r	Strickler particle roughness coefficient $= 26/d_{90}^{\frac{1}{6}}$;
t	time;
v	velocity;
x	a longitudinal coordinate;
x_n, x_{n+1} etc.	specific points along the channel;
$x_n - x_{n+1}$	the section of channel between points x_n and x_{n+1};
y	water surface elevation;
A	channel area;
D	flow depth;
G	bed-material discharge by weight;
G'	bed-material discharge by weight per second per unit width measured under water;
Q	discharge;
Q_s/Q	bank friction correction factor in the Meyer-Peter and Müller equation;
R	hydraulic radius;
S	sediment stored in the channel, by weight;
S_f	friction slope;
S_0	bed slope;
W	channel width;
α	velocity head coefficient;
γ_s	specific weight of sediment;
γ_w	specific weight of water.

G. H. DURY

Professor of Geography and Geology
University of Wisconsin, Madison

4

Peak Flows, Low Flows, and Aspects of Geomorphic Dominance

In an influential paper, Wolman and Miller (1960) demonstrated quantitatively that a large proportion of the work done by rivers is effected during events of modest magnitude and high frequency. They drew similar conclusions with respect to the work of winds and waves. Leopold, Wolman, and Miller (1964) reconfirmed that floods, although providing large discharges, occur so infrequently that they contribute less water than do more frequent lower flows. They drew similar conclusions with respect to the delivery of sediment (cf. also Knox and coworkers, 1975). Events of moderate frequency, rather than catastrophic events, account for a large percentage of the delivery both of suspended and of dissolved load.

By a kind of osmotic process, reliance on frequent events of modest magnitude seems to have diffused itself through much of geomorphic thought—doubtless with the assistance of a retrospect and prospect of an almost infinite timespan within which geomorphic processes can operate. At present rates of fluvial denudation, some 25 to 35 million years would be required to reduce the existing continental landmasses. For the most part, we seem inclined to take the long view, and to concentrate on the very long-term effect of processes which in the short term of a human life produce insignificant change.

Again, Leopold, Wolman, and Miller conclude that channel-forming discharge on meandering streams with floodplains seems not to be associated with infrequent or catastrophic events, even though great erosion can occur during exceptionally large flows. In point of fact, channel-forming discharge can be identified both conceptually and empirically with discharge at the most probable annual flood (Gumbel, 1958; Dury, 1974), which is an event of modest magnitude and certainly an event of high frequency, corresponding as it does to the 1.0-year event in the partial-duration series (Langbein, 1949). Although great erosion can occur on some channels during exceptionally large flows, on other channels even the 1000-year flood seems incapable of producing perceptible effects (Dury, 1974). From this grey area of information and understanding, we are driven back to safe reliance on the reasonably well-understood effects of events of low magnitude and high frequency.

A specific test of the effectiveness of events of great magnitude and low frequency is made possible by an analysis of the discharge records for the 1972 floods, caused by Hurricane Agnes in the northeastern U.S.A. In contrast to most previous studies, the test will be designed in terms of magnitude-frequency, rather than of proportional duration. The result of analysis is to reconfirm the concept of dominance of river action by events of low magnitude and high frequency. The reconfirmation holds, even though the dominant frequencies appear to be lower for discharge of water than for delivery of sediment. The analysis produces equations whereby total discharge of water and sediment in a flood wave can be retrodicted from peak flow. The equations can be speculatively applied to the former discharges of the ancestors of streams which are now underfit.

However, while the reconfirmation of the dominance of river action by events of low magnitude and high frequency enables that action to be regarded in a strictly uniformitarian light, some workers on mass-movement and on sedimentation appear to be attracted by what may be styled neocatastrophic thought. Although any apparent opposition of uniformitarianism to neocatastrophism may well be fallacious, it does seem likely that the magnitude-frequency relationships of geomorphic events on interfluves may differ strongly from the corresponding relationships for channelled surface flow.

The Hurricane Agnes Floods

In mid-June 1972, Hurricane Agnes developed in the Caribbean. The centre of the system crossed the panhandle coast of Florida on 19 June, moved northeastward overland to clear Cape Hatteras by 22 June, and then swung inland again across New York City. An associated extratropical low, stalled for a time west of Agnes, helped to promote unusually heavy rains and floods, especially in the states of New York, Pennsylvania, Maryland, and Virginia. Peak flows exceeded the recurrence interval of 100 years at many stations. Sediment deliveries were also unusually high. While no physical definition of the term *catastrophe* seems to exist (but cf. Dury, in press), the Agnes event was identified politically as a disaster.

The data employed are taken or derived from Bailey, Patterson, and Paulhus (1975; especially Table A8). They relate to basins in the Delaware, Potomac, and Susquehanna River systems. Basin areas range from 0.2 to 16,700 km^2. Recurrence intervals of the floods involved range from more

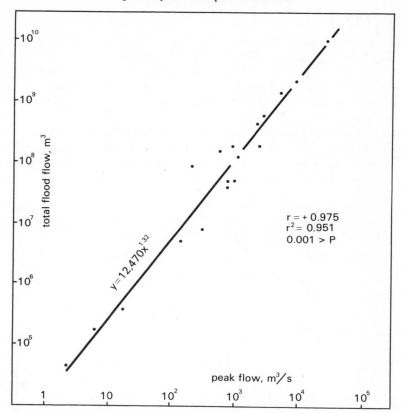

FIGURE 4.1 TOTAL FLOOD FLOW AS A FUNCTION OF PEAK FLOW, AGNES FLOODS

Each point plots, for an individual gauging station, total delivery of water in the flood wave against momentary delivery at the flood peak

than 100 to less than 2 years. Variation of recurrence interval seems completely irrelevant to the results obtained. Variation of basin size seems to be of minimal significance: its effect upon such characteristics as the form of flood hydrographs can be taken as integrated into the basic data.

Water Delivery

Total delivery of water (q, m^3/s) in The Agnes flood waves turns out to be a power function of peak flow (q_{max}, m^3/s, Figure 4.1). The equation is

$$q = 12.470 \, q_{max}^{1.32} \qquad (1)$$

With 95% of the variance in throughput explained, statistically speaking, by variance in peak flow, the equation provides a convenient means of retrodicting throughput of water from simple observations of peak discharges.

Sediment Delivery

Analyses under this subheading reject data on the Stave Run basin in the Potomac River system. This area of 0.2 km² was, at the time of the Agnes rains, undergoing urban development, and in consequence displayed highly aberrant relationships between delivery of sediment and delivery of water. The analyses retain data on two other small basins, respectively 0.8 and 2 km² in area.

Analysis indicates that total suspended sediment discharge (q_{sed}, tonnes) during the passage of a flood wave is close to a linear function of total delivery of water (q, m³) (Figure 4.2). A parallel relationship holds for mean annual totals. The relevant equations are:

$$q_{sed} = 0.002 \ q^{0.935} \tag{2a}$$

or

$$= 0.0006 \ q \tag{2b}$$

for flood waves, and

$$q_{sed} = 0.0008 \ q^{0.906} \tag{3a}$$

or

$$= 0.00013 \ q \tag{3b}$$

for mean annual totals.

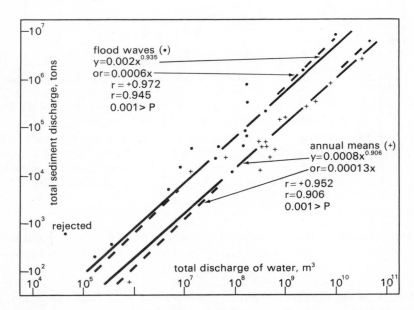

FIGURE 4.2 TOTAL SEDIMENT DISCHARGE AS A FUNCTION OF TOTAL DISCHARGE OF WATER, AGNES FLOODS, AND OF MEAN ANNUAL DISCHARGE OF WATER

The departure from unity in equations (2a) and (3a) is confirmed to some extent by regression of mean suspended-sediment concentration upon discharge of water (Figure 4.3). Both for flood waves and for annual discharges, concentration declines slightly as magnitude of flow increases. In neither case, however, is the decline statistically significant.

Direct regression of sediment delivery per flood wave ($\sum q_{sed}$) upon peak flow (q_{max}) explains, in the statistical sense, almost 97% of the variance in sediment delivery (Figure 4.4). In the same units as before, the relevant equation is

$$\sum q_{sed} = 8.36 \, q_{max}^{1.3} \qquad (4)$$

That is to say, throughput of sediment can be retrodicted from peak discharge of water.

There is no *a priori* reason to suppose that equations in the form of equation (4) cannot be constructed for other regions, although considerable regional variation is to be expected, especially in the value of the constant.

Application to the Magnitude-Frequency Scale

Equations (1) and (4) can be applied to a synthetic record of magnitude-frequency of streamflow. The essential requirement in constructing a synthetic record is the slope of a Gumbel graph of Type I probability. The slope is defined by the ratio between two floods of different recurrence interval. Calculations are simplified if a numerical value is assumed for discharge at one of the two recurrence intervals.

Leopold, Wolman, and Miller (1964) provide, for the eastern half of the United States, a dimensionless graph of depth ratio against discharge ratio. Discharge ratios read off for the recurrence intervals of 1.5, 5, 10, 25, and 50 years can be generalized by

$$q/q_{1.58} = 0.8427 \text{ r.i.} \qquad (5)$$

where r.i. is a selected recurrence interval (annual series) and $q/q_{1.58}$ is the ratio of discharge at that recurrence interval to discharge at the 1.5-year flood. However, the recurrence interval of 1.58 years is to be preferred as the basis of reference. Moreover, equation (5) produces a concave-upwards graph on Gumbel paper, indicating Type II probability. For purposes of the present enquiry, which assumes Type I probability and a straightline graph, the ratio of 4.4 between q_{50} and $q_{1.58}$, read from the graph of Leopold, Wolman, and Miller, has been adopted. It is of course based on the analysis of real streams. In addition, the ratio of 0.0175 between mean annual discharge q and natural bankfull discharge $q_{1.58}$, also read from the graph, has been retained. Calculations were based on an assumed value of $10 \text{ m}^3/\text{s}^{-1}$ for $q_{1.58}$, although, needless to say, the actual numerical value is immaterial to the general argument.

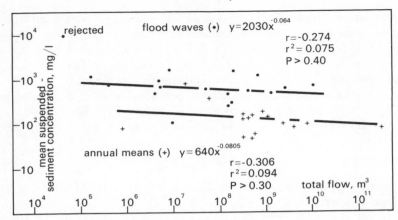

FIGURE 4.3 MEAN SUSPENDED-SEDIMENT CONCENTRATION AS A FUNCTION OF TOTAL DISCHARGE OF WATER, AGNES FLOODS AND ANNUAL MEANS

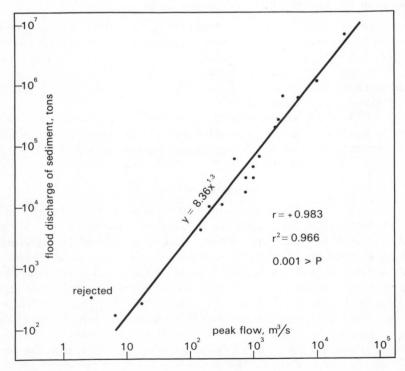

FIGURE 4.4 TOTAL DISCHARGE OF SEDIMENT AS A FUNCTION OF PEAK FLOW, AGNES FLOODS

The Synthetic Record: Simple Annual Series

The simple annual series permits no more than one overbank flow in a given water year. Since peak discharge in some years fails to exceed $q_{1.58}$, the number of years with overbank flow is less than the total of years in the series.

For synthetic series of record, respectively 4, 9, 24, 99, and 199 years in length, and with the indicated slope of the Gumbel graph, annual peak discharges have been read off against computed recurrence intervals. The lengths of record are selected to produce $n + 1 = 5, 10, \ldots, 200$. Summation of peak discharges at flows exceeding $q_{1.58}$ permits a comparison with total flow during a record period, as determined from the stated ratio of $\bar{q}$ to $q_{1.58}$.

The percentage of total delivery of water which is effected at overbank stages ($\% \sum q$) in the simple annual series can be expressed as

$$\% \sum q = 4.83t^{0.017} \tag{6}$$

where t is the length of record in years. As shown in Table 4.1, the percentage rises slowly as length of record increases, but is still below 10% for the 199-year record.

Application of equation (4) to $\bar{q}$ permits annual mean delivery of sediment to be determined. The percentage of sediment delivered by overbank flows ($\% \sum$ sed) becomes

$$\% \sum \text{sed} = 32.7t^{0.037} \tag{7}$$

rising from 34.4% for a 4-year record to 39.7% for a 199-year record (Table 4.1). At this rate, a record some 100,000 years in length would be

TABLE 4.1 SYNTHETIC RECORDS

Length of record, years	4	9	24	49	99	199
Simple annual series						
Number of overbank events	3	6	15	31	63	126
Percent overbank delivery of water	5.6	6.1	6.8	7.4	8.0	8.5
Percent overbank delivery of sediment	34.4	35.5	37.0	38.2	39.1	39.7
Partial-duration series						
Number of overbank events, including random additions: leads to equation (8a)	4	7	21	40	81	153
Number of overbank events, from equation (8c)	3	8	20	42	84	169
Percent overbank delivery of water	10.5	7.7	10.6	13.2	10.4	—

required, to ensure delivery of 50% or more of suspended sediment by overbank flows of the simple annual series.

The Synthetic Record: Partial-duration Series

A partial-duration series, with more than one overbank flow occurring in some years, is much more likely than is the simple series to resemble happenings in the real world. The problems of identifying number and magnitude for floods independent of, and smaller than, annual floods is precisely the opposite of that faced by Frost and Clarke (1973). These two authors undertook the extension of known partial-duration series. Here, a partial-duration series needs to be synthesized from the simple annual series.

Series were synthesized for the hypothetical situation where at least 50% of sediment delivery is effected by overbank flows. The object was to examine the magnitude-frequency relationships of discharge of water and sediment for the situation described. It was assumed that the required overbank flows, additional to those of the simple annual series, are randomly distributed through the range of annual floods, from the discharge next above $q_{1.58}$ to that of the second-ranked annual flood. For each record series from 4 to 99 years, additional flows were randomly selected from the indicated range, being added successively to the cumulative overbank total of the simple annual series, until the 50% mark of overbank sediment delivery was reached or passed.

A slight defect of the method, in practice, was the elimination of a once-selected flood from further consideration in the series in hand. Against this, the next higher and lower floods retained their own equal chances of selection. Trials for the periods of record 4 to 99 years produced the following power-functional relationship between number of overbank events ($n_{o/b}$) and length of record (t years):

$$n_{o/b} = 0.97 t^{0.956} \qquad (8a)$$

This result appears satisfactory, to the extent that the simple annual series leads to a similar relationship, namely:

$$n_{o/b} = 0.6236 t^{1.0004} \qquad (8b)$$

It may be that graphical determination of flood magnitude is responsible for the departure from unity of the exponent in equation (8b), and that a combination of such determination with the accidents of sampling for short series is similarly responsible in equation (8a). If so, the equations should be rewritten as

$$n_{o/b} = 0.85 t \qquad (8c)$$

for the partial duration series, and as

$$n_{o/b} = 0.63t \qquad (8d)$$

for the simple annual series.

For the 199-year partial-duration series, equation (8a) predicts 153 overbank flows. But since $q_{1.58}$ is not exceeded in 73 of the 199 years, 27 extra inundations need to be distributed among the 126 years in which $q_{1.58}$ is exceeded. On the average, two inundation years in nine must experience two independent inundations each. Equation (8c) calls for 43 extra inundations, or an average of two independent inundations in one inundation year in three.

This latter proposition may perhaps seem excessive, but the former is by no means impossible for some river systems. Calculations of the percentage overbank discharge of water for the floods which lead to equation (8a) produce no particular pattern in relation to length of record (Table 4.1). In the range from 4 to 99 years, this percentage averages about 10.5. Since the situation under discussion is the hypothetical, and possibly extreme, one where overbank flows effect 50% of sediment delivery, it may be reaffirmed that by far the greater part of total delivery of water is effected by flows at bankfull stage and below.

Because of the increase in suspended-sediment concentration at-a-station, commonly with the second to third power of discharge as flow increases, and of the relationship indentified in equation (4), the proportion of sediment delivered by overbank flows should be considerably greater than the proportion of water so delivered (compare also Table 4.1). The proportional durations given by Leopold, Wolman, and Miller (1964) for the delivery of given percentages of total sediment cannot be directly translated into the terms of the magnitude-frequency scale. However, durations as low as 4 days per year for the delivery of 50% of total sediment suggest that, on some streams at least, overbank flow can be responsible for a sizeable, if not a major, fraction of total delivery. As already noticed, some 40% is predicted as delivered overbank, in the region of the Agnes floods, by peak flows of the simple annual series. Nevertheless, we are still dealing with flows of modest magnitude and high frequency, even if not necessarily with bankfull or lower flows.

Figure 4.5 is drawn for the situation where the synthetic partial-duration series of 99 years in length results in the delivery of at least 50% of total suspended sediment by flow at overbank stage. For the indicated situation, 71% of total delivery is effected by flows (including below-bank flows) of 4 years recurrence interval or less, while only 25% is effected by flows of recurrence interval 5.26 years or greater. Flows in the top half of the overbank range effect no more than 35% of total sediment delivery. Discussion of comparative effectiveness in this hypothetical situation must obviously depend on precisely what is meant by *modest magnitude* and *high*

G. H. Dury

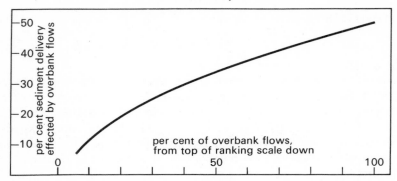

FIGURE 4.5 PER CENT OVERBANK DELIVERY OF SEDIMENT AS A FUNCTION OF PER-
CENTAGE OF OVERBANK FLOWS, FOR RANDOMLY EXTENDED 99-YEAR PARTIAL-
DURATION SERIES, WHERE 50% OF SEDIMENT IS DELIVERED OVERBANK

frequency. But it still seems unlikely that a major fraction of sediment delivery will be effected by major overbank flows, including catastrophic flows, as against flows which fall cumulatively within a modest range both of magnitude and of frequency.

Application to Former Rates of Fluvial Denudation

Some of the conclusions reached above can be applied, although in an admittedly speculative fashion, to the question of denudation rates at the time when the ancestors of presently underfit streams were flowing.

Refinement of the retrodiction of former discharges from the dimensions of former channels indicates that a ratio of 5/1 between former and present bedwidth and meander wavelength should correspond to a ratio of about 18/1 between former and present channel-forming discharge. A dimension ratio of 10/1 would imply a discharge ratio of about 66/1 (Dury, 1976). Equation (2a) predicts for these discharge ratios a sediment delivery some 15 to 50 times as great, at the bankfull stage, as that of today. But, as Judson (1973) has effectively urged, present-day denudation rates should be discounted by about half for former times, in order to allow for the influence of man on erodibility. We can thus arrive at a delivery of sediment, by former streams at their bankfull stage, some 7.5 to 25 times as great as existing rates.

It is however not possible to multiply mean annual delivery of sediment by the ratios applicable to delivery at bankfull stage, since nothing is known of the slope of former graphs of magnitude-frequency. But something can be done with mean annual runoff. I have reasoned elsewhere (Dury, 1965) that the former bankfull discharges can be accommodated in climates somewhat cooler and distinctly wetter than those of today. A temperature reduction of some 5.5 deg. C in combination with an increase in precipitation by a factor of 1.5 to 2.0, would cause mean annual runoff to increase, in many midlatitude climates, by a factor in the approximate range 5 to 10. Application of the exponent in equation (2a) retrodicts a mean annual sediment delivery in the approximate range 4.5 to 8.5 times as great as that of today, or in the range 2.25 to 4.25 at the half-rate discount.

Specific confirmation of considerable changes in mean annual runoff at the relevant time comes from the work of Webb and Bryson (1972). These authors have used the pollen record to reconstruct climatic change in the Upper U.S. Midwest between about 12,000 and 10,000 B.P. Their findings are expressed in quantitative terms. They concluded that the excess of precipitation over evapotranspiration was, at one site, more than twice as great at the beginning of the interval than at the end—380 mm against 160 mm. At a second site the difference is sevenfold—350 against 50 mm. A third site provides information on the interval 10,500 to 10,000 B.P., during which the precipitation excess was reduced from 240 to 130 mm. If the changes in precipitation excess are equated with changes in mean annual runoff, then runoff was up to seven times as much in 12,000 B.P. as it was in 10,000 B.P. Application of the exponent in equation (2a) retrodicts that mean annual sediment delivery can have been about six times greater at the beginning of the interval than at the end. Allowance for the discounting factor reduces the disparity into the range from present-day rates to rates about three times as great.

Although Schumm (1968, 1969) has attempted to demonstrate that, in some climatic conditions, an increase in precipitation could actually result in a decrease of sediment delivery, his conclusions have been challenged, for the field area and timespan concerned, by Bowler (1971). Schumm inferred that some large sandy paleochannels of the Riverine Plain of southeastern Australia carried lower discharges than those of smaller clay-lined channels, and reasoned accordingly about rates of sediment transport. However, Bowler has shown that former discharges, even in the sandbed channels, were greater than those of today.

The published record of the rate of deposition of terrigenous sediment includes marked variations, in the appropriate direction and at the appropriate time (Broecker, Turekian, and Heezen, 1958; Broecker, Ewing, and Heezen, 1960). For the Gulf of Mexico, the changes of rate identified by these authors can be separated in time from the effects of meltwater discharge, which had very greatly declined by 11,500 B.P. (Kennett, 1974). Thus, even though the magnitude-frequency relationships of former streams cannot be reconstructed, a considerable reduction in denudation-rate can be inferred, as associated with the reduction of streams to underfitness.

Problems of Geomorphic Dominance in Mass-movement

It is tempting to suppose that the magnitude-frequency analysis that works so well for discharge of water and sediment, for single rainstorms, and for monthly or annual precipitation, should work equally well for the mass-movements that deliver rock waste from slopes into streams. But there is developing in the background, mainly so far among geologists, what may be called a neocatastrophic school of thought (reviewed in some detail by Dury, in press). In some geomorphic contexts, a rethinking of the concept of dominance by events of low magnitude and high frequency is possibly in order (Dury, 1972; for a discussion of mechanics, see Scheidegger, 1975).

Selby (1974) finds that the dominant land-forming events in the hills of North Island, New Zealand, are rainstorms with a recurrence interval of 30 years on pasture and 100 years on forest. Major landslides and major eruptions of tephra can deliver more material into fluvial systems, in a single event, than rivers strip from an entire continent in a whole year (Dury, in press). Slow mass-movement, dominated by low magnitudes and high frequencies, may supply only some 10 to 20% of the material moved by rivers. The remainder comes from surface wash, throughflow, channel erosion (on which see Roehl, 1962), and rapid mass-movement. If this last process approaches or attains dominance in the delivery of sediment, even on the subcontinental regional scale, then the concept of the primacy of events of low magnitude and high frequency will not prove transferable from channels to interfluves. If storage is to be discounted, it must be supposed that rivers remove landslide debris almost at once. To the extent that storage is allowed for, the magnitude–frequency scale must be stretched to accommodate events of great magnitude and low frequency. Some kind of model is needed, of the impact of events (rapid mass-movements) that are assessed in one range of frequency, upon systems (stream channels) where a very different range of frequency is observed.

Summary and Conclusions

Analysis of the floods promoted by Hurricane Agnes in 1972 shows that total delivery of water and of sediment in a flood wave can be retrodicted from peak flood flow. The relationships are independent of basin area and of recurrence interval.

Synthetic annual and partial-duration series can be used, in conjunction with the equations derived from the analysis of the Agnes records, to derive proportional values for the delivery of water and sediment by overbank flow. It seems likely that, in the general case, major overbank flows fail to deliver a major fraction of total sediment, while discharge of water is effected mainly by flows at bankfull stage and below.

Application of the equations to the former discharges of streams that are now underfit suggests that former mean annual discharge of sediment can have been about 2 to 4 times what it is today.

The concept of dominance by events of low magnitude and high frequency may not be transferable from channels to interfluves, where events of very high magnitude and very low frequency include great mass-movements. Two contrasted systems, operating on highly contrasted magnitude–frequency scale, interact at the debris/channel interface.

References

Bailey, J. F. O., Patterson, J. L., and Paulhus, J. L. H. 1975. Hurricane Agnes rainfall and floods. *U.S. Geol. Surv. Prof. Paper*, 924.

Bowler, J. M. 1971. Pleistocene salinities and climatic change. In Mulvaney, D. J., and Golson, J. (Eds.), *Aboriginal Man and Climate in Australia*. Australian Nat. U. Press, Canberra, pp. 47–65.

Broecker, W. S., Ewing, M., and Heezen, B. C. 1960. Evidence for an abrupt change in climate close to 11,000 years ago *Amer. J. Sci.*, **258**, pp. 429–448.

Broecker, W. S., Turekian, K. K., and Heezen, B. C. 1958. The relations of deep-sea sedimentation rates to variations in climate. *Amer. J. Sci.*, **256**, pp. 503–517.

Dury, G. H. 1965. Theoretical implications of underfit streams. *U.S. Geol. Surv. Prof. Paper 452-C.*

Dury, G. H. 1972. Some current trends in geomorphology. *Earth-Sci. Rev.*, **8**, pp. 45–72.

Dury, G. H. 1974. Magnitude-frequency analysis and channel morphometry. In Morisawa, M. (ed.), *Fluvial Geomorphology*. State University of New York, Binghamton, pp. 91–121.

Dury, G. H. 1976. Discharge prediction, present and former, from channel dimensions. *J. Hydrol.*, **30**, pp. 219–245.

Dury, G. H. In press. Neocatastrophism? (expected to appear in *International Quaternary Symposium Volume*, Brazilian Academy of Sciences).

Frost, J., and Clarke, R. T. 1973. Estimating the T-year flood by the extension of records of partial duration series. *Hydrol. Sci. Bull.*, **17**, pp. 209–217.

Gumbel, E. J. 1958. Statistical theory of floods and droughts. *J. Inst. Water Engrs*, **12**, pp. 157–184.

Judson, S. 1973. Erosion of the land. In R. Tank, R. (Ed.), *Focus on Environmental Geology*. Oxford U.P., New York, pp. 167–180.

Kennett, J. P. 1974. Latest Pleistocene melting of the Laurentide ice sheet recorded in deep sea cores from the Gulf of Mexico. *Geol. Soc. Amer. Abstracts*. 1974 Annual Mtg., pp. 1015.

Knox, J. C., Bartlein, P. J., Hirschboeck, K. K., and Muckenhirn, R. J. 1975. *The Response of Floods and Sediment Yields to Climatic Variation and Land Use in the Upper Mississippi Valley*. University of Wisconsin-Madison, Institute for Environmental Studies Center for Geographic Analysis, I.E.S. Report no. 52.

Langbein, W. B. 1949. Annual floods and the partial-duration series. *Trans. Amer. Geophys. Union*, **30**, pp. 879–881.

Leopold, L. B., Wolman, M. G., and Miller, J. P. 1964. *Fluvial Processes in Geomorphology*. Freeman and Company, San Francisco.

Roehl, J. W. 1962. Sediment source areas, delivery ratios and influencing morphological factors. *Internat. Assoc. Sci. Hydrol. Pub.*, **59**, pp. 202–213.

Scheidegger, A. E. 1975. *Physical Aspects of Natural Catastrophes*. Elsevier, Amsterdam.

Schumm, A. A. 1968. River adjustment to Altered Hydrologic Regimen— Murrumbidgee River and Palaeochannels. *U.S. Geol. Surv. Prof. Paper 598.*

Schumm, S. A. 1969. River metamorphosis. *J. Hydraul. Div. Amer. Soc. civ. Engrs.*, **95**, pp. 255–274.

Selby, M. J. 1974. Dominant geomorphic events in landform evolution. *Bull. Internat. Assoc. Engng. Geol.*, No. 9, pp. 85–89.

Webb, T., and Bryson, R. A. 1972. Late- and postglacial climatic change in the Northern Midwest, U.S.A.—quantitative estimates derived from fossil pollen spectra by multivariate statistical analysis. *Quaternary Research*, **2,** pp. 70–115.

Wolman, M. G., and Miller, J. P. 1960. Magnitude and frequency of forces in geomorphic processes. *J. Geol.*, **68,** pp. 54–74.

E. MYCIELSKA-DOWGIAŁŁO

Institute of Geography
University of Warsaw

5

Channel Pattern Changes During the Last Glaciation and Holocene, in the Northern Part of the Sandomierz Basin and the Middle Part of the Vistula Valley, Poland

The northern part of the Sandomierz Basin and the Vistula Valley between Tarnobrzeg and Sandomierz (Figure 5.1) largely attained its present relief at the time of the Last Glaciation and during the Holocene. During the Riss Glaciation, the waters of the Vistula and its right bank tributaries were flowing north-eastward along the southern part of the area investigated, within the limits of the terrace-surface lying at a height of 10 to 15 m above the present bottom of the Basin.

The fluvial deposits preserved beneath the present floor of the Vistula Valley and beneath the northern part of the Sandomierz Basin (Figure 5.2) suggest that the deepest parts of the fossil valleys in this area were formed during the decline of the Riss Glaciation and during the Eems Interglacial. These valleys are cut in the Miocene Krakowiec clays, they extend to a depth of 20–24 m below the present bottom level of the valley, and in the base of the valleys there are several metres of a thick gravel series.

Two surface sulphur mines allow a detailed picture of the formation of alluvial series to be obtained. From the analysis of the deposits, their stratification, grain size distribution, and rounding of quartz grains of the sandy fraction, as well as of the petrographic and mineralogical composition, the palaeo-hydrological regime of the rivers was characterized in particular phases of development. Using the methods proposed by Dury (1965), Schumm (1965, 1968), and Falkowski (1965, 1967, 1970), the sequence of

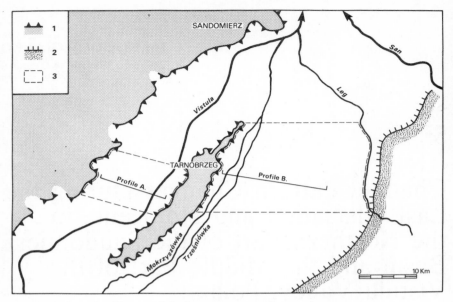

FIGURE 5.1. SKETCH OF THE STUDY AREA

1, Edge of the Sandomierz Upland and the Tarnobrzeg Monadnock; 2, edge of the accumulation terrace of the Riss Glaciation; 3, areas shown in the geomorphological maps (Figures 5.3 and 5.4). Profiles A and B are detailed in Figure 5.2

changes in the river channel pattern was deduced and related to changes of climate and of human activity of the last centuries.

The lowest gravel series, infilling the deepest valley form beneath the present floor of the Vistula Valley, is characterized predominantly by thick sets of tabular cross-stratification with only slight dispersion of laminae inclination, and with very poor sorting of the deposits (Mycielska-Dowgiałło, in press). The character of the deposit suggests that it was accumulated by high discharges and a braided channel pattern.

Numerous syngenetic frost structures of the ice wedge and involution types are present in the bottom gravel series. This indicates that the accumulation of the series took place during the presence of permafrost. The occurrence of sporadic, large blocks of rock up to 2.5 m in diameter, at a distance from the valley slopes, indicates a very thick seasonal ice cover which transported the above mentioned stone blocks during the spring thaw.

Petrographic analysis of the sandy fraction of the gravel deposits showed that crystalline-rock fragments were much more frequent in these deposits than in the deposits of the late Last Glaciation and the Holocene. This indicates that weathered material was supplied to river beds from the catchment area. Highly dynamic rivers affected by a short and violent spring thaw, were overloaded with the weathered materials, and these accumulated

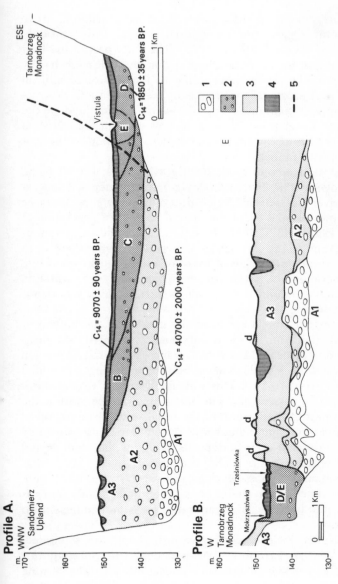

FIGURE 5.2 SYNTHETIC GEOLOGICAL CROSS-SECTIONS OF THE VISTULA VALLEY AND THE NORTHERN PART
OF THE SANDOMIERZ BASIN

The main stages of river channel development during the Last Glaciation and the Holocene are indicated. 1, Gravel series; 2, sandy-gravel series; 3, sandy and sandy-silty series; 4, mud series; 5, range limit of the Vistula Valley at the time of Last Glaciation. A₁, channel deposits of the early Pleniglacial; A₂, channel deposits of the late Pleniglacial; A₃, channel and high-water deposits from the early stage of the Late Last Glaciation; d, levees, with dune sands in their ceiling; B, channel deposits of the first meander generation from the Late Last Glaciation (Alleröd—Younger Dryas); C, channel deposits of the second meander generation from the beginning of the Holocene (Preboreal—Atlantic); D, channel deposits of the third meander generation, of the second part of Holocene (Subboreal—Subatlantic); E, channel deposits of contemporaneous rivers. Reproduced by permission of Wydawnictwa UW

as thick-grained deposits, badly sorted, with poorly rounded quartz grains within the sandy fraction.

In the area of Middle Poland, at the time of the Last Glaciation, arctic conditions prevailed only during the Pleniglacial (Dylik, 1966; Środoń, 1967; Jahn, 1975; Środoń, 1972; Goździk, 1973). The early part of the Last Glaciation was characterized by a cool and humid climate, which was hardly frosty (Kuydowicz-Turkowska, 1975). According to the nature of alluvial deposits filling the deepest erosional formations of the Vistula and Wisloka Valleys, one might suppose that these deposits were accumulated in the Pleniglacial of the Last Glaciation. Wisloka is now a rightsided Vistula affluent, which during the Last Glaciation was flowing separately on the east side of the Tarnobrzeg Monadnock.

The lack of erosional surfaces within the lower gravel series, as well as the occurrence of frost wedges within every successive overlying deposit, suggest that the accumulating river had a decidedly positive alluvial balance. Of basic importance was the rapid denudation of the upland, devoid of vegetation at that time (Falkowski, 1965, 1967, 1970). In spite of insignificant precipitation of the order of 200 mm yearly (Mycielska–Dowgiałło, 1969) there were high river flows on the spring as a result of thawing and of quick runoff due to the presence of permafrost. Rivers of important and very variable discharge carried enormous amounts of alluvial material, which was deposited in an unsorted state.

During a period of relative amelioration of the climate in the middle of the Pleniglacial (interstadials Hengelo and Denekamp), a partial stabilization of the basin level took place and there was an increase of lateral erosion activity. The initial phase of Vistula Valley bottom widening was dated 40,000 ± 2000 years B.P. (Figure 5.2). This was based upon ^{14}C analysis of organic materials, found at a level of 13 m below the surface, within the gravel series of the flat part of the bottom of the fossil valley. The period of relative increase in temperature and humidity of the climate is also shown by palynological analysis of the fine-grained deposits filling the flat, shallow and extensive forms—the remains of periodic channels (Krauss, Mycielska-Dowgiałło and Szczepanek 1965; Mycielska-Dowgiałło, 1967, 1969, 1972a, 1972b). The presence of ice wedges and involutions in this series is proof, however, that the increase in temperature was not sufficient to eliminate the permafrost.

The Vistula and Wisloka Valleys, which had hitherto flowed separately along opposite sides of the Tarnobrzeg Monadnock, fused during this period, most probably as a result of intensive lateral erosion. The Vistula Valley, together with the northern part of the Sandomierz Basin, was then covered with the braided net of the Vistula and Wisloka channels which flowed north-eastward towards the San Valley. They formed together, in the upper part of the Pleniglacial, one common, wide valley, about 20 km wide. Such an important extension of the valley at that time, the largest in the whole history of Pleistocene fluvial relief development, was probably connected with marked thermal erosion by the braided river channels when permafrost was present. This system of shallow, wide channels allowed for a higher absorption of thermal solar energy than in the case of deep but narrow meander channels. An important lateral development of river bottoms during the Last Glaciation is also noted in all the valleys of the Sandomierz Upland.

The period of the upper part of the Pleniglacial was marked in the Vistula Valley and in the Sandomierz Basin by the recurrence of the prevailing process of accumulation, together with further intensive active lateral erosion. The diminished significance of gravel grains (sandy gravel series) is ascribed by the author to the gradual widening of the valley bottom (Figure 5.2). This widening process greatly reduced the dynamics of flood waters and consequently the transportation capacity of rivers. Climatic conditions obtaining during accumulation of that series must have been generally much akin to those which prevailed at the time of deposition of bottom gravels.

In the literature, data on average temperatures at particular periods of the Last Glaciation generally suggest a higher average temperature for July in the lower part of the Pleniglacial (>5 °C) than for its upper part (<5 °C) (Hammen, Maarleveld, Vogel, and Zagwijn, 1967; Środoń, 1972). However, the nature of the gravel series and the sandy gravel series in the Vistula Valley does not support this generalization. The ice wedges in the upper sandy gravel series, less numerous and narrower than those found in the lower gravels, although they are at the same time of a much wider perpendicular span, suggest alternatively, that the climate ameliorated during the upper part of the Pleniglacial of the Last Glaciation. This is concordant with the thesis advanced by Dylik (1974) that the periods of formation of glaciers are characterized by a humid climate and cool only in so far as to give solid phases of precipitation. As is well known, the upper part of the Pleniglacial was the period of widest extent of the ice sheet of the Last Glaciation (Mojski, 1968; Geyh and Rohde, 1972).

However, a further point of view, recently advocated is that the period of largest spread of the Last Glaciation ice sheet took place at the time of transition of the climate from cool–humid to cool–dry (Gritchuk, 1969, 1973). If this hypothesis is to find its confirmation in further research, it is possible that the reduced-river activity in the upper part of the Pleniglacial may have been produced not only by extension of the valley bottoms, but also by the drought.

In the area studied, a gradual warming of the climate took place in the later part of the Last Glaciation (Wasylikowa, 1964), inducing a thickening of vegetation as well as a slow reduction of permafrost (Goździk, 1973). Therefore the water retention in the ground and in the vegetation was increasing, so that in spite of the probable increase in annual precipitation, the energy of rivers at high discharges was decreasing.

FIGURE 5.3 SKETCH OF THE VISTULA VALLEY FROM THE VICINITY OF TARNOBR-
ZEG, WITH INDICATION OF TRACES OF MEANDERS AND BRAIDED RIVER CHANNELS

I, Valley bottom formed during the Late Last Glaciation with indication of traces of
braided river channels; II, valley bottom formed during the period from the Late
Last Glaciation up to present times, with indication of meander ox-bows; II_a, zone of
first meander generation; II_b, zone of second meander generation; II_c, zone of third
meander generation; III, valley bottom formed by the present river of braided
pattern channel tendencies. 1, edge of valley; 2, dunes; 3, alluvial fans

The main denudation process characterizing the slopes and uplands was
slope wash (Kuydowicz-Turkowska, 1975). Traces of this activity are visible
in the diluvial surfaces occurring on the slopes of the Vistula Valley and in
the vicinity of the Tarnobrzeg Monadnock, as well as in the construction of
alluvial fans at the mouths of lateral valleys. The process of slope wash
supplied the river with fine grained material, which was later accumulated in
bars between the braided river channels (Figure 5.3). At this period, in the
northern part of the Sandomierz Basin extensive alluvial fans were formed
by rivers flowing from south to northeast, towards the San Valley (Figure
5.4 and 5.5). The type of deposit accumulated at that time and its stratification (well-sorted sands and muds frequently of horizontal lamination or
fine-set cross stratification) indicate a greatly diminished carrying power of
waters as compared with the Pleniglacial phase of the Last Glaciation. A
high percentage of well-rounded grains within the sandy fraction indicates
that intensive aeolian processes were active at the time on the uplands and

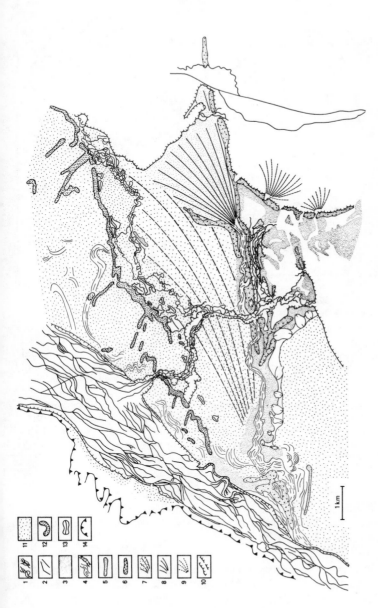

FIGURE 5.4 GEOMORPHOLOGICAL SKETCH OF THE MOKRZYSZÓWKA AND LĘG WATERSHED

1, Traces of river-channels preceding the river regulation; 2, regulated river channels; 3, valley bottoms with Holocene accumulation; 4, traces of river channels within the Pleistocene terrace; 5, levees; 6, levees with wind-blown sands in the ceiling; 7, alluvial fan of the oldest generation; 8, alluvial fan of the intermediate generation; 9, alluvial fan of the youngest generation; 10, edge of the Pleistocene terrace; 11, Pleistocene terrace of the Late Last Glaciation; 12, dunes; 13, depressions of deflation; 14, edge of the Tarnobrzeg Monadnock

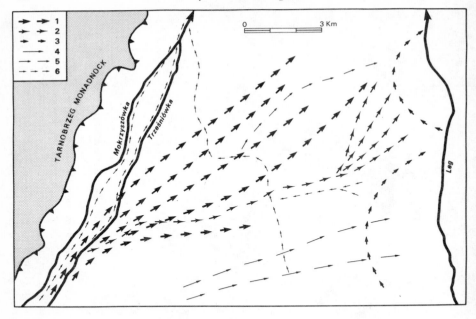

FIGURE 5.5 MAP OF THE MOKRZYSZÓWKA AND ŁĘG WATERSHED, INDICATING THE
SUCCESSIVE STAGES OF DEVELOPMENT OF THE VALLEY NETWORK

1, Flow of the waters of the eastern braided channels of the Vistula and Wisłoka
during the Pleniglacial of the Last Glaciation; 2, flow of the Mokrzyszówka and
Trześniówka waters operating since the Late Last Glaciation up to the beginning of
the present era; 3, flow of the San waters along the great arches of the meanders
(Alleröd–Younger Dryas); 4, flow of the left side tributaries of the San along the
southern limit of the Pleistocene terrace surface, still functioning to the eighteenth
century; 5, first stage of waterflow from the area of the terrace surface, eastward to
the San valley, with the meander channel pattern; 6, second stage of waterflow from
the terrace surface area, in the northward direction to the Mokrzyszówka Valley
(falling into Vistula), with the braided channel pattern. Reproduced by permission of
Wydawnictwa UW

higher terrace surfaces. This is confirmed by the character of the surface
texture of the quartz grains of the sand as analysed under the electron
microscope (Mycielska-Dowgiałło and Krzywoblocka-Laurow, 1975).

In the area of the Mokrzyszówka-Łęg watershed (Figure 5.5) there are a
few generations of alluvial fans from that time. These fans are younger and
smaller to the east and they occur on ever lower hypsometric levels. This is
the result of the tendency, also observed in the Vistula Valley as well as in
the northern part of the Sandomierz Basin, for the river channels to shift
eastward during the Last Glaciation and the Holocene.

The End of the Last Glaciation

At the end of the Last Glaciation, as a result of further changes in climatic conditions and associated alteration of vegetation density, there followed a slow transformation of valley bottoms with braided channel pattern into a system of meandering valleys. Contemporaneous with that transformation was a gradual incision of valleys and the lowering of valley bottoms. The ox-bows of the first generation of meanders were characterized by the largest radii of curvatures and channel widths against an insignificant depth and perpendicular reach of redeposition of alluvia in river channels. The largest meander forms are preserved in the San Valley (Figures 5.4 and 5.5). Szumański connects their development with the phase of erosion previous to the Alleröd (Szumanski, 1972; Falkowski and Szumański, 1975). As a result of the increasing vegetation density the supply of decayed rock material to the rivers diminished significantly, and this changed the balance of alluvia to a negative one (Falkowski, 1967). The less well loaded rivers started to incise into the substratum, and deepening their channels they produced a gradual lowering of valley floors. Although particular processes dominate particular periods other processes also occurred. Thus in the Holocene, during the meandering phase of river development, erosion processes were accompanied by accumulation, so that a 10-m thick series of alluvia occurs in the Vistula Valley.

Meanders of the oldest generation are very broad, up to 500 m, in the San Valley (Szumański, 1972). In the Vistula Valley there are no similarly extensive forms, though the period of development of the largest of them is also connected with the Alleröd and the younger Dryas (Figure 5.3; Mycielska-Dowgiałło, 1972). Meanders of that generation are characterized by shallow channels, of low perpendicular reach of alluvial redeposition. This is clearly seen on the walls of the sulphur surface mine in Piaseczno, within the bottom of the Vistula Valley, where fine-grained sandy alluvial series of that generation of meanders, 3–4 m thick, are inserted into sandy gravel deposits of the Pleniglacial. The radius of meander curvatures of the mentioned generation is also substantial, up to 1100 m in the San Valley (Szumański, 1972).

It is possible that this oldest generation of meanders constitutes a transitory form between a braided river pattern and a meandering one. This is also suggested by the type of alluvia, among which prevail deposits from the channel facies: fine-grained sand and sandy muds, with a large proportion of well rounded quartz grains, covered discontinuously by a layer of loamy muds (Szumański, 1972).

The ox-bows of the following meander generations became gradually narrower, with smaller radius of curvature and greater depth of alluvial redeposition—up to 10 m in the Vistula Valley. Based on the ^{14}C analysis of peat, infilling one of the Vistula ox-bows of that generation of meanders

(9070±90 years B.P.) it can be supposed that they started to form at the beginning of the Preboreal period (Figure 5.2). Undoubtedly this was connected with a marked improvement of climatic conditions at the time, and with the appearance of a more dense vegetation (Szczepanek, 1961).

Late Glacial and Holocene

The stages of river channel development since the decline of glaciation and during the Holocene are well preserved in the Vistula Valley (Figure 5.3). They are marked on the surface of the valley bottom in the design of river channels and in vertical profiles by the presence of flood and channel deposits. The channel facies deposits show a high directional dispersion of laminae inclination, a prevailing trough cross-stratification and, as a rule, a good sorting of deposits, a significant proportion of well rounded grains in the sandy fraction, a poor petrographic composition as well as a high admixture of vegetable remains. In the analysis of heavy minerals one may observe nearer to more recent series an increased incidence of garnet. Garnets, being particularly resistant to abrasion (Turnau-Morawska, 1955), constitute an indication of the frequent redeposition of alluvial deposits, without more important supply of fresh weathered material from the drainage area. This is also suggested by the considerably diminished admixture of crystalline rock fragments as compared with the lower series. The high amount of plant remains is a proof of the density of vegetation which then covered the valley floors.

The deposits of the flood facies are represented by muds which initially, in the first generation of meanders, contained more silt but gradually became more loamy. This is comparable to descriptions of the middle Vistula and San (Falkowski, 1967; Szumanski, 1972), and is the result of more complete vegetation cover in the early part of the Holocene.

The youngest stage of river-channel transformation induced by the effect of human activity within the drainage basins can be detected in the bottom of the Vistula Valley, as well as in the Mokrzyszówka Valley, by the change from a system of meandering channels to a braided pattern. This transformation was induced by an increased supply of denudation material from slopes and uplands from increased soil erosion. According to historical data, it may be supposed that this last stage of river-channel development occurred during the last 200 years. A similar date for the formation of similar braided river channels is given by Falkowski (1972, 1975) and Szumański (1972) for the Vistula and the San.

Between Mokrzyszówka and Łęg, this stage is clearly recorded in the change of course of river channels (Figures 5.4 and 5.5). As already mentioned during the whole period of the Last Glaciation and Holocene and until historical times, an eastern direction of water-flow to the San river predominated. In one channel an oak-trunk was discovered of an age proved to be 1795 ± 35 B.P.

Using data collected from old maps, one could suggest that the eastern direction of water-flow, with a prevailing meandering system of river channels, persisted in that area until about 200 years ago. It is only then that the Vistula started capturing the waters flowing from that area and that the course of the rivers changed from east- to north-flowing. This was undoubtedly the result of the already noted tendency of the San, as well as the Vistula channels, to shift eastwards. During this process the San gradually withdrew from the watershed area of Mokrzyszówka and Łęg, while from the west the Vistula channel was drawing nearer to that area. It is possible that the observed tendency of the Vistula and San channels to shift eastwards during the Last Glaciation and the Holocene occurred as a response to neotectonic movements, which raised the area of the Holy Cross Mountains together with the Sandomierz Upland and depressed the eastern part of the Sandomierz Basin.

Alluvial deposits from the period of the contemporaneous stage of development of river channels in the examined area are typically characterized by a sandy channels series, as well as by a sandy-mud, high-water series. A large accumulation of alluvia induced a gradual raising of valley bottoms and the immersion during floods of an important part of valley bottoms, developed in the previous stage of meandering river pattern. It is only when earth-walls were built along the rivers, about 100 years ago, that the expansion of floods became limited.

The sequence of events in the development of fluvial relief of the Last Glaciation and Holocene, should be referred to the scheme of glacial cyclicity of Trevisan (1949). Including the stage of bottom erosion and accumulation, this scheme discusses the variability of processes only in one, perpendicular, plane. According to the above results this model should be developed by the introduction of a component acting in horizontal plane— the process of lateral erosion. It seems this was inseparable from the braided channel pattern and from the permafrost in substratum.

Fluvial accumulation processes prevailed in the examined area during the whole of the Pleniglacial of the Last Glaciation, up to the initial stage of decline. They coincided with intensive lateral erosion. Bottom erosion predominated after the second part of the Last Glaciation and this tendency persisted during the Holocene up to the modern times. It is only human activity during the last 200 years that has changed the stabilized order of events and induced a reinforced accumulative action of rivers. This was the consequence of deforestation of the drainage basin, and the destruction of small retention pools, like ponds and mill-waters, induced by developing industrialization in the eighteenth century (Dembińska, 1972).

References

Dembińska, M. 1972. An attempt at periodization of the history of agriculture between the 7th century and middle of the 19th century. *Excursion Guide-Book. Symp. INQUA Comm. Stud. of the Holocene.* 2. PAN Poland, pp. 41–47.

Dury, G. H. 1965. Theoretical implications of underfit streams. *U.S. Geol. Surv. Prof. Paper 452-C.*

Dylik, J. 1966. Importance des éléments périglaciaires dans la stratigraphie du Pleistocène. (Résumé). *Czasop. Geogr.* **37,** pp. 131–151.

Dylik, J. 1967. Main elements of palegeography of Younger Pleistocene in Middle Poland. *Czwartorzęd Polski.*, PWN. pp. 311–352.

Dylik, J. 1974. Causes des changements climatiques dans le passé géologique. (Résumé). *Kwart. Geol.*, **18,** pp. 147–183.

Falkowski, E. 1965. History of Holocene and prediction of evolution of Middle Vistula valley from Zawichost to Solec. *Mat. Symp.*: '*Geologiczne problemy zagospodarowania Wisły środkowej od Sandomierza do Puław*'. SIT Górn. Zarząd Główny. Katowice, pp. 45–61.

Falkowski, E. 1967. Evolution of the Holocene Vistula from Zawichost to Solec with an engineering-geological prediction of further development. (Summary). *Biul. Inst. Geol.* **198,** pp. 57–150.

Falkowski, E. 1970. History and prognosis for the development of bed configurations of selected sections of Polish Lowland rivers. (Summary). *Biul. Geol.*, **12,** pp. 5–110.

Falkowski, E. 1972. Regularities in development of lowland rivers and changes in river bottoms in the Holocene. *Excursion Guide-Book. Symp. INQUA Comm. Stud. of the Holocene.* 2. PAN Poland, pp. 3–30.

Falkowski, E. 1975. Variability of channel processes of lowland rivers in Poland and changes of the valley floors during the Holocene. *Biul. Geol.*, **19,** pp. 45–78.

Falkowski, E. and Szumański, A. 1975. Problems of the engineering-geological mapping of the valley floors of lowland rivers under temperate climatic conditions. (Summary). *Aktualne problemy Geologii Inzynierskiej. Problem III— zastosowanie Geol. Inz. w Urbanistyce i Planowaniu Przestrzennym. Wyd. Geol.* Warszawa, pp. 23–36.

Geyh, M. A. and Rohde, P. 1972. Weichselian Chronostratigraphy [14]C Dating and Statistics. *Intern. Geol. Congress. 24. Section 12. Quaternary Geology.* Canada, pp. 27–36.

Goździk, J. 1973. Origin and stratigraphical position of periglacial structures in Middle Poland. (Summary). *Acta Geogr. Lodz.*, **31,** pp. 119.

Gritchuk, V. P. 1969. Les flores glaciales et leur classification, In La dernière glaciation au nord-ouest de la partie européene de L'URSS. *Congrès de L'INQUA VIII Paris.* '*Nauka*', Moscow, pp. 57–70.

Gritchuk, V. P. 1973. Vegetation in Europe during the maximum stage of the Upper Pleistocene (Valdai) glaciation. In *The Paleogeography of Europe during the Late Pleistocene. Reconstructions and models.* Inst. Georgr. USSR Academy of Sciences. Moscow. pp. 192–200.

Hammen, T., Maarleveld, G. C., Vogel, J. C., and Zagwijn, W. H. 1967. Stratigraphy, climatic succession and radiocarbon dating of the last glacial in the Netherlands. *Geol. Mijnb.*, **45,** pp. 79–95.

Jahn, A. 1975. Problems of the periglacial zone. *Polish Scientific Publ.* Warszawa.

Krauss, A., Mycielska-Dowgiałło, E., and Szczepanek, K. 1965. Preliminary data on the age of the deposits in the Vistula River valley, near Tarnobrzeg. (Summary). *Przegl. Geol.*, **6,** pp. 275–278.

Kuydowicz-Turkowska, K. 1975. Processus fluviaux périglaciaires sur le fond de la morphogenèse de la vallée de la Mroga. (Résumé). *Acta Geogr. Lodz.* **36,** pp. 122.

Mojski, J. E. 1968. Outline of the stratigraphy of North Polish Glaciation in North and Middle Poland. (Summary). *Prace Geogr. 74.* In *Ostanie zlodowacenie skandynawskie w Polsce.* IG PAN, pp. 37–64.

Mycielska-Dowgiałło, E. 1967. Frost fissures and involutions in sand and gravels of the Vistula valley near Tarnobrzeg. (Summary). *Biul. Peryglacjalny,* **16,** pp. 203–215.

Mycielska-Dowgiałło, E. 1969. An attempt at reconstructing the paleohydrodynamics of a river, based on sedimentological studies in the Vistula Valley near Tarnobrzeg. (Summary). *Przegl. Geogr.,* **41,** pp. 409–429.

Mycielska-Dowgiałło, E. 1972a. Holocene evolution of Middle Vistula Valley in the light of examinations made near Tarnobrzeg. (Summary). *Przegl. Geogr.,* **44,** pp. 73–83.

Mycielska-Dowgiałło, E. 1972b. Stages of Holocene evolution of the Vistula valley on the background of its older history, in the light of investigations carried out near Tarnobrzeg. *Excursion Guide-Book. Symp. INQUA Comm. Stud. of the Holocene. 2.* PAN Poland, pp. 69–82.

Mycielska-Dowgiałło, E. In press. Development of the fluvial relief in the Northern part of the Sandomierz Basin in the light of the sedimentological investigations. (Summary). *Prace IG UW.* Warszawa.

Mycielska-Dowgiałło, E. and Krzywobłocka-Laurow, R. 1975. Fluvial and dune sands from the Vistula and Łęg interfluve /the Sandomierz Basin/ in the light of an analysis of surface textures of quartz sand grains observed in electron microscope. (Summary). *Przegl. Geogr.,* **47,** pp. 567–576.

Schumm, S. A. 1965. Quaternary paleohydrology. In Wright, H. E. and Frey, D. G. (Eds) *Quaternary of the United States.* Princeton University Press, pp. 783–794.

Schumm, S. A. 1968. River adjustment to altered hydrologic regimen— Murrumbidgee River and palaeochannels, Australia. *U.S. Geol. Surv. Prof. Paper 598.*

Szczepanek, K. 1961. The history of the Late Glacial and Holocene vegetation of the Holy Cross Mountains. (Summary). *Acta Paleobot. 2.* Kraków, pp. 3–45.

Szumański, A. 1972. Changes in the development of the lower San's channel pattern in the Late Pleistocene and Holocene. *Excursion Guide-Book. Symp. INQUA Comm. Stud. of the Holocene. 2.* PAN Poland, pp. 55–69.

Srodoń, A. 1972. The Quaternary vegetation in Poland. *Szata roślinna Polski.* I. PWN, pp. 527–570.

Trevisan, L. 1949. Genese des terrassen fluviatiles en relation avec les cycles climatiques. *C.R. Congr. Inter. Geogr. Lisbone. 2.*

Turnau-Morawska, M. 1966. Significance of heavy mineral analysis in the solution of geological problems. *Acta Geol. Pol.,* **5,** pp. 363–388.

Wasylikowa, K. 1964. Vegetation and climate of the Late-glacial in Central Poland based on investigations made at Witów near Łęczyca. (Summary). *Biuletyn Peryglacjalny,* **13,** pp. 261–417.

Section II

CHANNEL GEOMETRY CHANGES

J. B. THORNES

Lecturer in Geography
London School of Economics

6

Hydraulic Geometry and Channel Change

For almost 25 years fluvial geomorphology has been dominated almost exclusively by the systematic investigation of the hydraulic geometry of stream channels. Following the seminal paper of Leopold and Maddock (1953), this work has highlighted and reiterated the essential uniformity of river cross-sections, planforms, and profiles, across a wide range of climatic, lithological, and (to a lesser extent) energy environments. This theme was reinforced by the publication of *Fluvial Processes in Geomorphology* in the mid-sixties and a series of exciting papers which sought to account for this uniformity. At the same time, several authors realized the potential scope of the hydraulic geometry equations for identifying, estimating, and verifying actual and suspected changes in environment, particularly those relating to climate. It also offered an opportunity to approach the frustrating difficulties of climatic geomorphology from a quantitative point of view and to add some substance to the speculative character of the previous work. Dury, in a series of papers dating from the mid-fifties, led in these studies of climatic change and channel geometry. Schumm (1960, 1968, 1969) evaluated the response of channels to changes in a wider variety of controls both spatially and temporally, drawing special attention to sediment and bank properties in relation to changing form.

The essential propositions of hydraulic geometry are a set of power function relationships between discharge and width, depth, velocity, slope and roughness, riffle and pool spacing, meander wavelength and sinuosity. Their application in channel-change studies originally arose from the assumption that these relationships represent or reflect (i) direct causal relationships operating in a system under dynamic equilibrium; (ii) reversible, continuous, and linear relationships; (iii) a stepped response between discharge, input changes, and the related variables changes in time, and (iv) a particular response which could be elicited by a particular change (i.e. one to one). The difficulty of the last assumption has long been expressed in the concept of indeterminacy of river channel morphology. Despite the efforts of the sixties, notably by Leopold, Langbein, Scheidegger, and others, acceptable solutions to this problem have not yet been obtained. However, the qualification of dependency or otherwise of variables in the system according to the time scale considered, as exemplified by Schumm and Lichty's paper (1963) has helped to constrain the difficulty to some degree and made the associated problems more manageable. Nonetheless, in the last few years investigators have progressively brought into question the character of the other assumptions, as well as highlighting a number of technical problems. In this paper, an attempt is made to clarify the nature of the problems, technical and conceptual, which relate to evaluating and forecasting channel change in terms of hydraulic geometry.

At the outset, one needs to be clear about what constitutes change. Generally, three types of change, or any combination of the three, are considered. Sometimes one is concerned with changes which are assumed to be spatially uniform across the entire catchment. That is, all points are assumed equally affected and the parameters of change are generally assumed uniform across the whole system. Presumably this is the type envisaged by the climatic change workers, or that supposed in regional diastrophism. Usually responses to change are spatially variable even though the parameters of change may be ubiquitously constant. This is because some of the variables controlling the response are themselves spatially variable. Thus, for example, whilst overall precipitation might increase, the response to this increase may reflect spatially varying infiltration capacities. Finally, there are those situations in which the parameters of change apply to a small area but effects are distributed throughout the system, as in the effect of urbanization in one part of a catchment on suspended solids further downstream. In short, the response reflects the spatial distribution of both the changing and non-changing control variables. Change also implies time scale and the magnitude, character, and direction of change only have meaning with reference to a particular time scale (Thornes and Brunsden, 1977). This is not to say that time explains change, rather that change can only be recognized over time and hence our perception of the change is time conditioned.

There have been many attempts to study channel changes. Most of these involve direct observation of the changes taking place, at best with some attempt to indicate the causative character of the changes. Characteristically, those have been observations of extreme change whose imprint is both swift in occurrence and dramatic in character. Gilbert's study of the changes resulting from the Californian mining debris load on the channels of the Sierra Nevada is of this type. Burkham's study of the response of the Gila river to flooding (1972), and Schumm and Lichty's (1963) careful recording of the changes along the Cimarron River are among the best studies of this type. Although referring to individual extreme events, they are most valuable for their emphasis on the transient response of the system rather than its steady-state behaviour. Unfortunately, there are few other systematic follow-up investigations of the long-term response to recent events of the type which would enable us to estimate the reaction and lag times of the systems involved.

The most common type of investigation has been the inference of past changes from morphology, a theme of long-standing in geomorphology. It is based on the principle of a fixed relationship between contemporary precipitation, discharge, and morphological attributes, and is achieved by moving 'along the graph' to higher or lower values of the changing variable and reading off the new value of the response variable. Alternatively, one asks what value of the control variable will satisfy this 'anomalous' value of a response variable? In essence this argument backs the work of the type done on misfit channels by Dury (1965), on expected response to future changes by Rango (1970), and in a multivariate, less climatically determined framework, on palaeochannels by Schumm (1968, 1969). A similar proposition is the use of spatial samples to infer sequential development using a transform of space for time. Faulkner's (1974) study of the geometry of growth-constrained and growth-unconstrained systems in Alberta badlands led her to suggest a model in which the same controls operated in time as available area for growth diminished and systems came into competition.

The third group of studies on channel changes in terms of hydraulic geometry have been those using direct laboratory experimentation. There have been surprisingly few of these studies relative to those investigating bedforms or sediment transport, presumably because of the difficulties presented by mobile channel boundaries. An outstanding group of such experiments were carried out by Ackers and Charlton (1970) into the meandering properties of alluvial channels, and in particular into the change in the meander properties during a phase of transient development. Likewise, Schumm and Khan (1971) made important observations in their experimental flume in terms of threshold values of slope.

Geomorphology has been less strong in using analytical models to evaluate change, and most of this work has been in the hands of hydraulic engineers. Even here there has not been a great deal of success, mainly because of the complexity of the phenomena under investigation. For obvious reasons the tendency has been to identify and analyse the stable section in a channel reach, such as in Engelund's (1974) analysis of the flow and bed topography in channel bends. The study of the stability conditions for transition from braided to non-braided conditions of the type considered by Engelund and by Kirkby (1972), are likely to assume considerable importance in future geomorphological work. Although little use has been made yet of Yalin's (1971) stochastic model of periodicities in channel geometry, it offers an exciting area of investigation not least because it has outcomes which are consistent with both early and recent empirical findings.

The problems which are created in attempts to explain channel changes arise from specific difficulties with hydraulic geometry itself, problems of comparisons of hydraulic geometry parameters in the evolving system, and conceptual difficulties involving matters such as steady-state and uniqueness of response (equifinality).

The technical difficulties arise from the problems associated with statistical analysis of the relationships involved. Benson showed in 1965 how the common use of a variable on both sides of the regression equation (for example of bankfull discharge and depth) led to spurious results in terms of the correlation coefficient. Since in many papers depth is obtained as the quotient of area and surface width, it is hardly surprising that it has a straight-line relationship of apparently significant proportions which correlated with discharge (which is the product of area and velocity). Many plots would show a far greater scatter of points and, perhaps, therefore allow of more curiosity, if this effect were taken into account. *A similar* problem is the assumption that the relationships are necessarily linear. Thornes (1970) observed sharp changes in the hydraulic geometry of small channels with increasing discharge, which were thought to represent threshold characteristics of bank resistance due to vegetation. More recently, Richards (1973) has drawn attention to non-linearities in several of the variables, for example it is proposed that the rate of increase of velocity with discharge is reduced when the rate of decrease of roughness becomes less. Since others, e.g. Culbertson and Dawdy (1964), have noted the break in the plot of velocity against hydraulic radius, ascribed to change in the flow regime, we may expect discontinuity as well as non-linearity in the hydraulic geometry relations. The spurious correlation problem mentioned above will tend to hide these effects.

Another technical problem of some significance is that the conventional hydraulic geometry ranks observations in order of increasing discharge in a network rather than increasing downstream direction. Given the complexity of response between area and discharge, despite the simplicity of the relationship between length along the main channel and area, the two meanings of 'downstream' are not the same. More to the point is the fact that in topological hydraulic geometry the tendency is to emphasize relatively low frequency relationships. These relate to macro-scale correlations, which we then attempt to explain in micro-scale processes, with some disregard for thresholds inbetween. For this reason, a truly downstream hydraulic geometry (hydraulic autogeometry) should give a better match with our present understanding of hydraulics. It is because of this that some workers have adopted observational schemes which emphasize the relationship between points in real channel space rather than in the topological space of the earlier studies (Melton, 1962; Speight, 1965; Bennett, 1976;

Richards, 1976). Even in the topological hydraulic geometry, there is some real danger of autocorrelation among the residuals if conventional regression is used. This is greatly enhanced when adjacent sections of channel are involved, and has implications for techniques of analysis as well as for the investigation of the real Euclidean hydraulic geometry. One of these is to question the basis of Langbein's (1964) minimum variance model.

These technical problems are small compared with the conceptual leaps required to use hydraulic geometry relations to infer past changes and predict those of the future. The most awesome of these is the indeterminacy problem and its converse, the equifinality problem. The first is a reflection of the fact that channels have many different ways of responding to change and, conversely, that a result (response) could be the product of several combined causes or the same result could be caused in different ways. This remains the interest and frustration of hydraulic geometry. The problem is well illustrated by the debate about meanders in the early seventies. In essence, the argument revolves around Dury's (1965) attribution of the differences between valley and stream meander wavelengths to precipitation differences at the time of origin of the valley meanders. In a subsequent paper Dury, Sinker, and Pannett (1972) described certain meanders in the upper Severn as showing Osage-type underfitness, i.e. the pool–riffle sequence in the non-meandering channel was at variance with the valley meanders. Tinkler (1971, 1972) has favoured a lithological explanation of underfitness, questioning the Dury thesis on the grounds that the lithological controls would induce the longer wavelengths where meanders are incised into bedrock. Kirkby (1972) criticized the Dury thesis on the grounds of its improbable hydrological implications, and Kennedy (1972) discussed the difficulty of the dominant discharge concept. The shortcomings of these conflicting arguments was expressed by Ferguson (1973) who pointed out that none of them adequately take account of the adjustment which could be made in terms of slope in relation to bank resistance. It seems reasonable to suggest, as Richards (1972) has done, that the river meanders in question can be regarded as being in equilibrium at the present time, but that it takes a different form. This debate, which is by no means at an end, reveals the way in which, despite the optimistic beginnings, the hydraulic geometry approach to solving the climatic change problems is no simpler or conceptually better founded, than for other morphological characteristics.

The second major difficulty of a conceptual nature is that channel changes are, by definition, transient in character, so that some kind of reference plane which defines the 'natural' state is difficult to achieve. We use the high correlation between variables, notwithstanding the technical problems alluded to earlier, to define the steady-state geometry of channels. It is for these states that analytical models have been sought (Smith, 1974). Chorley and Kennedy (1971) have drawn attention to the non-equilibrium behaviour of physical systems, and Allen (1974, 1976) has provided detailed and

explicit discussion of the implications of this non-uniform unsteady behaviour in terms of bed-form geometry. A conceptual approach to transient
behaviour in hydraulic geometry is now required. In several papers the
evolving nature of the relaxation paths has been empirically investigated for
major flood events (Burkham, 1970, 1972; Everitt, 1968; Schumm and
Lichty, 1963; Wolman and Eiler, 1968; Thornes, 1976). Something is also
known of channel response through time to the effects of reservoir construction (Komura and Simmons, 1967; Gregory and Park, 1974). However,
there has as yet been little investigation of, for example, the morphological
consequences of seasonal changes or sequences of wet and dry years. Yet if
the response to change of the basic variables is as immediate as Schumm and
Lighty (1963) suggest, the former at least should present excellent oppor-
tunities for investigating the causal mechanisms involved. Moreover, in these
disequilibrium studies emphasis has been almost exclusively on the response
to discharge. The time-lagged response of other variables which affect
channel changes are also important. Whilst reaction time to a flood is
instantaneous, because of the distributed nature of the system, the changes
in sediment yield and the hydraulic geometry response may be of a much
more complex character. In some respects, by moving away from the
hydraulic geometry of the wetted perimeter, to channel capacity perhaps
(Brown, 1971), more meaningful geometries will be observed which better
reflect the transient character of the phenomena under observation. A shift
from mean annual or dominant discharge to some ratio of a time-related
peak discharge to average annual peak as suggested by Stevens, Simons, and
Richardson (1975) will also highlight transient rather than steady behaviour.

The third area of difficulty lies in the existence of thresholds and discontinuities in the relationships involved. In using the standard power functions
it is generally assumed that they are continuous and relatively smooth. In
at-a-station relationships, where most of the investigations of change have
occurred, there is every reason to suppose that this is not the case. With a
change in flow regime, the fairly well documented changes in roughness
should be matched by responses in the hydraulic geometry; certainly there is
evidence to suggest that the shift from meandering to braiding may take
place in a rather rapid and discontinuous manner. In the true downstream
case the response to change must also be spatially variable and probably
discontinuous. The existence of floodplains is a measure of this spatial
discontinuity, and the way in which area increments downstream leads one
to suspect that the channel-geometry response function is also stepped. In
ephemeral channels the effect of transmission losses creates a hierarchy of
flow types so the effect of a particular flow of a given frequency and
magnitude has spatially varying effects on the channel (Thornes, 1977). By
analogy, we may expect that since flood peaks are attenuated in a downstream direction in perennial channels, as shifts in the flood magnitude and
frequency respond to changes in external variables, the response in the

channel will itself be spatially varying. Floods of a different magnitude will have peaks which go further or less far downstream. For a given time, the channel characteristics are a combination of the effects of the last major event, the intervening distribution of erosional (as opposed to flow) events, and spatial position within the network. In some situations the effects are irreversible over the range of time scales normally considered, a point made by Ferguson (1973) with reference to the meandering problem. In others, a hysteresis effect would be expected of time-related changes, especially given the depletion effects of sediment availability within the system. It is difficult to imagine that the path taken by the hydraulic geometry variables against rising discharge can be the same as that on the falling limb and during the subsequent relaxation period. However, the literature indicates that channel fill does restore depth to a position remarkably close to that obtaining before the event.

The three major problems outlined here, indeterminacy of response, disequilibrium relationships, and transient behaviour, and also discontinuity, thresholds and irreversibility, lie at the core of channel change in relationship to hydraulic geometry. It should be evident from what has been said that there is doubt as to whether the conventional hydraulic geometry can assist us much further in elucidating the causes of, and predicting the responses to, exogenous changes.

Three areas of investigation should help to improve our knowledge of both hydraulic geometry and the relationship of geometry to external changes. The first is a better understanding of the channel characteristics in relation to position in the network system as a whole (Chapter 19). Geomorphology is shifting away from the somewhat unproductive phase of network topology to a network description based on hydrological and sedimentological considerations. This is likely to yield a more meaningful set of network properties in relation to channel activity. In particular, the spatial ordering of stream power based on slope and discharge in trunk streams, will need to be described and formulated. Secondly, rather than focus attention on streams assumed to be in equilibrium of one type or another, more may be learned from streams characterized by dynamically unstable conditions, both at-a-station and in a spatial sense. The difficulty here, of course, will be that many models of sediment behaviour are developed for highly steady and uniform situations. Likewise, mathematical solutions to dynamically evolving systems are more difficult to obtain than for those in which an equilibrium form is to be developed. In recent years, however, work by Smith and Bretherton (1972), and Luke (1974), as well as by the kinematic modellers, have opened up possibilities for investigating dynamic behaviour in drainage basin systems. Besides investigating immediate responses of such channels, more needs to be known about the long-term response of channel systems to change of many different types. In this respect the systems response to unit impulses which can be observed in arid and semi-arid channels, as well as in laboratory situations, may be particularly useful. Likewise, a shift back towards historical studies of channel behaviour following the track so well defined by Schumm (1968), which have been unfashionable in recent years, or concern for the effects of man, could prove very productive.

Thirdly, it would appear that a change in the scale of analytical modelling is required if real progress is to be made. The difficulty in matching the field knowledge of hydraulic geometry and overall channel changes to the behaviour of the single, or even aggregates of, rather unnatural particles, described in countless publications by hydraulic engineers, seems at present almost insuperable. Geomorphologists have been both excited and encouraged by the work of Bagnold because it seems to offer a manageable model in terms of the scales at which they are working. The difficulties of such a match have been reflected by the rather empirical character of most hydraulic geometry work to date and by the failure to come to grips with the problems of sediment supply and transport, and those of channel macroroughness. There can be little doubt that sediment and roughness play a crucial role in the understanding and forecasting of change and that they have been somewhat neglected to date.

Channel changes are measured in terms of hydraulic geometry; the hydraulic geometry of channels are used to infer change; the causes of such change are usually to be found elsewhere. However, the study of temporal and spatial responses in terms of hydraulic geometry, particularly if the latter are linked to an understanding of sediment and channel macroroughness on a basin-wide scale, should assist in describing and discovering the causes of the changes.

References

Ackers, P. and Charlton, F. G. 1970. The geometry of small meandering streams. *Proc. Inst. Civil Engineers, Paper 73285*, pp. 289–317.

Allen, J. R. L. 1974. Reaction, relaxation and lag in natural sedimentary systems: general principles, examples and lessons. *Earth Sci. Reviews*, **10**, pp. 263–342.

Allen, J. R. L. 1976. Computational models for dune time-lag: general ideas, difficulties and early results. *Sedimentary Geol.*, **15**, pp. 1–53.

Bennett, R. J. 1976. Adaptive adjustment of channel geometry. *Earth Surface Processes*, **1**, pp. 131–150.

Benson, M. A. 1965. Spurious correlation in hydraulics and hydrology. *J. Hydraul. Div. Amer. Soc. Civ. Engrs.*, **91**, pp. 35–43.

Brown, D. A. 1971. Stream channels and flow relations. *Water Resour. Res.*, **7**, pp. 304–310.

Burkham, D. E. 1970. Precipitation, streamflow and major floods at selected sites in the Gila River basin above Coolidge Dam, Arizona. *U.S. Geol. Surv. Prof. Paper 655-B*.

Burkham, D. E. 1972. Channel changes of the Gila River in Safford Valley, Arizona, 1846–1870. *U.S. Geol. Surv. Prof. Paper 655-G*.

Chorley, R. J. and Kennedy, B. A. 1971. *Physical Geography: A Systems Approach*. Prentice-Hall, London.

Culbertson, J. K. and Dawdy, D. R. 1964. A study of fluvial characteristics and hydraulic variables, Middle Rio Grande, New Mexico. *U.S. Geol. Surv. Water Supply Paper 1498-F.*

Dury, G. H. 1965. Theoretical implications of underfit streams. *U.S. Geol. Surv. Prof. Paper 452-C.*

Dury, G. H., Sinker, C. A., and Pannett, D. J. 1972. Climatic change and arrested meander development in the River Severn. *Area,* **4,** pp. 81–85.

Engelund, F. 1974. Flow and bed topography in channel bends. *J. Hydraul. Div. Amer. Soc. Civ. Engrs.,* **100,** pp. 1631–48.

Everitt, B. L. 1968. Use of the Cottonwood in an investigation of the recent history of a flood plain. *Amer. J. Sci.,* **266,** pp. 417–439.

Faulkner, H. 1974. An allometric growth model for competitive gullies. *Z. Geomorph.,* Suppl. Vol. 20.

Ferguson, R. L. 1973. Channel pattern and sediment type. *Area,* **5,** pp. 38–41.

Gregory, K. J. and Park, C. 1974. Adjustment of river channel capacity downstream from a reservoir. *Water Resour. Res.,* **10**(4), pp. 870–873.

Kennedy, B. A. 1972. 'Bankfull' discharge and meander forms. *Area,* **4,** pp. 209–212.

Kirkby, M. J. 1972. Alluvial and non-alluvial meanders. *Area,* **4,** pp. 284–288.

Komura, S. and Simmons, D. B. 1967. River bed degradation below dams, *J. Hydraul. Div. Amer. Soc. Civ. Engrs.,* **93,** pp. 1–14.

Langbein, W. B. 1964. Geometry of river channels. *J. Hydraul. Div. Amer. Soc. Civ. Engrs.,* **90,** HY 2, pp. 301–312.

Leopold, L. B. and Maddock, T. 1953. The hydraulic geometry of stream channels and some physiographic implications. *U.S. Geol. Surv. Prof. Paper 252.*

Luke, J. C. 1974. Special solutions for non-linear erosion problems. *J. Geophys. Res.,* **79**(26), pp. 4035–4040.

Melton, M. A. 1962. Methods for measuring the effects of environmental factors on channel properties. *J. Geophys. Res.,* **67,** pp. 1485–1492.

Rango, A. 1970. Possible effects of precipitation modification on stream channel geometry and sediment yield. *Water Resour. Res.,* **6,** pp. 1765–1770.

Richards, K. S. 1972. Meanders and valley slope. *Area,* **4,** pp. 288–290.

Richards, K. S. 1973. Hydraulic geometry and channel roughness—a non-linear system. *Amer. J. Sci.,* **273,** pp. 877–896.

Richards, K. S. 1976. The morphology of riffle and pool sequences. *Earth Surface Processes,* **1,** pp. 71–88.

Schumm, S. A. 1960. The shape of alluvial channels in relation to sediment type. *U.S. Geol. Surv., Prof. Paper 352-B.*

Schumm, S. A. 1968. River adjustment to altered hydrologic regimen— Murrumbidgee River and Palaeochannels, Australia. *U.S. Geol. Surv. Prof. Paper 598.*

Schumm, S. A. 1969. River metamorphosis. *J. Hydraul. Div. Amer. Soc. Civ. Engrs.,* **95,** pp. 255–273.

Schumm, S. A. and Khan, H. R. 1971. Experimental study of channel patterns. *Nature,* **233,** pp. 407–409.

Schumm, S. A. and Lichty, R. W. 1963. Channel widening and flood plain construction along Cimarron river in south-western Kansas. *U.S. Geol. Surv. Prof. Paper 352-D,* pp. 71–88.

Schumm, S. A. and Lichty, R. W. 1965. Time, space and causality in geomorphology. *Amer. J. Sci.,* **263,** pp. 110–119.

Smith, T. R. 1974. A derivation of the hydraulic geometry of steady-state channels from conservation principles and sediment transport laws. *J. Geol.,* **82**(1), pp. 98–104.

Smith, T. R., and Bretherton, F. P. 1972. Stability and the conservation of mass in drainage basin evolution. *Water Resour. Res.*, **8**(6), pp. 1506–1529.

Speight, J. G. 1965. Meander spectra of the Angabunga river, *J. Hydrol.*, **3,** pp. 1–15.

Stevens, M. A. Simons, D. B., and Richardson, E. V. 1975. Non-equilibrium river form. *J. Hydraul. Div. Amer. Soc. Civ. Engrs.*, **101,** pp. 557–566.

Thornes, J. B. 1970. The hydraulic geometry of stream channels in the Xingu–Araguaia headwaters. *Geogr. J.*, **136,** pp. 376–382.

Thornes, J. B. 1976. *Semi-arid erosional systems: case studies from Spain.* Geography Department, London School of Economics.

Thornes, J. B. 1977. The autogeometry of semi-arid channels. *Proceedings of the 3rd Symposium on Inland Waterways for Navigation, Flood Control and Water Diversions,* Fort Collins, Colorado, 9–12 August, 1976, II, pp. 1715–1726.

Thornes, J. B. and Brunsden, D. B. 1977. *Geomorphology and Time.* Methuen, London.

Tinkler, K. J. 1971. Active meanders in south central Texas and their wider implications. *Geol. Soc. Amer. Bull.*, **81,** pp. 1873–1899.

Tinkler, K. J. 1972. The superimposition hypothesis for incised meanders—a general rejection and specific test *Area*, **4,** pp. 86–91.

Wolman, M. G. and Eiler, J. P. 1968. Reconnaissance study of erosion and deposition produced by the flood of August 1955 in Connecticut. *Trans. Amer. Geophys. Union*, **39**(1), pp. 1–14.

Yalin, M. S. 1971. On the formation of dunes and meanders. *International Assoc. for Hydraulic Res., 14th Congress, Paris, Proc.*, **3,** paper C 13, pp. 1–8.

A. D. KNIGHTON

Lecturer in Geography
University of Sheffield

7

Short-term Changes in Hydraulic Geometry

Hydraulic geometry is an empirical model devised by Leopold and Maddock (1953) to provide a quantitative description of stream behaviour either at a particular cross-section or along a particular stream. Only at-a-station hydraulic geometry is considered here. The application of the model in a wide range of physiographic environments (for example, Wolman, 1955; Leopold and Miller, 1956; Fahnestock, 1963; Lewis, 1969; Heede, 1972) indicates the extent to which it has become accepted as a suitable means of analysing the behavioural characteristics of natural streams. However, the model does contain certain assumptions. Firstly, discharge is regarded as the dominant independent variable to which the dependent variables adjust, all other independent variables being controlled to a greater or lesser extent. Secondly, simple power functions are considered to be a suitable expression of the relationships between the dependent variables and discharge, the three basic relationships being

$$w = aQ^b \qquad d = cQ^f \qquad v = kQ^m \qquad (1)$$

where w is width, d is mean depth, v is mean velocity, and Q is a discharge. Since discharge is the product of width, depth, and velocity, it follows that

$$b + f + m = 1 \qquad (2)$$

Richards (1973) has seriously challenged this assumption on the basis that, since resistance variation is not necessarily linear, neither depth nor velocity can be expected to vary linearly with discharge. He suggested that higher order functions are more appropriate than simple log-linear functions in describing the relationships. Finally, it is assumed that a given set of relations represents the mean condition of stream behaviour at a particular cross-section, with the result that little or no scatter should be present about regression lines. Essentially this is a corollary to the assumption of linearity. However, Knighton (1975) has shown that even when measurements are made over a short time period the resultant graphs can have a wide scatter of points. One aim of this paper is to consider the possible causes of this internal inconsistency which, in effect, represents unexplained short-term variation in the hydraulic parameters. In addition, it is intended to show how that variation associated with changes in channel form can be usefully examined within a Euclidean space context. For this purpose the simple log-linear model is retained as a first approximation.

Sources of Short-term Variation

Wolman (1955) has examined small-scale spatial variation in hydraulic geometry by comparing measurements made at several stations within a single reach, but no attempt has been made to consider the possible causes and significance of temporal variation in hydraulic geometry at individual stations. The measurements made at a cross-section in order to obtain a set of hydraulic relations are a discrete sample through time of a continuous sequence of events. Hence any temporal variation should be reflected on the resultant graphs by a wide scattering of the data points which may in some way be correlated with time. The following analysis of the causes of scatter uses examples from a study of 12 stations on the Bollin-Dean in Cheshire, a fifth-order stream which has generally cohesive banks and a bed composed largely of pebble-sized grains. The stations were located at sections of different cross-sectional and planimetric form, and at points along the stream to represent the range of flow conditions in the upper, middle, and lower parts of the stream's course. At each station regular measurements were made of streamflow and channel characteristics for a range of discharges up to bankfull. The examples used in this text are all drawn from the middle reaches where bankfull discharge ranges from $2 \, m^3 \, s^{-1}$ to $6 \, m^3 \, s^{-1}$, and suspended sediment load concentrations rarely exceed 400 p.p.m. A detailed discussion of the general results of this study can be found elsewhere (Knighton, 1975).

The main sources of scatter are classified under four headings: measurement error, analytical error, random variation, and systematic variation.

Measurement Error

This is always present to some extent and results from failure to make observations at precisely the same cross-section each time, and from inaccuracy of measurement. The former, considered by Wolman (1955) to be a major cause of scatter, can be completely controlled, whilst the latter can be controlled within acceptable bounds. Stream width is the parameter which can be measured with greatest accuracy, although the boundary of the flow zone is sometimes difficult to define precisely. Errors in depth data are produced by incorrect measurement, changes in depth during measurement, and the method used to calculate mean depth. Mean depth is usually obtained by dividing cross-sectional area by width, yielding a value which is only indirectly related to measured depths. However, mean velocity is probably the variable most subject to error. The calculation of mean velocity at a cross-section is based on the assumption that velocity changes logarithmically with distance from the bed and that therefore the mean velocity in a vertical can be obtained by averaging the velocities measured at 0.8 and 0.2 of the depth or by using the velocity measured at 0.6 as an estimate of the mean. Such an assumption is not valid for all verticals over the complete range of flow conditions. In addition, point velocity observations often produce erratic results because of surges in the flow which affect the velocity distribution, particularly at high discharges. However, provided adequate precautions are taken and a consistent sampling procedure is followed, measurement error should make only a small contribution to scatter about regression lines.

Analytical Error

The fact that certain hydraulic variables may not have a simple power functional relationship to discharge, as previously assumed, provides an additional explanation of scatter. Richards (1973) has suggested that a log-quadratic rather than a log-linear function may be more appropriate for the depth-discharge and velocity-discharge relationships in particular. In many of the cases which he studied, the addition of a quadratic term significantly increased the explained variance and, by implication, reduced the amount of scatter. From a mathematical standpoint this result is hardly surprising since, given a set of n data pairs, a polynomial of degree $n-1$ can be fitted that will pass through all the plotted points. The argument does, however, have a physical basis in that resistance variation is considered to be non-linear. Thus, whilst the suitability of linear power functions in hydraulic geometry remains an unresolved issue, the method of analysis must be regarded as a possible source of scatter on traditional plots.

Random Variation

The more important sources of variation are, however, those which can be associated with particular physical conditions or events whose effect on the dependent hydraulic parameters varies over a short period of time. The magnitude of a channel variable at a given discharge may vary depending on whether the measurement was made on the rising or falling stage during the passage of a flood. Leopold, Wolman, and Miller (1964, p.230) have provided an example in which the same discharge produced a higher velocity and smaller depth on the rising limb of the hydrograph than on the falling limb. Clearly, where hydraulic parameters describe hysteresis loops during floods, the scatter on the resultant graphs may be quite large. In a similar way temporary scour and fill, and temporary channel widening, may cause perturbation in the depth–discharge and width–discharge relationships. Changes in the relative influence of other factors are also important. Thus, for example, local resistance elements may have a variable effect over a short time period, producing inconsistency in the relation of velocity to discharge. Variation in the amount of material supplied to the stream from the extra-channel area may result in different sediment loads being transported at equivalent discharges, leading to an unequal response in those variables which are directly or indirectly affected by sediment load. All of these effects can be regarded as random in the sense that they represent variation about some ill-defined mean state, with complete or partial return to initial conditions over some unspecified time period. In a system characterized by simultaneous adjustment amongst a large number of variables, a degree of indeterminacy in the magnitude of certain hydraulic parameters is to be expected.

Figure 7.1 illustrates the amount of scatter which can be produced by changes of this kind. The graphs indicate variable stream behaviour over a period of 18 months at a cross-section sited in the top limb of a meander in which a central bar had been deposited. During the measurement period small changes to channel morphology occurred at the cross-section itself (Figure 7.1), but of greater importance were the changes which occurred immediately upstream of the section. The bar grew laterally and vertically to such an extent that all but the higher flows were excluded from the left channel during the second part of the study period, with the result that stream width decreased and mean velocity increased at low and intermediate discharges. Towards the end of the study period the bar had regained its former shape and initial hydraulic conditions had been re-established.

Figure 7.2 indicates the effect on hydraulic variables of minor modifications to stream bed topography produced during a single high flow of magnitude $2.4 \, \mathrm{m^3 \, s^{-1}}$ and recurrence interval 2.5 years. Segmentation of the original data set into two phases representing conditions before and after this flow event, respectively, served to explain most of the scatter remaining after the initial analysis. In addition to explaining residual variance, this form of analysis provides a useful means of identifying the principal changes through time. In this case bed scour caused an increase in mean depth which, in order to maintain flow continuity at approximately constant width, required a corresponding decrease in mean velocity. Thus, whilst the rates of change of depth or velocity did not differ significantly between the two phases, the magnitudes of those variables were altered in the short term, notably at low and intermediate discharges.

These two examples demonstrate how small changes in local conditions

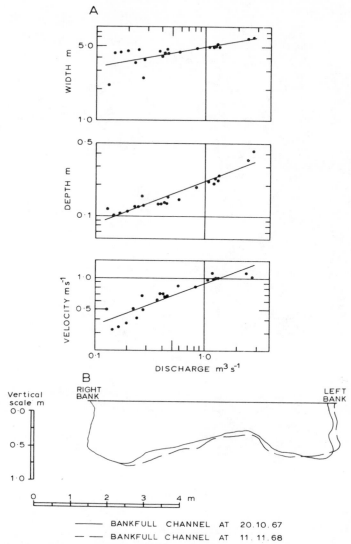

FIGURE 7.1 RELATION OF WIDTH, MEAN DEPTH AND MEAN VELOCITY TO DIS-
CHARGE (A), AND CHANGES IN CROSS-SECTIONAL FORM (B), RIVER DEAN, LOW-
ERHOUSE MILL, BRAIDED SECTION

can produce considerable scatter on at-a-station graphs. Hydraulic geometry
is sensitive to the variable influence of local factors which may remain
undetected if the data set is small or if observations are made at widely
spaced time intervals. There is a danger also that random variation may be
confused with the kind of analytical error which Richards (1973) has
suggested.

A. D. Knighton

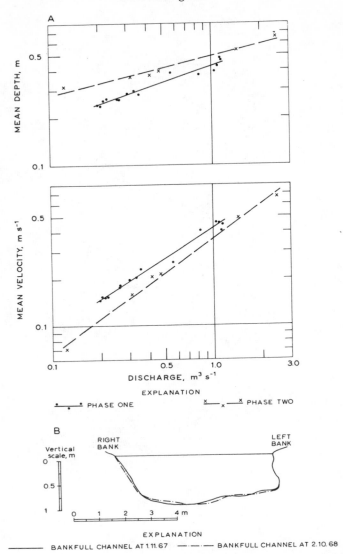

FIGURE 7.2 CHANGES IN HYDRAULIC GEOMETRY (A) AND CROSS-SECTIONAL FORM (B), RIVER DEAN, LOWERHOUSE MILL, STRAIGHT REACH SECTION

Systematic Variation

The final source of variation is that associated with progressive change in channel condition. Fundamental changes to channel form occur as the stream develops an alternative state, precluding the possibility of a return to initial conditions at least in the short term. Hydraulic relations obtained during a period when substantial change is taking place are characteristically

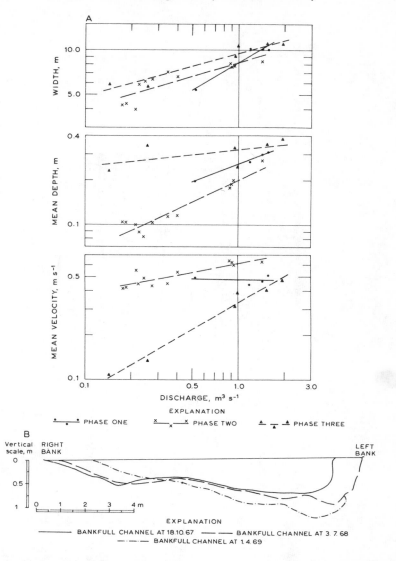

FIGURE 7.3 CHANGES IN HYDRAULIC GEOMETRY (A) AND CROSS-SECTIONAL FORM (B), RIVER DEAN, ADLINGTON HALL, MEANDER SECTION (DAM)

inconsistent, the respective graphs showing a wide scattering of points. In order to identify the changes in hydraulic geometry, the original sample of measurements can be subdivided into distinct temporal phases associated with different channel morphologies and new sets of relations calculated, one for each phase. One of the problems in the subsequent interpretation is that the data subsets may cover different discharge ranges.

A. D. Knighton

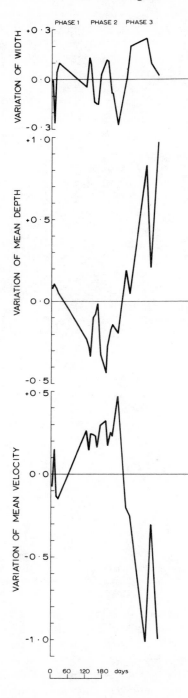

FIGURE 7.4 PLOT OF RESIDUAL SCATTER ABOUT REGRESSION LINE CALCULATED FOR ALL DATA AGAINST TIME FOR THE VARIABLES, WIDTH, MEAN DEPTH, AND MEAN VELOCITY, AT RIVER DEAN, ADLINGTON HALL, MEANDER SECTION (DAM)

Three phases were identified at a station located at the apex of a well-developed meander which was migrating both laterally and downstream (Figure 7.3). Small changes in channel morphology occurred between the first two phases, but the major changes occurred between phases two and three as the direct result of a high discharge with a recurrence interval of 2.5 years. The width and depth of the channel were significantly increased which, to maintain continuity, required a decrease in velocity. This was brought about by the enlargement of wetted perimeter which intensified the effect of bed and bank roughness, and by the increase in channel curvature which further distorted the flow pattern. Thus the resistance properties of the channel altered with cross-sectional form. The scale of these changes can be appreciated from Figure 7.4 in which the amount of variation of each variable about the mean state calculated for all data is plotted against time. Also, the fact that the rates of change of depth and velocity changed significantly between phases two and three (Table 7.1) indicates the extent to which hydraulic geometry can be modified by a flow of sufficiently large magnitude.

Changes to channel form and hydraulic geometry do not always take place so erratically. Figure 7.5 describes the variable flow behaviour at a cross-section where the changes were not simply the result of a single high flow but represented the cumulative effect of a wide range of discharges, including flows with a frequency of at least 15%. Although distinct phases were more difficult to recognize because the changes occurred gradually, tripartition of the original data set did help to explain much of the scatter remaining after the overall analysis. Initially the left channel carried the larger discharge. However, point-bar and island deposition towards the left bank and increase in the cross-sectional area of the right channel led to the dominance of that channel by phase three, and to the abandonment of the left

TABLE 7.1 CHANGES IN THE HYDRAULIC EXPONENTS THROUGH TIME
AT THREE MEASUREMENT STATIONS

Station	Phase	Hydraulic exponents		
		b	f	m
R. Dean, Adlington	1	0.61	0.40	−0.01
Hall, Meander	2	0.31	0.49	0.20
Section (DAM)	3	0.29	0.11	0.60
R. Dean, Adlington	1	0.44	0.23	0.34
Hall, Braided	2	0.12	0.47	0.42
Section (DAB)	3	0.38	0.37	0.24
R. Bollin, Mottram	1	0.06	0.44	0.50
Bridge, Pool	2	0.32	0.25	0.42
Section (BMP)	3	0.20	0.38	0.42

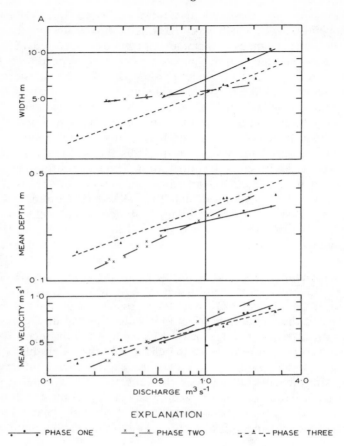

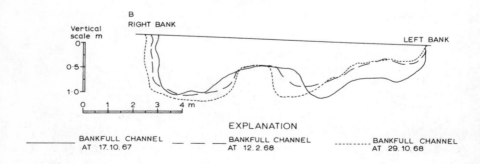

FIGURE 7.5 CHANGES IN HYDRAULIC GEOMETRY (A) AND CROSS-SECTIONAL FORM
(B), RIVER DEAN, ADLINGTON HALL, BRAIDED SECTION (DAB)

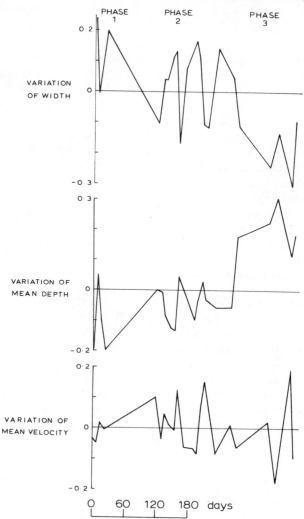

FIGURE 7.6 PLOT OF RESIDUAL SCATTER ABOUT REGRESSION LINE CALCULATED FOR ALL DATA AGAINST TIME FOR THE VARIABLES, WIDTH, MEAN DEPTH, AND MEAN VELOCITY, AT RIVER DEAN, ADLINGTON HALL, BRAIDED SECTION (DAB)

channel by all but the higher discharges. A recent visit showed that this trend has continued since the original measurements were made, for no trace of the left channel now remains. Variation in the relative amounts of erosion and deposition affected both the width and depth of the channel (Figures 7.5 and 7.6) and their respective rates of change with discharge (Figure 7.5, Table 7.1). The plot of residual variance against time (Figure 7.6) clearly indicates a trend in width and depth variation, with phase two being transitional between the first and third phases. There was no abrupt change in velocity conditions despite the large modifications to channel form, although the rate of velocity increase did change significantly between phases two and three. These measurements reflect conditions during the gradual development of a unichannel flow pattern and, although the changes were not as discontinuous as in the previous example, there was nevertheless considerable variation in hydraulic geometry.

This discussion has identified some of the major sources of scatter on at-a-station graphs. Without detailed knowledge of particular physical situations it would be difficult to determine the relative contributions of the various sources to total scatter, which has obvious implications on subsequent interpretation. If measurement error and analytical error are disregarded, then scatter represents short-term variation in hydraulic geometry. As the examples have shown, without any basic change in environmental conditions or any exceptionally high flow events, variation about the mean due to random or systematic effects can be quite considerable. Clearly, any attempt at synthesizing available information in order to identify basic patterns of at-a-station stream behaviour must take account of this fact, for a single set of relations may not adequately describe conditions at a cross-section, particularly where channel form is subject to change. The analysis raised a further problem regarding the physical significance of associated systematic changes in channel morphology and hydraulic geometry. The final two examples demonstrated that different types of adjustment exist within natural streams, and that adjustment can occur relatively rapidly under favourable physical conditions, but the question remains of whether the development of an alternative channel state represented an attempt by the stream to approach some form of dynamic or quasi-equilibrium, if such a concept is relevant in this context. One approach to the solution of this problem is considered in the next section.

Changes in Channel State

The state of stream behaviour at a channel cross-section for a given time period can be described by the ordered triple (b, f, m), the components of which represent respectively the rates of change of width, depth, and velocity with discharge. (b, f, m) can be regarded as a vector $\boldsymbol{\alpha}$ belonging to the three-dimensional Euclidean space R^3, with the restriction that $b, f, m \in [0, 1]$. At time t_0 the channel has state $\boldsymbol{\alpha}_0 = (b_0, f_0, m_0)$ and at time t_1 it has changed to state $\boldsymbol{\alpha}_1 = (b_1, f_1, m_1)$, where $b_1 = b_0 + b'_0$, $f_1 = f_0 + f'_0$, and $m_1 = m_0 + m'_0$. Since $b + f + m = 1$, it follows that $b' + f' + m' = 0$. Three further definitions are required. If $\boldsymbol{x}$ and $\boldsymbol{y}$ are two vectors in R^3 such that $\boldsymbol{x} = (x_1, x_2, x_3)$ and $\boldsymbol{y} = (y_1, y_2, y_3)$, then the scalar or inner product $\boldsymbol{x} \cdot \boldsymbol{y}$ is given by

$$\boldsymbol{x} \cdot \boldsymbol{y} = (x_1, x_2, x_3) \cdot (y_1, y_2, y_3) = (x_1 y_1 + x_2 y_2 + x_3 y_3) \tag{3}$$

the length of a vector, denoted by $\|x\|$ is

$$\|x\| = (x \cdot x)^{\frac{1}{2}} \tag{4}$$

and the distance, $d(x, y)$, between two vectors is

$$d(x, y) = \|x - y\| = [(x - y) \cdot (x - y)]^{\frac{1}{2}} \tag{5}$$

The application of these ideas makes the following assumptions. Firstly, the channel is assumed to be in an unstable state, with the result that its form and associated hydraulic geometry are subject to change. The previous analysis demonstrated that even in the absence of exceptionally high flows, at-a-station hydraulic geometry can alter appreciably in the short term. The exponents b, f, and m are conservative quantities, so that a change from one state to another may be physically significant without the two sets of hydraulic relations being significantly different in a statistical sense. Secondly, change is assumed to take place in such a way that the channel approaches a kind of quasi-equilibrium characterized by stability in the b, f, and m values and regarded here as a limit state $\alpha = (b, f, m)$. By definition of vector convergence, therefore, we can write

$$n \xrightarrow{\text{lim}} \infty \{\alpha_n\} = \alpha \tag{6}$$

where $\{\alpha_n\}$ signifies the sequence of vector states produced during channel modification.

A natural stream can be regarded as a self-regulating system, one property of which is a tendency to counteract the effects of external changes in such a way that the amount of internal adjustment required is minimized. Approach to quasi-equilibrium is characterized by a decrease in the magnitude of adjustment through time. Within the Euclidean space context considered here, this implies that the distance between the ith and $(i+1)$th states is greater than the distance between the next two consecutive states, α_{i+1} and α_{i+2}. Hence, with the equality sign introduced for completeness to signify the possibility of no change,

$$d(\alpha_i, \alpha_{i+1}) \geq d(\alpha_{i+1}, \alpha_{i+2}) \tag{7}$$

which yields

$$[(\alpha_i - \alpha_{i+1}) \cdot (\alpha_i - \alpha_{i+1})]^{\frac{1}{2}} \geq [(\alpha_{i+1} - \alpha_{i+2}) \cdot (\alpha_{i+1} - \alpha_{i+2})]^{\frac{1}{2}} \tag{8}$$

Neglecting the square root since $d(x, y) \geq 0$ for all $x, y \in R^3$ and $d(x, y) = 0$ if and only if $x = y$, we have

$$(b_i - b_{i+1})^2 + (f_i - f_{i+1})^2 + (m_i - m_{i+1})^2$$
$$\geq (b_{i+1} - b_{i+2})^2 + (f_{i+1} - f_{i+2})^2 + (m_{i+1} - m_{i+2})^2 \tag{9}$$

Using the initial definitions $b_{i+1} = b_i + b'_i$, $b_{i+2} = b_i + b'_i + b'_{i+1}$, etc., we obtain

$$(b'_i)^2 + (f'_i)^2 + (m'_i)^2 \geq (b'_{i+1})^2 + (f'_{i+1})^2 + (m'_{i+1})^2 \tag{10}$$

Approach to a limit state under this condition is therefore characterized by

$$(b')^2 + (f')^2 + (m')^2 \to \text{minimum} \tag{11}$$

Whilst the converse argument that minimization of $(b')^2 + (f')^2 + (m')^2$ implies attainment of quasi-equilibrium does not immediately follow from this derivation, it should be noted that, by definition, $b' + f' + m' = 0$ when no further change occurs. Using this last equation, statement (11) could be replaced by any of the equivalent statements

$$\left. \begin{array}{c} -2(b'f' + b'm' + f'm') \\ 2(b')^2 + b'f' + 2(f')^2 \\ 2(b')^2 + b'm' + 2(m')^2 \\ 2(f')^2 + f'm' + 2(m')^2 \end{array} \right\} \to \text{minimum} \tag{12}$$

One further development of these ideas is considered. If adjustment is towards a quasi-equilibrium condition represented by the limit vector $\boldsymbol{\alpha} = (b, f, m)$, then the $(i+1)$th state will be closer to that limit than will the ith state, whence

$$d(\boldsymbol{\alpha}_i, \boldsymbol{\alpha}) \geqslant d(\boldsymbol{\alpha}_{i+1}, \boldsymbol{\alpha}) \tag{13}$$

This is in fact a property of vector convergence itself since (6) can also be written

$$n \xrightarrow{\ \lim\ } \infty \ \|\boldsymbol{\alpha}_n - \boldsymbol{\alpha}\| = 0 \tag{14}$$

If the limit $\boldsymbol{\alpha}$ exists, then for some real number $\varepsilon > 0$ an integer N can be found such that

$$\|\boldsymbol{\alpha}_n - \boldsymbol{\alpha}\| < \varepsilon \tag{15}$$

for all $n > N$, where N depends on ε. Approach to the limit state is therefore characterized by a decrease in the distance between $\boldsymbol{\alpha}_n$ and $\boldsymbol{\alpha}$ as n increases. Neglecting the square root sign again, development of (13) produces

$$(\boldsymbol{\alpha}_i - \boldsymbol{\alpha}) \cdot (\boldsymbol{\alpha}_i - \boldsymbol{\alpha}) \geqslant (\boldsymbol{\alpha}_{i+1} - \boldsymbol{\alpha}) \cdot (\boldsymbol{\alpha}_{i+1} - \boldsymbol{\alpha}) \tag{16}$$

whence

$$(b_i - b)^2 + (f_1 - f)^2 + (m_i - m)^2 \geqslant (b_{i+1} - b)^2 + (f_{i+1} - f)^2 + (m_{i+1} - m)^2 \tag{17}$$

Multiplying out and substituting $b_{i+1} = b_i + b'_i$, etc., we have

$$b_i^2 + f_i^2 + m_i^2 + 2(b'_i b + f'_i f + m'_i m) \geqslant b_{i+1}^2 + f_{i+1}^2 + m_{i+1}^2 \tag{18}$$

Now, if

$$\boldsymbol{\alpha}'_i \cdot \boldsymbol{\alpha} = b'_i b + f'_i f + m'_i m \leqslant 0 \tag{19}$$

which requires that

$$f_i' \leqslant -\lambda m_i' \tag{20}$$

where $\lambda = (m - b/f - b)$ can be regarded as positive since the rates of change of depth and velocity are usually larger than the rate of change of width, then (18) reduces to

$$b_i^2 + f_i^2 + m_i^2 \geqslant b_{i+1}^2 + f_{i+1}^2 + m_{i+1}^2 \tag{21}$$

Approach to the limit state under these conditions would be characterized therefore by

$$b^2 + f^2 + m^2 \rightarrow \text{minimum} \tag{22}$$

This is essentially a restatement of the simpler version of Langbein's (1964, 1965) minimum variance hypothesis obtained without recourse to the statistical assumptions which are contained therein and which have been seriously challenged elsewhere (Kennedy, Richardson, and Sutera, 1964; Richards, 1973). Statement (18) can be used instead of statement (22) if (19) is considered to be invalid, but then a knowledge of the limit state $\alpha = (b, f, m)$ is required. For the purpose of subsequent analysis, the components of α are taken to be 0.16, 0.43, and 0.41 respectively, these values being the average rates of change obtained from 280 sets of at-a-station relations.

Statements (11), (18), and (22) constitute a means of testing whether adjustment at a channel cross-section is towards a limit state. In addition, they can be used to indicate the position of a given channel relative to a limit state, where a higher value of, for example, $(b')^2 + (f')^2 + (m')^2$ or $b^2 + f^2 + m^2$ represents a greater departure from that condition. The theoretical minima associated with the three statements are respectively 0, 1/3, and 1/3. In order to test the model, detailed measurements of changes in hydraulic geometry through time are required. The relevant information is available for only three stations, two of which (DAM and DAB) were considered in the previous section and one of which (BMP) has been analysed in detail elsewhere (Knighton, 1975, p. 214–216). At each station three phases were defined. The relevant data expressed in vector terms together with the requisite calculations are given in Table 7.2. With respect to the calculations corresponding to statements (11) and (18), there is a tendency for the sum of squares terms to decrease through time at DAB and BMP but not at DAM. This tendency is repeated to a lesser extent in the values of $b^2 + f^2 + m^2$. The increase in this term between phases one and two at DAB probably reflects the transitory nature of the second phase identified previously. It should be noted also that the sums of squares are lower for DAB and BMP than they are for DAM, and that the final values for the former are close to the theoretical minima. These results suggest firstly that adjustment was towards a limit state at DAB and BMP but not at DAM, and secondly that DAB and BMP were closer to that limit than was DAM.

TABLE 7.2 CHANNEL STATES IN VECTOR TERMS AND CALCULATIONS OF STATEMENTS (11), (18), AND (22)

Station	Channel state vectors, α_i ($i = 1, 2, 3$)	Changes in channel state vectors, α_i'($i = 1, 2$)	Statement (11): $\alpha_i' \cdot \alpha_i'$($i = 1, 2$)	Statement (18): $\alpha_i \cdot \alpha_i + 2(\alpha_i' \cdot \alpha_i)^*$, $\alpha_{i+1} \cdot \alpha_{i+1}$($i = 1, 2$)	Statement (22): $\alpha_i \cdot \alpha_i$($i = 1, 2, 3$)
R. Dean, Adlington Hall, Meander Section (DAM)	$\alpha_1 = (0.61, 0.40, -0.01)$	$\alpha_1' = (-0.30, 0.09, 0.21)$	0.142	0.584, 0.376	0.532
	$\alpha_2 = (0.31, 0.49, 0.20)$	$\alpha_2' = (-0.02, -0.38, 0.40)$	0.304	0.372, 0.456	0.376
	$\alpha_3 = (0.29, 0.11, 0.60)$				0.450
R. Dean, Adlington Hall, Braided Section (DAB)	$\alpha_1 = (0.44, 0.23, 0.34)$	$\alpha_1' = (-0.32, 0.24, 0.08)$	0.266	0.447, 0.412	0.362
	$\alpha_2 = (0.12, 0.47, 0.42)$	$\alpha_2' = (0.26, -0.10, -0.18)$	0.110	0.337, 0.339	0.412
	$\alpha_3 = (0.38, 0.37, 0.24)$				0.339
R. Bollin, Mottram Bridge, Pool Section (BMP)	$\alpha_1 = (0.06, 0.44, 0.50)$	$\alpha_1' = (0.26, -0.19, -0.08)$	0.110	0.374, 0.341	0.447
	$\alpha_2 = (0.32, 0.25, 0.42)$	$\alpha_2' = (-0.12, 0.13, 0)$	0.031	0.378, 0.361	0.341
	$\alpha_3 = (0.20, 0.38, 0.42)$				0.361

* $\alpha = (0.16, 0.43, 0.41)$

They support a classification of these three stations based on physical grounds for, whilst the changes at DAM were irregular and related to single high flows, adjustments at DAB and BMP were less discontinuous and reflected the cumulative effect of a far wider range of discharges.

A larger body of data collected over a longer period of time at many more stations is required to provide an adequate test of the model but these results are sufficiently encouraging to warrant its further development. With a suitable increase in Euclidean space dimension, variables additional to width, depth, and velocity could be considered. The method is not necessarily restricted to regarding stream behaviour in terms of Leopold and Maddock's (1953) simple log-linear model but is flexible enough to take account of the higher order model proposed by Richards (1973). Various constraints could be imposed on the hydraulic exponents in order to define the theoretical limit state. Thus, for example, minimization of $b^2 + f^2 + m^2$ under the condition $b = 0.33f$, which is the relation between the rates of change of width and depth in a channel with a perfect semicircular form at bankfull, yields a limit state of (0.15, 0.45, 0.40). However, probably the greatest benefit to be gained from such an approach is that linear algebra and vector space methods can be applied to problems of stream behaviour. To illustrate the point, Langbein's (1964, 1965) minimum variance hypothesis is considered. The hypothesis, represented by statement (22), has been derived here from the initial assumption that a change in channel state can be written in the form $\boldsymbol{\alpha}_{i+1} = \boldsymbol{\alpha}_i + \boldsymbol{\alpha}'_i$. A more general form of the relation between consecutive channel states is, however, $\boldsymbol{\alpha}_{i+1} = P\boldsymbol{\alpha}_i$ where

$$P = \begin{bmatrix} p_{11} & p_{12} & p_{13} \\ p_{21} & p_{22} & p_{23} \\ p_{31} & p_{32} & p_{33} \end{bmatrix}$$

Thus $\boldsymbol{\alpha}_{i+1} = \boldsymbol{\alpha}_i + \boldsymbol{\alpha}'_i$ can be regarded as a special case of $\boldsymbol{\alpha}_{i+1} = P\boldsymbol{\alpha}_i$ in which $p_{ii} = 1$ and $\boldsymbol{\alpha}'_i = (p_{12}f_i + p_{13}m_i, \ p_{21}b_i + p_{23}m_i, \ p_{31}b_i + p_{32}f_i)$. The form of $\boldsymbol{\alpha}'_i$ indicates that, with respect to the original model, the amount of change in any one hydraulic exponent is linearly dependent upon the previous values of the other two exponents. This modest generalization suggests that Langbein's minimum variance hypothesis is one aspect of a more general problem, and that a more profitable approach might be to concentrate on this extended version rather than the original formulation.

Conclusion

The most important physical causes of scatter on at-a-station graphs are random variation about some modal state with partial or complete return to initial conditions after an appropriate time lag, and systematic variation related to changes in channel form. The presence of large amounts of scatter may invalidate any attempt to define a mean hydraulic geometry for a group of stations or even for a single station. In effect, scatter represents unexplained short-term variation in the hydraulic parameters. Provided detailed information on local conditions is available, explanation of that variation associated with systematic change can be achieved by segmentation of the original data set into distinct temporal phases and calculation of new sets of relations, one for each phase. With respect to the examples cited in the text, the main differences between phases were at low and intermediate discharges, the values of the variables at high flows showing less interphase difference. On the basis of this limited evidence, greater consistency in the relations may be achieved if hydraulic geometry is defined over a narrower range of discharges, excluding those flows which are incapable of performing work.

At cross-sections undergoing systematic change there is a problem of determining whether adjustment is towards some form of quasi-equilibrium. The potential for adjustment in the short term is related to flow regime and channel boundary conditions. As the examples in the text have shown, marked changes to channel form and associated hydraulic geometry can occur over a short period in the absence of exceptionally high flows and in a channel with high boundary resistance, suggesting that the approach to quasi-equilibrium or establishment of a new equilibrium position is relatively rapid. In recognition of this apparently high potential for short-term changes in channel state, a model based on Euclidean space concepts has been developed in an attempt to determine whether adjustment at a cross-section is towards a quasi-equilibrium or limit state and whether, in the case of two sections, one is closer to that state than is the other. Preliminary results indicate firstly that where change is taking place gradually rather than as the result of single high flows there is a greater likelihood of development towards quasi-equilibrium, and secondly that Langbein's (1964, 1965) minimum variance hypothesis can be derived without recourse to the statistical assumptions contained within the original derivation. Indeed, the hypothesis can be shown to be a special case of a more general problem. Further development of these ideas may yield valuable information on the nature of stream behaviour and the adjustment characteristics of river channels.

Acknowledgements

Financial support was provided initially by the Natural Environment Research Council and latterly by the University of Sheffield Research Fund.

References

Fahnestock, R. K. 1963. Morphology and hydrology of a glacial stream—White River, Mount Rainier, Washington. *U.S. Geol. Surv. Prof. Paper 422-A.*
Heede, B. H. 1972. Flow and channel characteristics of two high mountain streams. *U.S. Dept. Agric. Forest Service Research Paper RM-96.*

Kennedy, J. F., Richardson, P. D., and Sutera, S. P. 1964. Discussion of 'Geometry of river channels' by W. B. Langbein. *J. Hydraul. Div. Amer. Soc. Civ. Engrs.*, **90,** pp. 332–341.

Knighton, A. D. 1975. Variations in at-a-station hydraulic geometry. *Amer. J. Sci.,* **275,** pp. 186–218.

Langbein, W. B. 1964. Geometry of river channels. *J. Hydraul. Div. Amer. Soc. Civ. Engrs.,* **90,** pp. 301–313.

Langbein, W. B. 1965. Closure of 'Geometry of river channels' by W. B. Langbein. *J. Hydraul. Div. Amer. Soc. Civ. Engrs.,* **91,** pp. 297–313.

Leopold, L. B. and Maddock, T., Jr. 1953. The hydraulic geometry of stream channels and some physiographic implications, *U.S. Geol. Surv. Prof. Paper 252.*

Leopold, L. B. and Miller, J. P. 1956. Ephemeral streams—hydraulic factors and their relation to the drainage net, *U.S. Geol. Surv. Prof. Paper 282-A.*

Leopold, L. B., Wolman, M. G., and Miller, J. P. 1964. *Fluvial processes in geomorphology.* W. Freeman, San Francisco and London.

Lewis, L. A. 1969. Some fluvial geomorphic characteristics of the Manati Basin, Puerto Rico. *Ann. Ass. Amer. Geog.,* **59,** pp. 280–293.

Richards, K. S. 1973. Hydraulic geometry and channel roughness—a non-linear system. *Amer. J. Sci.,* **273,** pp. 877–896.

Wolman, M. G. 1955. The natural channel of Brandywine Creek, Pennsylvania. *U.S. Geol. Surv. Prof. Paper 271.*

C. C. PARK

Lecturer in Geography
St. David's University College
Lampeter

8

Man-induced Changes in Stream Channel Capacity

Recent years have witnessed a growing awareness of the need for increased understanding of the various effects which man's activities have upon the landscape. The importance of this 'environmental-impact' orientation to geomorphology is evidenced in the recent growth of Environmental Geomorphology (Coates, 1971; Cooke and Doornkamp, 1974), in the increasing number of geomorphological contributions to recent symposia concerned directly with environmental modification (e.g. Oglesby, Carlson, and McCann, 1972; I.A.S.H., 1974), and in the appearance of papers dealing with local examples of man's impact on landscape forms (e.g. Hammer, 1972) and processes (e.g. Walling and Gregory, 1970).

Interest in man as a geomorphological agent is not, however, an entirely new phenomenon. Marsh's *Man and Nature* (1864), and more recently Sherlock's *Man as a Geological Agent* (1922), signposted early advances in understanding man's significance, whilst the popular interdisciplinary volume *Man's Role in Changing the Face of the Earth*, edited by Thomas (1958), provides a striking milestone in the history of the subject. The orientation of much geomorphological research, within the last two decades, towards man-induced landscape changes, has been stressed in several reviews such as those offered by Jennings (1965), Brown (1970), and Gregory (1976). This redirection of some research effort was conditioned by several stimuli, including the impetus offered by the International Hydrological Decade (1965–1974)—one of the aims of which was to increase understanding of man's role in the hydrological cycle (Keller, 1968); by the increasingly interdisciplinary nature of drainage basin research (Rodda, 1971) and a growing awareness of the potential contribution of hydrological research to planning (McCulloch, 1972; Painter, Rodda, and Smart, 1972); by the recent logistical advances offered, for example, by the development of continuous monitoring schemes (e.g. Walling and Teed, 1971), and by the increasing application of remote sensing techniques (Painter, 1973), as well as a general increase in interest in environmental problems.

The nature and magnitude of man's influences upon the physical landscape in terms of rates of erosion (e.g. Douglas, 1967; Judson, 1968) and drainage basin dynamics (Gregory and Walling, 1973) have attracted a considerable amount of attention, particularly within the last decade, although many aspects of man-induced landform changes have hitherto not received the attention which their potential significance and magnitude seem to warrant. This is true, in particular, of stream channel changes. The extent of human impact on channel changes is highly variable, and will depend largely upon the activities involved and the proportion of the drainage basin which is directly affected. Man-induced channel changes are of two basic types—direct and indirect.

Types of Channel Change Induced by Man

Direct changes are those brought about by some direct and generally purposeful human action upon the stream channel, and these are generally related to engineering schemes designed to alleviate existing or impending threats of flooding, sedimentation or erosion. Channel stabilization schemes, for example, involving upper bank paving, subaqueous mattressing, dikes, and jetties (Stanton and McCarlie, 1962; Task Committee, 1965) directly modify the stream channel in cross-section, profile, and possibly also planform, although in many cases the stabilized channel conforms to the channel morphology dimensions which existed before intervention (Winkley, 1972). Such schemes are often essential for the provision of maintained or increased channel capacities, and the prevention of bed or bank erosion (Task Committee, 1972). However, whilst alternative watershed management schemes would often affect the stream channels substantially less, in circumstances such as those facing Ferrell (1959) in the San Gabriel Mountains of Los Angeles County, for example, stabilization provides the only practical method of reducing excessive sediment volumes. Channel improvement schemes also assist in the stabilization of streambanks and removal of excessive sedimentation (Erichsen, 1971). Whilst many channel equilibrium problems may be solved in individual reaches (Jefferson, 1965), subsequent channel changes may be induced in downstream reaches by sedimentation and increased flooding directly related to upstream activity (Emerson, 1971). Channel realignment schemes are often necessary, for example, to prevent the undermining of roadways adjacent to laterally migrating natural channels, but such schemes often result in channel degradation upstream and sedimentation downstream (Yearke, 1971).

Indirect changes are those brought about by the effects of human activity upon the processes which control stream channel form. In alluvial rivers, channel form adjusts to maintain hydraulic relationships between the channel boundary and the water, and sediment discharge through the channel reach. The peak flow magnitude and frequency characteristics of discharge have been shown to condition the overall channel size (Leopold and Maddock, 1953), slope (Hack, 1957), and planform (Carlston, 1965) properties, and Harvey (1969) has demonstrated the significance of hydrologic regime to channel cross-sectional form adjustment. Both channel perimeter sediment and sediment transport through the reach influence channel form—the perimeter sediment directly controlling channel shape (Schumm, 1960) and also reflecting sediment transport (Schumm, 1963), whilst suspended sediment (Leopold and Maddock, 1953) and bedload (Gilbert, 1914; Wilcock, 1971) influence channel form via their effects on channel roughness and hydraulics. Langbein and Leopold (1964), moreover, maintain that the equilibrium channel represents a state of balance between two opposing tendencies—towards a minimum rate of energy expenditure, and towards an equable rate of energy expenditure along the channel. Man-induced changes in one or more of these factors should, therefore, logically lead to channel changes. A guide to the general nature of such channel changes is offered in the set of relationships derived by Schumm (1969) on the basis of results of his own research on alluvial channels in Australia and North America.

Indirect man-induced channel changes may be the result of many different causes, and amongst the various human environmental impacts on drainage basins—which include afforestation and deforestation, precipitation modification, road construction, and interbasin water transfers—two in particular have attracted attention of late. These are, first, the effects of reservoir construction and, second, the effects of urbanization.

Reservoir Construction

The changes in water and sediment discharge in stream channels brought about by reservoir construction, and the attendant channel changes, have attracted the attention of a number of workers in the present century. The deposition of sediment in reservoirs, for example, was considered by Harris (1901) as far back as the turn of the present century, and methods of reducing sedimentation (Kenyon, 1938) and the factors controlling rates of deposition (Gottschalk, 1947) have subsequently commanded considerable attention. As a consequence of this sedimentation, the waters issued to the channels downstream from dams are largely devoid of much of their sediment load, and a tendency for substantial degrees of bed scour to occur immediately below dams—often threatening engineering works—has been widely documented (e.g. Lane, 1934; Stanley, 1972). The presence of the reservoirs effectively reduces the magnitude and frequency of peak streamflow response in the downstream channels by impounding flood waters, and the general channel response in many cases will be a reduced channel capacity. Such reductions in channel size have been observed in several areas, such as the Republican–Kansas River system in the United States (Wolman, 1967b), in Devon (Gregory and Park, 1974; Gregory, 1976), in Yorkshire (Gregory and Park, 1976a), and in the Pennines (Chapter 9).

Urbanization

Undoubtedly the most extensively documented man-induced drainage basin modifications are those associated with urbanization, both in terms of the effects of the construction activities, and also the contrasts between rural and urban environments. Changes in both discharge and sediment transport, related to construction and post-construction phases of urbanization, have been measured by a number of workers, and many of the early results have been summarized by Leopold (1972). Changes in runoff during construction include decreases in hydrograph lag times (Carter, 1961; Anderson, 1968), increased peak discharges (Wilson, 1967; Hollis, 1975), increased flood volumes, and increased percentages of runoff (Kinosita and Sonda, 1969; Gregory, 1974), although the amount of change appears to be related to local construction practices and the relative areas affected. Changes in sediment supply and transport are related to the changed runoff characteristics and the extensive areas which lie unvegetated during construction, and sediment loads may be increased between five- and tenfold (Guy and Ferguson, 1962; Keller, 1962; Walling, 1974; Walling and Gregory, 1970). The most striking changes in sediment loads have been observed during phases of active construction (Wolman, 1967a; Dawdy, 1967; Walling, 1974; Leopold, 1973). Bull and Scott (1973) have summarized the different effects upon water and sediment yield which different stages of the urbanization process may have. The nature and magnitude of channel change induced by urbanization have been examined in Wolman's (1967a) deductive model, by sequential observation of channel changes in Watts Branch, Maryland over a 20-year period during which urbanization progressively encroached into the basin (Leopold, 1973), and by spatial interpolation of channel morphometric properties by Hammer (1972) and, more recently, by Gregory and Park (1976a).

There are hence many ways in which the effects of man may induce stream channel changes. The number of studies which have been made of the actual nature and magnitude of such channel changes remains relatively small. One reason for this may be the apparent difficulty of identifying and of quantifying, the changes which might be anticipated.

Possible Ways for the Identification of Channel Change

There appear to be three approaches to the identification of man-induced channel changes. Firstly, the ideal method is actually to monitor the channel changes at monumented sites, through time, in streams subjected to modification (e.g. Leopold, 1973). The procedure is time consuming, however, and records have to be collected over many years. Such an approach also requires a period of calibration observations, before the modification begins, to provide background data on the nature of channel changes through time on that stream. As an alternative, if the changes in discharge and sediment load which accompany the man-induced effects are known or can be estimated with reasonable accuracy, then the nature and magnitude of the channel changes could be predicted either through the use of Regime Theory relationships (Blench, 1972), or through similar relationships established between channel form, and water and sediment yield (Schumm, 1969; Rango, 1970). The necessity to know the character and magnitude of the process changes, however, creates problems, and accounts for the lack of attempts to adopt this approach. Thirdly, channel changes may be identified by some form of spatial interpolation technique in which channel-form properties observed under modified conditions can be compared with estimates of the form properties at the same sites under natural conditions (e.g. Hammer, 1972). Two basic variations of this approach exist; either the comparison of adjacent streams, one of which is modified and one not, or a comparison along individual streams where the upper portions are natural and the downstream reaches are believed to have been modified to some degree. The ease with which these spatial interpolation approaches can be applied, their relative flexibility, and the fact that they do not require any previous knowledge of the magnitude of changes in streamflow and sediment yield, commend the approach.

Feasibility Study of Channel Morphometric Relationships

Before an attempt can be made to identify and quantify any induced channel changes by employing a spatial interpolation approach, it is clearly necessary to determine which of the two alternative approaches is the more suitable, with regard to the nature and magnitude of channel form variations along and between essentially 'natural' channels. Channel cross-sections were surveyed in the field by establishing a horizontal datum across the bankfull channel, and measuring down to the channel perimeter at a large number of points (see Park, 1975, 1976b), at a number of sites along 16 'natural' channels in Devon. Adopting drainage area as a measure of spatial location within the drainage basin, as well as a surrogate for discharge, relationships have been established between bankfull channel capacity (i.e. channel cross-sectional area, Brown, 1971) and drainage area in each stream (Table 8.1). Although sample sizes vary between 7 and 84, the significance of each relationship exceeds the 95% level, and in over half of the streams the coefficient of determination exceeds 85% (Figure 8.1C). What is most revealing from the table is the fact that the detailed form of the individual

TABLE 8.1 RELATIONSHIPS BETWEEN CHANNEL CAPACITY AND DRAINAGE AREA
FOR SAMPLE STREAMS IN DEVON (FROM PARK, 1976b)

Stream	Constant (a)	Exponent (b)	n	Correlation coefficient r	r^2
Kenn	0.7379	0.5300	28	0.8432	71.10%
Dart (Exe)	1.1890	0.4790	50	0.8133	66.15%
Burn	0.8155	0.8433	7	0.9870	97.41%
Shobrooke	0.7365	0.4100	10	0.9065	82.17%
Barle	0.7340	0.7000	84	0.8347	70.00%
Quarme	0.5147	0.6382	20	0.9690	93.90%
West Dart	0.4312	0.8954	18	0.9639	92.90%
East Dart	0.3775	1.0226	19	0.9741	94.90%
Main Dart	0.5504	0.8186	49	0.9808	96.19%
W. Okement	0.5129	0.7081	16	0.8308	69.03%
Meavy[a]	0.3486	0.9993	23	0.9251	85.60%
Hawkerland	0.5754	0.4435	16	0.9853	97.10%
Woodbury[b]	0.6693	0.3302	10	0.9836	96.74%
Gorhuish	1.4709	0.4143	43	0.8400	70.56%
Tone[c]	0.4400	0.5780	14	0.8637	74.60%
Deer[d]	1.0028	0.5438	20	0.9837	96.77%

NOTE: all relationships are of the form $y = aX^b$, where y is channel capacity in m^2, and X is the drainage area in km^2

[a] Natural channel above Burrator Reservoir
[b] Natural channel above Woodbury village
[c] Natural channel above Clathworthy Reservoir (Somerset)
[d] Natural channel above Holsworthy town

relationships, as characterized by both the exponent and the constant values, differs considerably between the sample streams in Devon. The exponent values quantify the rate of increase of channel size relative to increases in drainage area, and in the Devon streams these vary between 0.33 and 1.02. The constant values provide an index of the channel size at sites with a drainage area of 1 km^2 in the individual streams, and these vary between 0.35 and 1.47. It has been shown that the constant values in this form of relationship may be related to rainfall characteristics as surrogates for discharge (Gregory and Park, 1976a). Despite the considerable variations in both the constant and the exponent values, some signs of regularity in the relationships are suggested in the generally inverse correlation between the constant and the exponent values (Figure 8.1B). The variability of the relationships summarized in Table 8.1, however, suggest that it is more meaningful to base spatial interpolation on variations in channel form properties *along* individual streams, rather than *between* streams, even between two streams with essentially similar environmental conditions. Hawkerland and Woodbury streams, in east Devon, for example, are similar in their rock types, bed and bank material, and rainfall regimes, and both

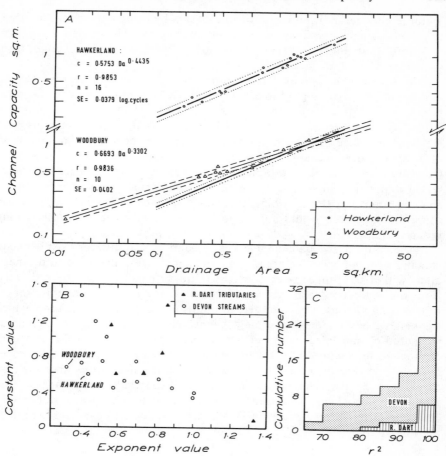

FIGURE 8.1 CHANNEL MORPHOMETRIC RELATIONSHIPS IN DEVON STREAMS

Relationships between bankfull channel capacity (*c*) and drainage area (Da) in Hawkerland and Woodbury streams are illustrated in A. The Woodbury regression relationship shows the Hawkerland regression line for comparison. Variations in the constant and exponent values from relationships between channel capacity and drainage area for a sample of Devon streams (Table 8.1), and for six tributaries of the River Dart (Park, 1976a) are plotted in B. A cumulative histogram of the coefficients of determination in these relationships appears in C, drawn as the cumulative number of cases with r^2 less than the stated value

flow intermittently in their upper reaches, although differences are apparent in the relationship between channel size and drainage area (Figure 8.1A) between the two streams, Over only a restricted range of drainage areas (*c.* 0.7–5 km^2) do the 68% confidence limits fitted to the relationships overlap. At all other points it would clearly not be advisable to estimate 'natural' channel morphometry for one stream by using data from the other, using only drainage area as an independent variable. The evidence suggests, therefore, that spatial interpolation be based on downstream channel form variations along individual streams.

Whilst channel morphometric relationships are, on the whole, well defined along individual streams (Table 8.1; Figure 8.1A), differences in the relationships arise both between streams (Table 8.1), and between tributaries within individual stream systems (Park, 1976a). Since significant regression relationships have been demonstrated between drainage area and a variety of channel form variables, characterizing channel form in cross-section (capacity, width, depth, width/depth ratio, hydraulic radius, wetted perimeter), profile (map slope, field slope), and planform (meander wavelength, channel sinuosity) (Park, 1975), then the approach of spatial interpolation using drainage area would appear to be equally applicable to determining induced changes in any of these channel form properties.

In the 'natural' stream channels clearly defined relationships exist between channel size and drainage area, so that channel sizes, at sites with drainage areas within the range covered by the 'natural' channels, can be estimated from the regression relationships (Table 8.1). It should be possible, therefore, having measured channel size at points along the stream channel which we believe has undergone some 'channel change', to compare the field-measured channel sizes with the channel sizes predicted from the upstream channel form relationships, on the basis of drainage area. The differences between the observed and the predicted channel dimensions may then be considered as the 'channel change'. To illustrate this approach to the identification and quantification of induced channel changes, a number of case studies based on streams in Devon are outlined and evaluated employing this approach.

Examples of Man-induced Changes in Stream Channel Capacity

Channel Changes Below Burrator Reservoir

One valuable situation in which man-induced channel changes may be examined is below a reservoir, where channel adjustments to reduced peak flow magnitude and frequency conditions may be anticipated. Reductions in channel size below reservoirs have been identified by Wolman (1967b) on the basis of observations of channel cross-sectional changes through time, although suitable data is rarely available for chronologically comparative studies of this sort. Gregory and Park (1974) therefore adopted a spatial interpolation approach to identify channel changes below Clatworthy Reservoir on the River Tone in Somerset. A convenient illustration of this approach is offered by Burrator Reservoir, constructed in 1893 on the River Meavy, which flows in a southwesterly direction from Dartmoor (Figure 8.2), and which has been documented by Gregory (1976).

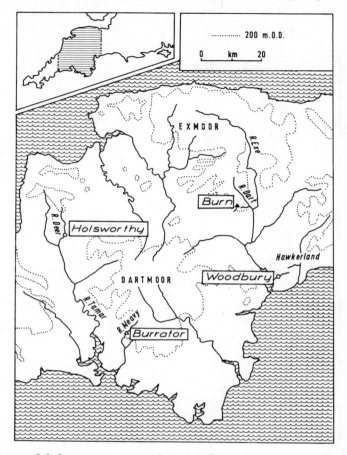

FIGURE 8.2 LOCATION OF THE STREAMS REFERRED TO IN THE TEXT

Channel cross-sections were surveyed in the field at 23 sites on 'natural' channels draining into Burrator Reservoir, in collaboration with Gregory. Bankfull channel capacities, widths, and depths at each site were measured from scale-plotted cross-sections based on these field measurements, and drainage areas were measured from the appropriate 1 : 25,000 Ordnance Survey maps. Correlations between channel capacity, width and depth, and drainage area, were established for the 'natural' channels above the reservoir (Figure 8.3) and the regression lines were calculated by the method of least squares (Table 8.2). Clearly, channel size increases relative to drainage area in a relatively systematic manner, the correlation coefficient exceeds the 99.9% significance level, and both width and depth show orderly and significant relationships with drainage area. Confidence limits, for the estimates of the regression constants, were calculated in the standard manner

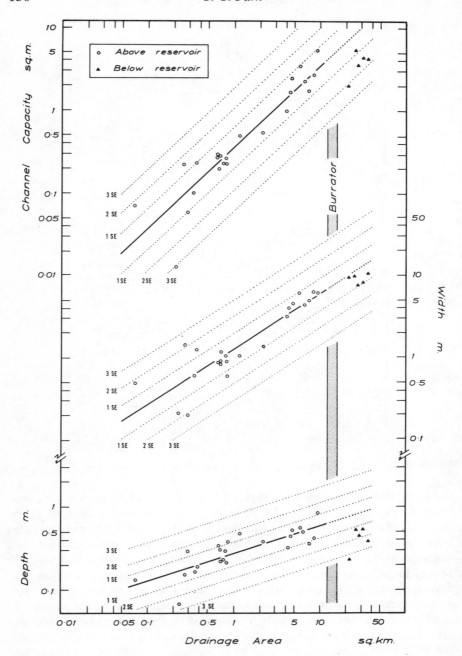

FIGURE 8.3 BURRATOR CHANNEL CHANGES

TABLE 8.2 SUMMARY OF BURRATOR CHANNEL DATA

(a) *Upstream Channel Relationships*

Relationship	n	r	$r^2(\%)$	$p(\%)$	$\log_{10}$ (s.e.)
$c = 0.3486\ Da^{1.0000}$	23	0.9251	85.59	>99.9	0.2455
$w = 1.2540\ Da^{0.6677}$	23	0.8892	79.06	>99.9	0.2055
$d = 0.2780\ Da^{0.3323}$	23	0.8029	64.46	>99.9	0.1474

(b) *Downstream Channel Changes*

Site	Drainage area (km²)	Capacity (m²) Predicted	Observed	%[a]	Width (m) Predicted	Observed	%[a]	Depth (m) Predicted	Observed	%[a]
a	36.839	12.84	4.20	33[1]	13.94	10.75	77	0.92	0.39	42[2]
b	32.045	11.17	4.40	39[1]	12.70	8.07	64	0.88	0.55	63[1]
c	27.385	9.95	3.60	38[1]	11.43	7.73	68	0.84	0.45	53[1]
d	21.832	7.61	2.02	27[2]	9.83	8.60	88	0.77	0.24	30[3]
e	25.876	9.02	5.57	62	11.01	9.73	88	0.82	0.57	70[1]
			$\bar{x}$ 39.62			$\bar{x}$ 76.85			$\bar{x}$ 51.72	
			δ 13.35			δ 11.28			δ 15.75	

[a] Observed value as % of predicted value. Significance levels of reductions, based on standard error values: 1, 68%; 2, 95%; 3, 99.7%.

(Gregory, 1963) and are shown in Figure 8.3. Channel cross-sections at five sites along the mainstream River Meavy below Burrator were subsequently surveyed, and estimated bankfull levels were identified on these cross-sections employing the same criteria as were applied in the above-dam sections, i.e. pronounced breaks of bank slope and dominant bench levels. It is clear from Figure 8.3 that the channel cross-sections below the reservoir are considerably smaller than would be predicted from upstream channel relationships, this difference occurring mainly in the channel depth dimension. In order to determine the extent of these channel reductions in greater detail, estimates were made of the channel dimensions which might be expected to exist at 'natural' sites with the same drainage areas as the below-dam sites, by the application of the upstream regression relationships (Table 8.2). The ratio between the observed and predicted channel properties at each site, expressed as a percentage, provides a convenient 'reduction ratio' index (Gregory and Park, 1976a). It is clear from Table 8.2, for example, that the channel at site (d)—immediately below the reservoir—is merely 27% of what would be predicted. The channel depth here is only 30% and the width 88%, of the predicted levels. As might be expected, the reduction ratios tend, in general, to decrease in the downstream direction (Figure 8.3), since the relative reduction in peak discharge levels below the reservoir will tend to decrease downstream with the contribution of flood waters from major tributaries. Nonetheless, on average, the channel sizes below the reservoir represent only 40% of what would be predicted from upstream (Table 8.2). It is possible to assess the significance of the reductions of channel dimensions, with reference to the confidence limits shown in Figure 8.3. At four of the five sites observed, channel capacity is less than the predicted capacity by an amount which exceeds the 68% significance level, and at all five sites the channel mean depth reduction exceeds the 68% level. It is unlikely that the observed differences in channel dimensions above and below Burrator Reservoir could have occurred entirely by chance, so that the hypothesis that reduced channel size is an adjustment to reduced peak flow characteristics seems to be acceptable. The general channel size reductions noted elsewhere below reservoirs (Wolman, 1967a; Gregory and Park, 1974, 1976a) thus appear to have occurred on the River Meavy below Burrator Reservoir and the effect appears to have been brought about principally in the channel depth dimension. The reductions in channel size generally exceed the 68% significance level, and the approach appears suitable for identifying, and for quantifying, channel changes of this magnitude and nature.

Channel Changes Associated with Holsworthy Urban Area

The town of Holsworthy, in west Devon, is located on the River Deer, a left-bank tributary of the River Tamar (Figure 8.2). Whilst Holsworthy is not a large urban area, its size relative to the size of the River Deer, and the fact that the greater part of the town has been developed around the river and on the adjacent valley side, lend substance to the hypothesis that channel changes on the River Deer below Holsworthy may be related to the modification in steamflow and sediment yield associated with the presence of the town (Gregory, 1974; Hollis, 1974, 1975).

Channel cross-sections were levelled in the field at 18 sites along the mainstream River Deer above Holsworthy, and at two sites on the River Tamar into which it flows (one above and one below the confluence). Clearly, from Figure 8.4, the relationships between the channel dimensions and drainage area which exist above the town, continue to apply in the two 'natural' sections on the River Tamar at the confluence. Thus the channel morphometric relationships in the 'natural' sections can be defined by regression [Table 8.3(a)] and confidence limits placed on the regression intercept values (Figure 8.4). For comparison with the 'natural' sections, cross-sections were levelled at ten sites on the River Deer between Holsworthy and the confluence with the River Tamar, at roughly equally spaced distances along the mainstream. The channel dimensions at these sites are shown in Figure 8.4 and although channel mean depth and channel capacity for the 'below urban' channel sections are in general larger than might be expected in the 'natural' channels, the differences appear to

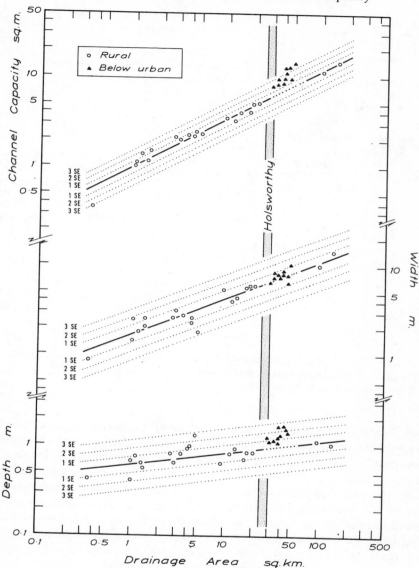

FIGURE 8.4 HOLSWORTHY CHANNEL CHANGES

be relatively minor. Channel dimensions, for sites with similar drainage areas to the 10 downstream sites, were predicted from the 'natural' channel morphometric relationships [Table 8.3(b)] and the ratio between the observed and predicted channel dimensions at each site provide an 'enlargement ratio' index (Hammer, 1972; Gregory and Park, 1976a). It is clear

C. C. Park

TABLE 8.3 SUMMARY OF HOLSWORTHY CHANNEL DATA

(a) *Upstream Channel Relationships*

Relationship	n^a	r	$r^2(\%)$	$p(\%)$	$\log_{10}$(s.e.)
$c = 1.0028\ Da^{0.5438}$	20	0.9837	96.77	>99.9	0.0652
$w = 1.6459\ Da^{0.4145}$	20	0.9415	88.66	>99.9	0.0974
$d = 0.6094\ Da^{0.1293}$	20	0.6718	45.12	99	0.0936

(b) *Downstream Channel Changes*

Site	Drainage area (km²)	Capacity (m²) Predicted	Observed	%[b]	Width (m) Predicted	Observed	%[b]	Depth (m) Predicted	Observed	%[b]
a	29.600	6.33	8.00	126[1]	6.70	6.50	97	0.94	1.23	130[1]
b	31.120	6.50	8.25	127[1]	6.84	7.25	106	0.95	1.14	120
c	33.430	6.76	9.75	144[2]	7.05	8.50	121	0.96	1.15	120
d	36.170	7.06	8.75	124[1]	7.28	8.00	110	0.97	1.09	113
e	37.700	7.22	9.75	135[2]	7.41	7.70	104	0.97	1.27	130[1]
f	39.900	7.44	11.00	148[2]	7.58	8.65	114	0.98	1.27	129[1]
g	43.000	7.75	13.00	168[3]	7.82	8.00	102	0.99	1.63	164[2]
h	43.840	7.84	13.00	166[3]	7.89	8.00	101	0.99	1.63	164[2]
i	46.110	8.05	9.75	121[1]	8.05	6.50	81	1.00	1.50	150[2]
j	48.260	8.26	14.50	176[3]	8.21	10.50	128[1]	1.01	1.38	137[1]
		$\bar{x}$	143.46%			$\bar{x}$	106.36%		$\bar{x}$	135.67
		δ	20.17%			δ	13.05%		δ	18.01

[a] Includes two mainstream sites on River Tamar, below confluence
[b] Significance levels as in Table 8.2(b)

from Table 8.3(b) that downstream channel capacities demonstrate an average enlargement ratio of nearly 145%, although the range of ratios is between 121% and 176%. Most of the channel enlargement has occurred in the depth dimension [Figure 8.4; Table 8.3(b)], where the average enlargement ratio of 135% contrasts markedly with an average width enlargement ratio of 106%. Changes in channel width have thus been relatively minor. Indeed at only one of the ten sites studied did the observed channel width differ from the 'natural' channel relationship by an amount significant at the 68% level. Of the channel depth changes, seven exceeded the 68% significance level and three of these exceeded the 95% level.

The channel cross-sections of the River Deer below Holsworthy therefore appear to be significantly larger (at the 68% level) than predicted 'natural' channel sections, representing increases generally in the channel depth dimension. This evidence thus demonstrates the nature of the channel changes which might be attributed to the influence of the 'urban' area on water and sediment yield into the channel system.

Channel Changes Associated with Woodbury 'Urban' Area

It is logical to expect the channel changes to be larger as the proportion of urbanized drainage basin increases. It may therefore be revealing to consider very small streams. A convenient east Devon stream for such a study is a stream referred to as Woodbury stream. It flows westward from Woodbury Common, through the built-up area of Woodbury which is large in comparison with the drainage area (Figure 8.2).

Ten cross-sections in the upstream 'natural' channel were levelled in the field, and the 'natural' morphometric relationships illustrated in Figure 8.5

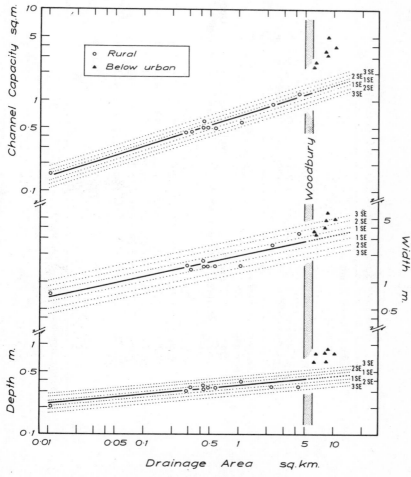

FIGURE 8.5 WOODBURY CHANNEL CHANGES

TABLE 8.4 SUMMARY OF WOODBURY CHANNEL DATA

(a) *Upstream Channel Relationships*

Relationship	n	r	r^2(%)	p(%)	$\log_{10}$(s.e.)
$c = 0.6693 \, Da^{0.3302}$	10	0.9836	96.74	>99.9	0.0402
$w = 1.9293 \, Da^{0.2333}$	10	0.9269	85.91	>99.9	0.0626
$d = 0.3470 \, Da^{0.0969}$	10	0.8720	76.04	99%	0.0361

(b) *Downstream Channel Changes*

Site	Drainage area (km²)	Capacity (m²) Predicted	Observed	%[a]	Width (m) Predicted	Observed	%[a]	Depth (m) Predicted	Observed	%[a]
a	6.109	1.22	2.33	191	2.94	3.65	124[1]	0.41	0.64	154[3]
b	6.492	1.24	2.70	218[3]	2.98	3.35	112[1]	0.42	0.81	194[3]
c	8.230	1.34	3.65	272[2]	3.15	4.00	127[1]	0.43	0.84	197[3]
d	8.532	1.36	3.13	230[3]	3.18	4.80	151[2]	0.43	0.65	152[3]
e	8.690	1.37	5.20	380[3]	3.20	5.80	182[3]	0.43	0.90	210[3]
f	10.388	1.45	4.00	276[3]	3.33	5.00	151[3]	0.44	0.80	184[3]
			$\bar{x}$	261.16%		$\bar{x}$	140.93%		$\bar{x}$	181.72%
			δ	66.86%		δ	25.06%		δ	23.55%

[a] Significance levels as in Table 8.2(b)

are summarized in Table 8.4. Channel capacity, width, and depth are each related significantly to drainage area. Six cross-sections were levelled in the mainstream below Woodbury, and it is clear from Figure 8.5 that the observed channel sizes are substantially larger than would be predicted from upstream. Channel capacity below Woodbury is nearly 200% of the predicted level [Table 8.4(b)], and increases to some 380% further downstream. At each of the six sites the increased channel capacity exceeds the 99.7% significance level (Figure 8.5) by a substantial extent. The contributions of channel width and mean depth to these increased channel sizes follows the general pattern outlined for the River Deer below Holsworthy. Whilst channel depth increases are all substantial, average 182%, and exceed the 99.7% significance level, the width increases are notably less and average 141%, although three exceed the 95% significance level (Figure 8.5).

Both studies associated with 'urban' areas, have several conclusions in common. Channel sizes below the 'urban' areas are generally substantially larger than predicted from the 'natural' channel cross-sections upstream, and the increase in channel size has apparently occurred almost entirely in the channel depth dimension. Changes in channel width have been minor. Most of the channel changes differ from the 'natural channel morphometric relationships by. statistically significant amounts and it seems likely that in both cases the approach has identified channel changes representing the adjustment of the stream channels to the increased streamflow associated with urbanization.

Channel Changes Associated with a Gully

The logic of focusing attention upon small streams where the magnitude of channel changes relative to the size of stream is relatively large, may be taken further, and examples sought of different forms of man-induced channel change on small headwater streams. In central Devon, the development of an incised gully (Burn in Figure 8.2) has recently been documented by Gregory and Park (1976b), and the available evidence demonstrates that the gully development was a result of road drainage installation associated with the metalling of a roadway which runs down the eastern watershed of the formerly small stream. Estimates of the time period required for gully development, based on bedload movement measurements, suggest an age of the feature of approximately 29 years, which appears to be in accord with the fact that road metalling appears to have occurred about 1949, with the personal evidence of local farmers who have observed the progressive increases of size of the gully over 30 years, and with the fact that neither gully nor stream appear on the Ordnance Survey 1 : 10,560 maps of the area which were surveyed before 1930.

The original study concentrated upon the processes in, and evolution of the gully channel, and the morphometry of the feature has subsequently been examined by the spatial interpolation approach. Since the feature is small, total length of the order of 500–600 metres, it was difficult to delimit and measure the drainage areas contributing to a number of sample sites along the channel, and so distance downstream was applied as the 'independent' variable.

Channel cross-sections were levelled at a total of 50 sites, spaced at 10-metre intervals, between the uppermost point along the channel at which a distinct channel form was apparent, and the confluence of the Burn Gully

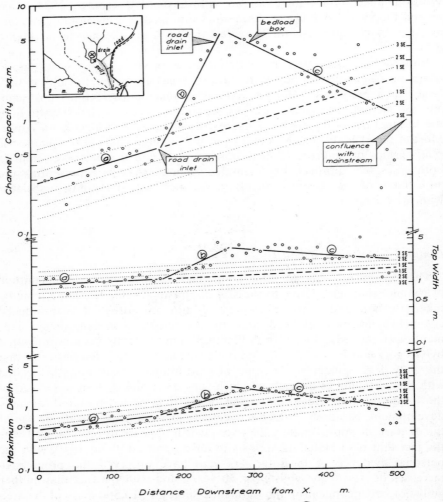

FIGURE 8.6 BURN GULLY CHANNEL CHANGES

with the mainstream Burn stream, a tributary to the River Dart of mid-Devon (Park, 1975). Two road drain inlets lead storm water from the road directly into the channel (see inset in Figure 8.6) and so the channel reach may conveniently be considered in three segments—one upstream from the upper inlet (a), one between the inlets (b), and one downstream from the lower inlet (c). From the field measurements, scale cross-sections were drawn, and the 'bankfull' level at each site was delimited at the discrete morphological break of slope between the nearly vertical channel banks and the adjacent relatively 'flat' surface of the fan into which the gully has been incised (Gregory and Park, 1976b). Channel capacity, bank top width, and maximum channel depth were measured and variations in these three variables relative to distance downstream are plotted in Figure 8.6.

Along segment (a) the channel size increases in the downstream direction, although the 'independent' variable accounts for only 40% of the variations in channel capacity (Table 8.5). These changes occur principally in the depth dimension (Figure 8.6), and bank width is seen to vary relatively little over the reach. Downstream from the first road drain inlet, however, in segment (b), channel size increases considerably, both in terms of bank width and maximum depth, and it is clear from Figure 8.6 that these increases are relatively systematic, with over 90% of the variations in channel capactiy being 'accounted for' by distance downstream. Along segment (c), downstream from the lower drain inlet, channel size is seen to decrease considerably (Figure 8.6), in both width and depth, this decrease being both systematic and significant (p greater than 99.9%, Table 8.5). It emerges from Figure 8.6 that at the lower sections of this reach (c), near to the confluence, the channel dimensions of the gully are of a similar order, if not lower, than would be predicted from the upstream 'natural' channel relationships, characteristic of the upper segment (a).

Thus the Burn Gully exhibits significantly systematic spatial adjustments of form, reflecting the man induced modifications brought about by the

TABLE 8.5 SUMMARY OF BURN GULLY CHANNEL DATA

Channel segment[a]	Relationship	n	r	r^2(%)	$\log_{10}$(s.e.)
a	$\log c = 0.0018$ Dist $- 0.5615$	16	0.6327	40.03	0.1024
a	$\log w = 0.00048$ Dist $+ 0.0091$	16	0.3086	9.52	0.0696
a	$\log d = 0.001245$ Dist $- 0.3313$	16	0.5865	34.40	0.0819
b	$\log c = 0.0114$ Dist $- 2.2163$	10	0.9548	91.16	—
b	$\log w = 0.00517$ Dist $- 0.8198$	10	0.8448	71.37	—
b	$\log d = 0.00327$ Dist $- 0.6586$	10	0.7490	56.10	—
c	$\log c = 1.5767 - 0.0031$ Dist	20	0.8890	79.03	—
c	$\log w = 0.8076 - 0.00089$ Dist	20	0.5814	33.80	—
c	$\log d = 0.7357 - 0.001513$ Dist	20	0.9126	83.28	—

[a] See Figure 8.6 for details.

presence of the road drain inlets. The channel morphometry of the upper segment reflects the 'natural' morphometry of the Burn channel. During times of rainfall, however, segment (b) receives both the runoff contributed from segment (a) and the possibly substantial amounts of direct runoff from the upper portions of the adjacent roadway (inset in Figure 6.8), fed into the channel via the road drain inlet. The natural response to this increased discharge has been the substantially increased channel size evident along the reach. Since further increments of direct roadway runoff will have been added to the stream at the lower drain inlet, it is logical that the maximum channel change should have occurred here (Figure 8.6), both because of the direct effects of the influx of water, and the tendency towards headward erosion of the gully form—a tendency whose effects have clearly been magnified in the middle reach of the gully by the contributions of storm water from the upper road drain. Expansion of gully form along this reach appears to have been assisted by the trampling effects of cattle on the bank tops, which also have access to the channel floor at several points along the reach.

The reductions in channel size along segment (c), in particular at the lower end of the reach, may be related to the slope of the channel in this reach, because the height of the lower portion of the gully is conditioned by the height of the confluence, and the height of the upper portion of segment (c) is conditioned by the depth of incision of the gully around the region of the lower drain inlet. It is possible that the overall nature of the gully form adjustment along the entire length of the channel has been conditioned by slope, for this would largely determine the maximum depth of the feature, whilst the bank width is largely related to slumping of the incohesive bank material. The channel size reductions along segment (c) may also be related to deposition in the downstream sections of the reach, conditioned by rapid gully development upstream which is currently providing bedload sediment in the order of $1164 \, \text{kg} \, \text{km}^{-2} \, \text{year}^{-1}$ (Gregory and Park, 1976b). Sedimentation appears to be responsible for the small channel capacities along this reach, which are substantially smaller than would be predicted from the upstream 'natural' channel relationships (Figure 8.6) and for the fact that, with the exception of the three most downstream sites, many of the sections along the reach are wider than predicted from upstream, though less deep. The higher than predicted width/depth ratios along this reach suggest channel adjustments to predominantly bedload sediment movement (cf. Schumm, 1963), which is consonant with the observed high rates of bedload movement through the gully.

The morphometric channel changes along the Burn Gully can thus be related quite clearly to the effects upon the fluvial system of the two road drain inlets, and the spatial interpolation approach would appear to be well suited to the identification and quantification of such man-induced channel changes.

Conclusion

Whilst a relatively large number of studies of the effects of human impacts upon drainage basin processes have been made in recent years, it is to some extent paradoxical that studies of such impacts on drainage basin forms have hitherto been few. It seems likely that the main reason for this is the difficulty of identifying such changes in the field. Considerable promise in the identification and quantification of man-induced changes in stream channel morphometric properties is offered by a spatial interpolation approach, based upon the evidence that channel form dimensions vary along 'natural' stream channels in a systematic and quantifiable manner. Field evidence from Devon illustrates that significant channel morphometric relationships can be established, for example, between channel size and drainage area. Whilst it is shown that the detailed form of these relationships varies considerably between streams, along individual streams it is possible to predict channel morphometric properties at downstream sites on the basis of upstream relationships, using either drainage area or distance upstream as 'independent' variables. The application of such an approach to identification of channel changes below a reservoir, below 'urban' areas, and related to inputs of direct runoff via road drainage systems, is illustrated. It seems likely that channel changes associated with a large variety of man-induced drainage basin process changes may eventually be delimited by the use of such an approach. The basic assumption behind the approach is that the downstream variations in channel form along individual streams are regular and continuous, and that discrete breaks in this continuity relate to the man-induced change. The evidence available from the 16 Devon streams (Park, 1976b) shows that this regularity of downstream variations is found along each of the streams studied, and no abrupt or major 'breaks' in the continuity of the variations can be related to differences in adjacent bank vegetation or underlying rock type. This is important, because one might anticipate that lithologically strong rocks could outcrop in the vicinity of a dam and these might constrain channel capacity below the dam and thus influence the below-dam channel reductions. Whilst the possibility of some lithological control must remain, in particular below the reservoir, the available evidence does strongly suggest that the channel changes have occurred in response to changes in water and sediment yield occasioned by the drainage basin modifications.

Acknowledgements

A special debt of thanks is owed to Professor K. J. Gregory, who, as mentor and friend has proved to be a most willing source of inspiration, illumination, and encouragement. Dr D. N. Wilcock, Dr V. Gardiner, Dr D. A. Davidson, Mr G. N. Sumner, and my wife Brenda, kindly read and commented on an earlier draft of the paper, and offered many valuable suggestions for its improvement. The field work was undertaken during the tenure of a N.E.R.C. Research Studentship at the University of Exeter, for which grateful thanks are offered.

References

Anderson, D. G. 1968. Effect of urbanization on floods in Northern Virginia. *U.S. Geol. Survey. Prof. Paper 475A.*

Blench, T. 1972. Morphometric Changes. In Oglesby, R. T., Carlson, C. A., and McCann, J. A. (Eds.), *River Ecology and Man.* Academic Press, New York, pp. 287–308.

Brown, D. A. 1971. Stream channels and flow relations. *Water Resour. Res.,* **7,** pp. 304–310.

Brown, E. H. 1970. Man shapes the Earth. *Geog. Journ.,* **136,** pp. 74–84.

Bull, W. B. and Scott, K. M. 1974. Impact of mining gravel from urban stream beds in the Southern United States. *Geology,* **2,** pp. 171–174.

Carlston, C. W. 1965. The relation of free meander geometry to stream discharge; and its geomorphic implications, *Amer. J. Sci.,* **263,** pp. 864–885.

Carter, R. W. 1961. Magnitude and frequency of floods in suburban areas. *U.S. Geol. Surv. Prof. Paper 424B,* pp. 9–11.

Coates, D. R. 1971. *Environmental Geomorphology,* State University of New York, Binghamton.

Cooke, R. U. and Doornkamp, J. C. 1974. *Geomorphology in Environmental Management: An Introduction.* Clarendon Press, Oxford.

Dawdy, D. R. 1967. Knowledge of sedimentation in urban environments. *J. Hydraul. Div. Amer. Soc. Civ. Engrs.,* **93,** pp. 235–245.

Douglas, I. 1967. Man, nature and the sediment yield of rivers. *Nature,* **215,** pp. 925–928.

Emerson, J. W. 1971. Channelization—a case study. *Science,* **173,** pp. 325–326.

Erichsen, F. P. 1971. Let's take a look at Channel improvement. *Proc. Symp. on Inter-disciplinary aspects of watershed management.* Montana State University. A.S.C.E. pp. 261–267.

Ferrell, W. R. 1959. Mountain channel treatment in Los Angeles County. *J. Hydraul. Div. Am. Soc. Civ. Engrs.,* **85,** pp. 11–20.

Gilbert, G. K. 1914. The transportation of debris by running water. *U.S. Geol. Survey. Prof. Paper 86.*

Gottschalk, L. C. 1947. A method of estimating sediment accumulation in stock ponds. *Trans. Amer. Geophys. Union,* **28,** pp. 621–625.

Gregory, S. 1963. *Statistical Methods and the Geographer.* Longman, London.

Gregory, K. J. 1974. Streamflow and building activity. In Gregory, K. J. and Walling, D. E. (Eds.) *Fluvial Processes in Instrumented Watersheds.* Inst. Brit. Geog. Spec. Publ. **6,** pp. 107–122.

Gregory, K. J. 1976. Drainage basin adjustments and Man. *Geographica Polonica,* **34,** pp. 155–173.

Gregory, K. J. and Park, C. C. 1974. Adjustment of river channel capacity downstream from a reservoir. *Water Resour. Res.,* **10,** pp. 870–873.

Gregory, K. J. and Park, C. C. 1976a, Stream channel morphology in north west Yorkshire. *Rev. de Geom. Dyn.,* **25,** pp. 63–72.

Gregory, K. J. and Park, C. C. 1976b. The development of a Devon Gully and Man. *Geography,* **61,** pp. 77–82.

Gregory, K. J. and Walling, D. W. 1973. *Drainage basin form and process.* Edward Arnold, London.

Guy, M. P. and Ferguson, G. E. 1962. Sediment in small reservoirs due to urbanization, *J. Hydraul. Div. Amer. Soc. Civ. Engrs.*, **88**, pp. 27–37.

Hack, J. T. 1957. Studies of longitudinal stream profiles in Virginia and Maryland. *U.S. Geol. Survey. Prof. Paper 294-B*, pp. 45–97.

Hammer, T. R. 1972. Stream channel enlargement due to urbanization. *Water Resour. Res.* **8**, pp. 1530–1540.

Harris, E. G. 1901. Effects of dams and like obstructions in silt-bearing streams. *Engineering News*, **47**, pp. 110–111.

Harvey, A. M. 1969. Channel capacity and the adjustment of streams to hydrologic regime. *J. Hydrol.*, **8**, pp. 82–98.

Hollis, G. E. 1974. The effect of urbanization on floods in the Canon's Brook, Harlow, Essex. In Gregory, K. J. and Walling, D. E. (Eds.), *Fluvial Processes in Instrumented Catchments.* Inst. Brit. Geog. Spec. Pub. **6**, pp. 123–139.

Hollis, G. E. 1975. The effect of urbanization on floods of different recurrence intervals. *Water Resour. Res.*, **11**, pp. 431–435.

Jefferson, P. O. 1965. Performance of channel changes. *Alabama Highway Research H.P.R.* Report 9.

Jennings, J. N. 1965. Man as a geological agent. *Australian J. Science*, **28**, pp. 150–156.

Judson, S. 1968. Erosion of the lands; or what's happening to our continents? *Amer. Scientist*, **56**, pp. 356–374.

Keller, F. J. 1962. Effect of urban growth on sediment discharge, North West Branch Anocostia River Basin, Maryland. *U.S. Geol. Survey Prof. Paper 450-C*, pp. 129–131.

Keller, R. 1968. The role of geography within the International Hydrological Decade. *I.G.U. Commission on Internat. Hydrol. Decade*, **6**, pp. 7–14.

Kenyon, E. C. 1938. The functions of debris-dams and the loss of reservoir-capacity through silting. *Trans. Amer. Geophys. Union* **1**, pp. 16–21.

Kinosita, T. and Sonda, T. 1969. Changes in runoff due to urbanization. *Int. Assoc. Sci. Hydrol.*, Publ. 85. pp. 787–796.

Lane, E. W. 1934. Retrogression of levels in riverbeds below dams. *Eng. News Record*, **112**, p. 836.

Langebein, W. B. and Leopold, L. B. 1964, Quasi-equilibrium states in channel morphology *Amer. J. Sci.*, **262**, pp. 782–794.

Leopold, L. B. 1972. Hydrology for urban land planning—a guidebook on the hydrologic effects of urban land use. *U.S. Geol. Survey. Circular*   554, pp. 1–18.

Leopold, L. B. 1973. River channel changes with time—an example. *Geol. Soc. Amer. Bull.*, **84**, pp. 1845–1860.

Leopold, L. B. and Maddock, T. 1953. The hydraulic geometry of stream channels, and some physiographic implications. *U.S. Geol. Survey Prof. Paper 252.*

Marsh, G. P. 1864. *Man and Nature.* Scribner, New York.

McCulloch, J. S. G. 1972. Hydrology and government in the United Kingdom. *Trans. Amer. Geophys. Union*, **53**, pp. 741–743.

Oglesby, R. T., Carlson, C. A., and McCann, J. A. (Eds.) 1972). *River Ecology and Man.* Academic Press, New York.

Painter, R. B. 1973. Potential applications of satellites in river regulation. *Water & Water Engng.*, **77**, pp. 487–490.

Painter, R. B., Rodda, J. C., and Smart, J. D. G. 1972. Hydrological research and the planner. *Surveyor*, **1**,

Park, C. C. 1975. Stream channel morphology in Devon. *Trans. Devonshire Assoc.*, **107**, pp. 25–41.

Park, C. C. 1976a. The relationship of slope and stream channel form in the River Dart, Devon. *J. Hydrol.*, **29**, pp. 139–147.

Park, C. C. 1976b. *Variations and controls of stream channel morphometry.* Unpublished Ph.D. thesis, University of Exeter.

Rango, A. 1970. Possible effects of precipitation modification on stream channel geometry and sediment yield. *Water Resour. Res.*, **6**, pp. 1765–1770.

Rodda, J. C. 1971. Why Hydrology? *Nature*, **232**, pp. 301–303.

Schumm, S. A. 1960. The shape of alluvial channels in relation to sediment type. *U.S. Geol. Survey. Prof. Paper 352 B.*

Schumm, S. A. 1963. A tentative classification of river channels. *U.S. Geol. Survey Circular 477.*

Schumm, S. A. 1969. River metamorphosis. *J. Hydraul. Div. Amer. Soc. Civ. Engrs.*, **95**, pp. 255–273.

Sherlock, R. L. 1922. *Man as a Geological Agent.* Witherby, London.

Stanley, J. W. 1972. Retrogression on the lower Colorado River after 1935. *Trans. Amer. Soc. Civ. Engrs.*, **116**, Paper 2453. pp. 943–957.

Stanton, C. R. and McCarlie, R. A. 1962. Streambank stabilization in Manitoba. *J. Soil & Water Conservation*, **17**, pp. 169–171.

Task Committee for Preparation of Manual on Sedimentation. 1965. Channel stabilization of alluvial rivers. *Proc. Amer. Soc. Civ. Engrs. Waterways & Harbours Div.*, **91**, WW1 pp. 7–35.

Task Committee for Preparation of Manual on Sedimentation. 1972. Sediment control methods. B. Stream Channels. *J. Hydraul. Div. Amer. Soc. Civ. Engrs.*, **98**, pp. 1295–1326.

Thomas, W. L. 1958. *Man's Role in Changing the Face of the Earth.* Univ. Chicago Press, Chicago.

Walling, D. E. 1974. Suspended sediment production and building activity in a small British basin. *Proc. Paris Sympos. on Man's Influence on Interface of Hydrological Cycle—Physical Environment*, IASH-UNESCO-WMO, pp. 137–144.

Walling, D. E. and Gregory, K. J. 1970. The measurement of the effects of building construction on drainage basin dynamics. *J. Hydrol.*, **11**, pp. 129–144.

Walling, D. E. and Teed, A. I. 1971. A simple pumping sampler for research into suspended sediment transport in small catchments. *J. Hydrol.* **13**, pp. 325–337.

Wilcock, D. N. 1971. Investigation into the relations between bedload transport and channel shape. *Geol. Soc. Amer. Bull.*, **82**, pp. 2159–2176.

Wilson, K. V. 1967. A preliminary study of the effect of urbanization on floods in Jackson, Mississippi. *U.S. Geol. Survey. Prof. Paper 575D*, pp. 259–261.

Winkley, B. R. 1972. River regulation with the aid of nature. In *Factors Affecting River Training and Flood plain Regulation.* Internal. Commission on Irrigation and Drainage. **29**, pp. 433–457.

Wolman, M. G. 1967a. A cycle of sedimentation and erosion in urban river channels. *Geog. Annaler.*, **49A**, pp. 385–395.

Wolman, M. G. 1967b. Two problems involving river channel changes and background observations. *Northwest University Studies in Geography*, **14**, pp. 67–107.

Yearke, L. W. 1971. River erosion due to channel changes. *Civil Engineering*, **41**, pp. 39–40.

G. E. PETTS
Department of Geography
Dorset Institute of Higher Education

9

Channel Response to Flow Regulation: The Case of the River Derwent, Derbyshire

Man first attempted dam construction for the purpose of river regulation some 5000 years ago in Egypt (Smith, 1971). Increasingly, dams are being used for a variety of water resource development schemes so that the number built since 1900 has more than equalled those already standing before the turn of the century. Despite the rapid increase in geomorphological research over the past two decades, and the awareness of the need to evaluate environmental impacts on the landscape (Emmett, 1974), on continental waters (Unesco, 1972), and more specifically on river channels (Leopold and Maddock, 1954), the effects of dam construction upon stream channel form have been virtually neglected. This paper outlines the effects of stream regulation, identifies five methods for the detection of changes in river channels, and describes the effects of reservoir construction upon the channel of the River Derwent, Derbyshire.

Causes and Consequences of an Altered Flow Regime

The retention of water behind a dam and its gradual release downstream results in the reduction of peak discharges and regulation of the flow regime. The variety of impacts that a single reservoir may have upon streamflow has been summarized by Rutter and Engstrom (1964). Reduction of peak discharges is achieved not only by the storage of floodwaters in the reservoir volume proper, responsible for a reduction of 98% in peakflow when no overflow occurs, but also through the storage provided by the rise in water-level above the overflow weir. Routing through a reservoir with no available storage may reduce peak discharges by over 50% (Moore, 1969).

Reduction in the magnitude of the mean annual flood by 60% has been recorded on the Colorado River below Hoover Dam (Dolan, Howard, and Gallenson, 1974). The total volume of flow may be reduced by the increase in time during which seepage and evaporation losses occur, possibly responsible for a loss of 25% (Gilbert and Sauer, 1970). Base flow tends to be increased through seepage and the provision of a compensation flow to the channel below the dam.

Reservoirs having a large storage capacity will trap and store in excess of 95% of the sediment load transported by the river (Leopold, Wolman, and Miller, 1964), the actual percentage depending upon the ratio between storage capacity and inflow. Although sluicing and venting (Brown, 1944) may reduce the trap efficiency by as much as 10%, the reduction of the water storage capacity of reservoirs by over 1% per year has been recorded in Europe (Gvelesiani and Shmalkmzel, 1971) and America (Frickel, 1972).

Effects of dams upon the sediment load must not be considered in isolation but in relation to changes in competence, in regime, in channel morphometry, and with regard to major tributaries. The problems of tributaries transporting large quantities of sediment into a regulated mainstream with reduced flows and which has lost the ability to flush sediments, have been frequently documented (Lane, 1955; Makkaveyev, 1970). These effects have been shown to involve continued aggradation, increase in bed slope, and trenching of the deposit to form a channel that is once again in quasi-equilibrium with the flow regime (King, 1961). The rapids of the Colorado River were mostly created following the construction of Hoover Dam (Dolan, Howard, and Gallenson, 1974). Sediment introduced into reaches below a dam during construction may affect the roughness and form of the channel, forming benches along the channel sides even after flushing (Eustis and Hillen, 1954).

Reduction in sediment supply is usually greater than that in sediment carrying capacity so that the channel below the dam experiences erosion. The literature published over the past 50 years is replete with reports of such channel degradation. Several theories have been advanced relating the rate of sediment transport to the many parameters involved, but tremendous difficulties have been encountered when predicting the total amount and the rate of degradation (Livesey, 1963). Degradation occurs where the outflow from a reservoir has sufficient tractive force to initiate the movement of sediment in the channel below the structure (Gottschalk, 1964). Early research concentrated upon homogeneous bed material but numerous interrelated hydraulic, sedimentologic, and biotic factors (Tinney, 1962) complicate the formulation of a theory of channel degradation which would apply to all natural streams. In the last decade, differential equations have been employed to include the effect of heterogeneous bed materials and armouring (Komura and Simons, 1967). Stabilization of the river-bed by armouring requires only a surface layer, one grain thick, of non-movable material (Livesey, 1963). Once the bed has become stabilized, either by armouring or by the exposure of bed-rock, then the banks, that usually consist of finer material than the bed, begin to fail and the channel will widen. Maximum degradation usually occurs in the tail-water of the dam but may occur up to 69 channel widths downstream (Wolman, 1967). Rates of degradation of up to 15 cm per year have been observed both in the United States (Leopold, Wolman, and Miller, 1964) and Europe (Shulits, 1934). The longitudinal extent of degradation with time, which may be predicted from a mathematical–graphical procedure advanced by Hales, Shindala and Denson (1970), is of several kilometres per year in lowland streams and tens of kilometres in mountain streams (Fedorev, 1969). Channel adjustment to bed degradation and the associated reduction in bed slope has been observed for nearly 250 km below Elephant Butte Dam. U.S.A. (Stabler, 1925). In contrast, Hammad (1972) demonstrated that the River Nile may attain an armoured condition before an appreciable change of bed slope occurs.

Changes in stream channel morphometry below dams would be expected to be associated with changes in aggradation and degradation. Where an armoured layer or an outcrop of bedrock occurs, preventing erosion, a simple adjustment will occur whereby the water discharge will be accommodated within the predetermined channel.

Several studies from the semi-arid southwest United States have documented the reduction in channel width (Frickel, 1972) associated with a trend towards a well-defined channel (Kennon, 1966) downstream from detention reservoirs in response to reduced peak flows and prolonged low flows. Furthermore, the change from an ephemeral to a perennial stream due to reservoir seepage has led to the encroachment of riparian vegetation (Fraser, 1972), stabilizing the bed and bank materials, and increasing channel roughness, resulting in a higher stage for a given discharge. Vegetation incursion caused a loss of 66% in operational capacity below Trenton and Harlan County Dams (Northup, 1965). In Britain, Gregory and Park (1974), and Gregory (1976a) have observed a reduction in channel capacities below reservoirs due to the modification of the flow regime.

Although it is difficult to model the effects of river regulation mathematically, empirical equations have been developed to indicate the direction in which changes will occur (Schumm, 1969). These indicate that for any change in water and/or sediment discharge, channel metamorphosis will occur involving an adjustment of channel width, depth, slope or planform. Whether a channel aggrades or degrades appears to be related to the ratio of mean annual discharge before and after construction (Wolman, 1967), and to the amount of sediment discharged from tributaries (Lawson, 1925).

The River Derwent, Derbyshire

Regulation of the headwaters of the River Derwent was authorized in 1899, and involved the formation of the Howden and Derwent reservoirs which were completed in 1912 and 1914 respectively. Intakes were later

G. E. Petts

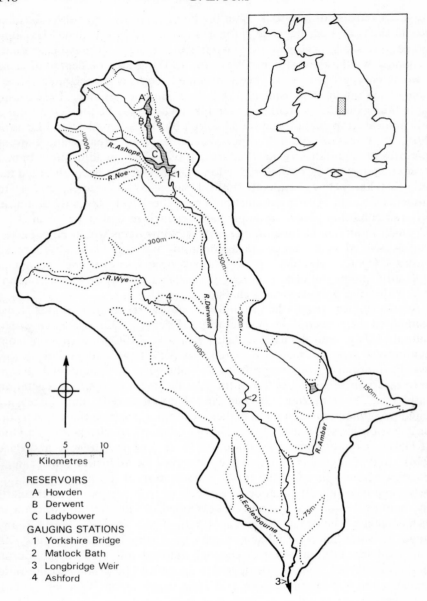

FIGURE 9.1 THE RIVER DERWENT CATCHMENT AREA

Location of reservoirs and gauging stations

constructed on the neighbouring Ashope and Alport rivers in order to supplement supply to the Derwent Reservoir. Construction of the earthfill, bellmouth overflow, Ladybower dam was begun in 1935 and contrasted with the masonry, crest-weir overflow dams of Howden and Derwent. Tunnels, later to be used for the bellmouth overflows, were effectively employed for dealing with floods during construction (Hill, 1949). The supply to Ladybower, completed in 1943, was later augmented by the construction of diversions from the River Noe and Jaggers Clough.

The catchment of the River Derwent (Figure 9.1) extends for 74 km upstream of the gauging station at Longbridge weir, Derby, encompassing an area of 1120 km². The River Derwent rises at an elevation of over 575 m O.D. on the northeast side of Bleaklow Hill, falls rapidly to below 200 m O.D. within the first 30 km of its length, and subsequently at a decreasing gradient to 44.4 m O.D. at Longbridge Weir.

Geologically, the area may be divided into three main units (Figure 9.2a). The most extensive is the Millstone Grit Series of alternating sandstones and shales, which provides a broad area of outcrop in the north, narrowing southwards to separate the Carboniferous Limestone from the Coal Measures. Although the grits are often permeable, the frequent beds of shale ensure the impermeability of the Millstone Grit Series as a whole and results in the formation of perched water tables which are responsible for frequent springs along the valley sides. In the north, the occurrence of peat overlying a weathered, fractured, gritstone surface provides some regulation of runoff and thus tends to balance the seasonal irregularities in the rainfall.

The west of the catchment is dominated by the Carboniferous Limestone Series of the Derbyshire Dome, a series composed predominantly of limestones with interbedded basaltic lavas and tuffs. The pervious limestone typically has little channelized flow in contrast to the remainder of the catchment (Figure 9.2b). In the south and east the mudstones and sandstones of the Coal Measures have an undulating topography with summits up to 180 m O.D. Restricted outcrops of Permian limestones with mudstones and Bunter sandstones occur in the south of the area.

Rock outcrops at several points along the river channel give temporary base levels for reaches upstream and, although the valley is generally wide, at Matlock the river enters a narrow gorge cut in Carboniferous Limestone. The valley fill is composed of terrace, alluvial fan, and head deposits, and glacial material occurs in rare, scattered patches on the higher slopes (Straw and Lewis, 1962). The river channel is often tree lined with birch, ash, oak, elm, and willow which, together with the several weirs and mills, tend to stabilize the slope and planform of the river.

Mean annual rainfall generally decreases in relation to altitude (Figure 9.2c) from over 1600 mm on the High Moors around Bleakflow to under 800 mm in the extreme south and east, so that the catchment average is 1080 mm. Runoff from the 127 km² headwater area is impounded in the

G. E. Petts

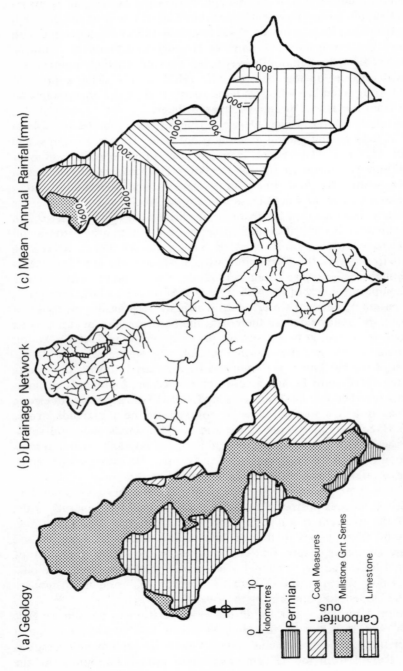

FIGURE 9.2 PHYSICAL CHARACTERISTICS OF THE DERWENT BASIN

reservoirs and a compensation flow for the channel downstream is maintained by the Ladybower Reservoir.

According to discussions in the literature, it may be expected that the reservoirs of the River Derwent will trap a large percentage of the sediment load and significantly alter the discharge hydrograph of flows downstream. Indeed, the significant percentage of reservoired surface area, some 2.7% of the regulated catchment area, indicates that there would be an appreciable reduction in the rate of outflow to reaches below the reservoirs (I.C.E., 1933). Nixon (1962) estimated that successive flood peaks in 1960 at Yorkshire Bridge were reduced by over $50 \, m^3/s$. Although streamflow records have been maintained for a number of years at the three main gauging stations on the River Derwent (Figure 9.1), the length of record available prior to reservoir construction is insufficient to enable comparison with the present flow regime. Furthermore, gross, 'naturalized' discharges recorded on a daily basis can only be compared with daily mean gauged flows. Consideration of channel form and process requires information on instantaneous peak discharges.

With regard to the sediment trap efficiency of the reservoirs, deposition by tributaries has been observed in all three cases. The capacity-inflow ratio of 0.239 indicates a high trap efficiency of the order of 80–90% (Brune, 1953), and with a storage capacity of $37.2 \times 10^4 \, m^3/km^2$ a reduction of reservoir capacity of significantly less than 0.3% per annum may be anticipated (Brown, 1944).

The reduction of peak flows and the virtually complete abstraction of the sediment load may be expected to initiate an adjustment of channel capacity downstream from the dams. Between 4 and 8 kilometres below Ladybower Dam a flat bench, bounded by marked breaks of slope, has developed within the main channel. The bench surface is generally well vegetated and bears evidence of overtopping at high flows. This feature (p. 157) appears to be related to the new flow regime imposed by the reservoirs.

Channel Response to Flow Regulation

Channel adjustments consequent upon reservoir construction may be recognized from a comparison of data collected from field surveys and values of channel capacity predicted from a regional relationship. Four other techniques may be employed to corroborate this evidence; comparison of surveys undertaken before and after construction, derivation of bankfull capacities from flow equations, sedimentological evidence, and dating of the deposits.

Channel Capacities Compared with a Regional Relationship

It has been demonstrated that the mean form, or 'hydraulic geometry', of river channels is adjusted to the dominant discharge (Ackers and Charlton,

1970), resulting from the interaction between water discharge, quantity and character of sediment discharge, and the composition of the bed and bank materials (Leopold, Wolman, and Miller, 1964). The discharge at bankfull stage has been employed in some studies to approximate the dominant discharge from the viewpoint of both sediment transport (Wolman and Miller, 1960) and flow resistance (Hey, 1972).

The problems of identifying the bankfull capacity of river channels, including along- and between-channel variation in channel form, have been reviewed by Gregory (1976b). Sedimentological (Nunally, 1967), vegetational (Leopold and Skibitzke, 1967; Gregory, 1976c), morphological (Brush, 1961; Kilpatrick and Barnes, 1964), and morphometric (Wolman, 1955; Riley, 1972) criteria have previously been advanced for the determination of the bankfull capacity of river channels. For the purpose of this study, bankfull capacity was consistently observed as the major break of slope separating a well-defined channel from a floodplain or bench, which may be as little as 1 metre in width. For compound sections the lower limit of non-aquatic vegetation was used as an additional indicator.

Measurements of channel cross-sectional area at the bankfull stage were made at 110 sites on the River Derwent, above and below the reservoirs, and on the major tributaries. Bankfull cross-sectional form was measured by stretching a tape horizontally across the channel and measuring verticals with a calibrated staff at frequent intervals. To avoid unnecessary error, sections greater than 10 metres in width were surveyed using a Hilger and Watts Quickset level. Sections were surveyed immediately upstream from riffles, and on the non-regulated channels three sections were surveyed at each location to indicate the magnitude of natural variation. A sample of bank sediment was collected at each site in order to obtain the proportion of silt-clay in the banks (Schumm, 1960) and a measure of channel slope was obtained for each site from 1:25,000 Ordnance Survey (Provisional Edition) maps as proposed by Park (1976).

As a simple regional relationship should exist between channel capacity and bankfull discharge (Nixon, 1959) it is possible to suggest the pre-dam bankfull capacity of the River Derwent below Ladybower dam. However, due to the lack of detailed discharge records describing the variation of discharge downstream, the need has arisen to find a surrogate for discharge. Drainage area has often been employed for this purpose (Gregory and Walling, 1973; Park, 1975). However, this assumes that the rate of runoff is uniform over the catchment, and that a linear relationship exists between the increase in drainage area and increase in bankfull discharge downstream. The attenuation of flood peaks due to channel storage and the spatial variation in the rate of runoff will tend to reduce the rate at which peak discharges increase downstream. The assumed linear relationship may therefore be invalid.

Drainage areas, measured from 1:25,000 Ordance Survey (Provisional

TABLE 9.1 SUMMARY OF REGRESSION ANALYSIS OF CHANNEL CAPACITY ON DRAINAGE AREA AND TOTAL STREAM LENGTH

Data source	Independent variable	Number of data points	Correlation coefficient	Two standard errors	Regression constant	Regression coefficient
Upper Derwent and Ashope	Drainage area	31	0.98	0.1752	−0.1507	0.8860
Upper Derwent and Ashope	Total stream length	31	0.97	0.2083	−0.7010	0.9870
All non-regulated streams	Drainage area	77	0.81	0.3837	−0.0040	0.5746
All non-regulated streams	Total stream length	77	0.94	0.2242	−0.5360	0.8532

Edition) Maps using a polar planimeter, and channel capacity at bankfull stage were related by simple regression analysis of log-transformed values. Although a significant relationship exists for the data from the non-regulated Upper Derwent and Ashope rivers (Table 9.1), it is statistically unsound to extend the regression beyond the extent of the data. Although incorporation of the Ecclesbourne, Amber, Noe, and Wye data (Figure 9.3) significantly increases the range of the regression to twice the drainage area of the Ladybower dam, the standard error of the regression equation is increased considerably (Table 9.1). The scatter of points reflects the presence of varied lithologies within the Derwent catchment. Furthermore, neither the addition of channel slope, nor of the percentage of silt-clay of the banks, into a multiple regression achieved a significant improvement in the explanation.

Not only does the rate of runoff vary markedly between the limestone and sandstone–slate areas (Table 9.2) but there is also difficulty in determining drainage divides. Several square kilometres of limestones to the north of Peak Forest Village in the Wye basin drain to Castleton in the Noe catchment (Fearnsides, 1932). As lithology is an important control governing both the amount and timing of surface runoff (Gregory and Gardiner, 1975), and drainage density is adjusted to the most efficient removal of flood runoff (Carlston, 1963), total stream length may be a more pertinent surrogate for discharge than is drainage area.

Total stream length was measured as the 'blue lines' on 1:25,000 Ordnance Survey (Provisional Edition) maps (Gregory, 1976). Although researchers have called attention to the limitations of measuring stream lengths from maps (Chorley and Dale, 1972; Drummond, 1972), for this purpose the accuracy of stream delineation was considered less important

G. E. Petts

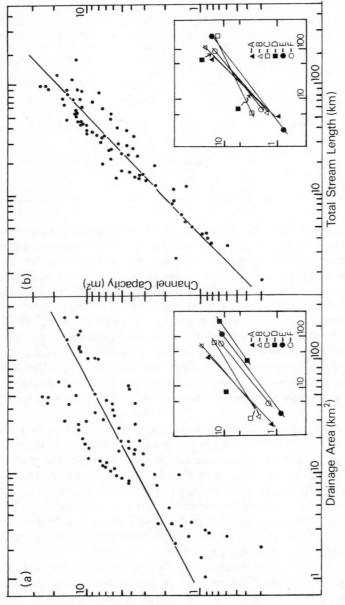

FIGURE 9.3 THE REGIONAL REGRESSION BETWEEN DRAINAGE AREA (a) AND TOTAL STREAM LENGTH (b) AND CHANNEL CAPACITY

Inserts show regression lines for the individual rivers; A, Upper Derwent; B, Ashope; C, Noe; D, Wye; E, Amber; F, Ecclesbourne

TABLE 9.2 COMPARISON OF THE DISCHARGE RECORDS OF THE SANDSTONE-SHALE
AND LIMESTONE REGIONS

River, gauging station and geology	Drainage area (km^2)	Average rainfall (mm)	Gauged flows		
			Mean	Maximum	Minimum
Derwent, Yorkshire Br. sandstone and shales	127	1220	3.80	150.60 (9.12.65)	0.47 (often)
Wye, Ashford limestone	154	1150	3.96	37.8 (9.12.65)	1.05 (2.11.69)

than the use of a consistent mapping convention. The Ordnance Survey shows streams at their 'normal winter level' on 1:25,000 maps.

The linear regression relating total stream length to channel capacity decreases the standard error (Table 9.1), but the range of the data is insufficiently large to enable the prediction of channel capacities downstream of the Derwent dams. Nevertheless, the variation of peak discharge downstream may be illustrated by relating total stream length, which at any point reflects the characteristics of the drainage basin upstream, to drainage area. Examination of the scatter of points above a drainage area of 100 km^2 suggests that more than one regression line would give an improvement in explanation. A test employed by Thornes (1970) was utilized to determine points of inflection minimizing the combined error sum of squares. A sequence of regressions was calculated by grouping the data, beginning at the lowest value for drainage area. The two points of inflection recognized (Figure 9.4a) give a significant (0.5% level) improvement in the standard error from the 'F' statistic, reflecting the occurrence, increase, and decrease of the percentage of catchment area underlain by Carboniferous limestone. The points of inflection may now be applied to the relationship between channel capacity and drainage area downstream from Ladybower dam.

Upon first examination the wide scatter of points (Figure 9.4b) shows only a poor degree of correlation. However, classification of the sections according to the nature of the site is revealing. Erosional, compound, obstructed (influenced by man-made structures), and normal sections (simple, stable sections, isolated from structures) were identified. Consideration of the 'normal sites', assumed to represent 'pre-dam' conditions, indicates that the inflection points derived above appear to be valid for this relationship (Figure 9.4b). Channel capacities derived from compound sections appear to be reduced so that they lie outside two standard errors of the 'normal' relationship. Hence, even this conservative estimate of pre-dam conditions indicates a marked change in channel form prevailing along a reach between 4 and 8 km downstream from the Ladybower Reservoir.

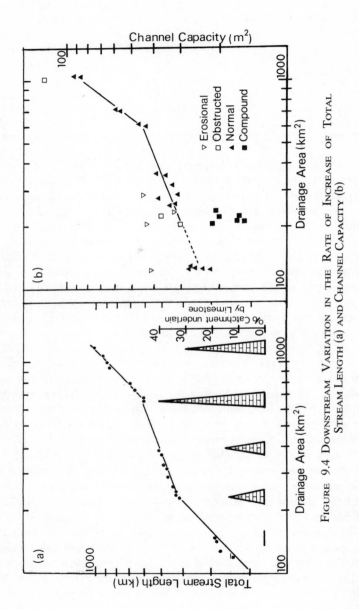

FIGURE 9.4 DOWNSTREAM VARIATION IN THE RATE OF INCREASE OF TOTAL STREAM LENGTH (a) AND CHANNEL CAPACITY (b)

Interpretation of the Compound Sections

To identify change, a comparison of surveys of channel form before and after reservoir construction is desirable, but suitable cross-sectional surveys or large scale maps are not available for the River Derwent. Furthermore, the occurrence of trees bounding long reaches of the channel inhibit the use of available air photographs. Nevertheless, comparison between air photographs taken during 1972 at a 1:12,000 scale, and the 1:2500 Ordnance Survey (2nd edition) maps surveyed in 1879 for a short reach some 8 km downstream from the Ladybower Reservoir indicates the development of a marked bench since this time.

The use of a flow equation facilitates the estimation of bankfull discharge from field measurements of channel form, and Manning's equation was employed for this purpose. Calculated bankfull discharges were compared with the discharge having a return period of 1.5 years on the annual series, representing the average bankfull discharge for stable, gravel bed rivers in the U.K. (Hey, 1975). The 1.5 year flood at each gauging station was determined from instantaneous peak flow data, whilst values for intermediate reaches were estimated from a linear relationship developed between the 1.5-year flood and drainage area. The results (Table 9.3) indicate that although a close similarity exists between the calculated bankfull discharge and the 1.5-year flood for three of the reaches cited, the channels having compound sections are either adjusted to a more frequent event or have not had sufficient time to reach equilibrium with the new flow regime.

Samples collected from the compound sections show that the depositional benches are formed of coarse sand and gravels, often with a coating of finer sand and silt, contrasting with the terrace materials which have between 16

TABLE 9.3 COMPARISON OF BANKFULL DISCHARGES DERIVED FROM MANNINGS EQUATION AND FLOW RECORDS

Reach	Drainage area (km^2)	Range of channel capacities	Value of Manning n	Bankfull discharge derived from Manning's equation	1.5-year flood determined from flow records
Yorkshire Bridge	127	22.2	0.037	28.79	38.70
to Bamford	132	28.2		49.47	
Noe Confluence	205	16.4		17.29	
to			0.035		50
Hathersage	235	21.4		29.37	
Upstream	617	43.5		71.90	
of			0.033		76
Matlock	625	45.8		72.63	
Allestree,		84.3		145.96	
Derby	1104		0.030		133.83
		91.3		185.80	

and 45% silt-clay. The terrace deposits are similar to the bank materials of simple sections both upstream and downstream. The benches, however, are not always depositional, a certain amount of bank collapse is evident and on occasion this material has become stabilized and incorporated into a bench form.

Techniques available for dating deposits include radiocarbon dating (Goede, 1973), dendrochronology (Helley and La Marche, 1973), lichenometry (Gregory, 1976c), the study of historical records (Leopold and Snyder, 1951), and archaeological evidence such as pottery (Miller and Wendorf, 1958) and other artefacts, including beer cans (Beaumont and Oberlander, 1973), and car number plates (Costa, 1975). The lack of historical records has been referred to earlier but the numerous trees lining the channel of the River Derwent allows the use of dendrochronology to determine the minimum age of the bench features.

Dendrochronology involves a count of annual rings exposed in cores taken with an increment borer. The differential seasonal growth rates of new wood produced by the cambial layer enables the recognition of distinct annual layers appearing as rings in transverse section. Although this technique has a high potential accuracy, variations in growth rates may cause problems of resolution necessitating the collection of a large sample of cores representing a variety of environmental conditions.

Some 200 trees were studied along a reach between 5 and 7 km downstream from the Ladybower dam. Each tree was classified according to situation as terrace top, degraded slope, bench form, or stranded. The latter category comprised trees whose roots were clearly exposed. An inferred ground surface level related to height above water level (low flow was observed throughout) was recorded for each tree and related to dates obtained from cores (Table 9.4). For 'stranded' forms the ground surface elevation was taken as the level of the highest root, not as the point of inflection of the basal flare which has commonly been misidentified as representing the original ground level. The inflection point will migrate

TABLE 9.4 SUMMARY OF TREE-RING EVIDENCE

Form	Height above water level (m)	Maximum age (years)
Bench	1.0 < 1.5	51
Stranded	1.75 < 2.25	65% greater than 70 years
Degraded slope	1.5 <	82
Terrace top	Generally 2.5 +	101 +

TABLE 9.5 INTERPRETATION OF COMPOUND SECTIONS

Summary of the evidence for channel change consequent upon dam construction at a single channel section

Details of the channel section:	
Post-dam capacity from Field Survey	21.4 m^2
Pre-dam capacity from Field Survey and	
identified from level of stranded trees	43.9 m^2
Ratio of pre-dam to post-dam capacity	0.49
Drainage area	210 km^2
Total stream length	430.2 km

Age of post-dam channel (River Derwent first regulated in 1914):
From historical evidence—maximum age 97 years
From dendrochronological evidence—minimum age 51 years

Comparison of the post-dam channel capacity with the pre-dam capacity predicted from regression analyses of regional data:

Source of data	Independent variable	Ratio of predicted capacity to the post-dam capacity	Direction of change
Upper Derwent and Ashope	Drainage area	0.27	Reduction
	Total stream length	0.55	Reduction
All non-regulated channels	Drainage area	1.00	No change
	Total stream length	0.42	Reduction
Segmented regression	Drainage area	0.64	Reduction

Comparison of bankfull discharges estimated from flow records and Manning's formula

1.5-year flood estimated from flow records	50 m^3/s
Ratio of 1.5-year flood to the pre-dam bankfull discharge	1.20
Ratio of 1.5-year flood to the post-dam bankfull discharge	0.59
Conclusion: post-dam channel is adjusted to a more frequent event.	

upward with respect to even a static ground surface as a consequence of increasing stem and root diameter (La Marche, 1968).

The results outlined in Table 9.4, show two points of interest. Firstly, all 'bench' forms of up to 1.5 m a.w.l. (above water level) are less than 51 years old and secondly, a former bank level at approximately 2.0 m a.w.l. is suggested by the 'stranded' forms. Furthermore, although the evidence is sparse and growth curves of individual species and climatic data need to be considered, cores taken from 'terrace top' forms subjected to massive root exposure show a marked reduction in growth rate, in each case between 52 and 40 years ago. It would be expected that rapid exposure of a large proportion of a tree's root system due to bank erosion would lead to a suppression of growth (La Marche, 1968).

Compound sections having a flat bench within the main channel occur for 5 km below the Noe confluence. Above this the channel bed is composed of gravel and small boulders which provide a natural armoured layer preventing degradation; also, the absence of sediment input results in an accommodation of the water discharge within the pre-dam channel. Downstream of the Noe confluence even the most conservative estimate of the pre-dam channel form indicates that the deposition of a bench has reduced the channel capacity by nearly 40% (Table 9.5). That a bench is not evident until the tributary has joined the mainstream implies that the introduction of sediment by the River Noe into the regulated mainstream may be the significant factor controlling bench formation. The post-dam channel, adjusted to a flow event of greater frequency than the 1.5-year flood, may be adjusted to the more efficient transport of sediment at moderate flows. The chronology of the compound sections indicates that the bench has been formed certainly since 1879 and most probably its present form is a response to river regulation. Furthermore, if the level of the 'stranded' trees is assumed to represent a former bankfull stage, then a reduction of the 1.5-year flood by 20% is indicated by estimates using Manning's equation. Thus the post-dam channel may be a response to the regulation of the flow regime and change in the nature of sediment transport consequent upon dam construction.

Conclusion

Analysis of the effects of flow regulation consequent upon dam construction indicates that changes in channel morphometry must be interpreted not only in relation to changes in flow regime and competence, but also considering the reduction in sediment load and the input of water and sediment from tributaries. Below the Derwent reservoirs the formation of a bench has reduced channel capacity by 40%, although this reduction is by no means consistent along the length of the channel. Although the potential for an adjustment of channel form is provided by the reduction of peak discharges, the introduction of sediment by a tributary is necessary for the physical adjustment of channel form to occur.

Five methods facilitate the recognition of changes of channel form and these may be widely applicable not only to determine the direction of change, but also the magnitude and rate of adjustment. Man's dependence on dams for flood control and water storage is such that more dams are being planned and more rivers are being controlled. Storage is constantly being lost due to the deposition of sediment and hence the flood probability distribution is constantly changing. The predictive techniques employed are dependent upon simplifying the relationships. In reality adjustments are complex, so that the rate, magnitude, and direction of response may vary both with regard to time and to different parts of the stream channel. Therefore, to understand the complex response of a fluvial system, involving the crossing of geomorphic thresholds initiating negative feedback mechanisms, it is necessary not only to explain variations between river channels (Schumm, 1973) but also to evaluate the significance of spatial patterns in the magnitude, rate, and direction of response along a single river.

Acknowledgements

The author wishes to acknowledge the encouragement and advice generously offered by Professor K. J. Gregory, and the Natural Environment Research Council for providing a research studentship and the help provided by the Severn Trent Water Authority.

References

Ackers, P. and Charlton, F. G., 1970. The geometry of small meandering streams. *Proc Inst Civ. Eng.*, 73285, pp. 289–317.

Beaumont, P. and Oberlander, T. M., 1973. Litter as a geomorphological aid—Death Valley, California. *Geography*, **58**, pp. 136–141.

Brown, C. B. 1944. Sedimentation in reservoirs. *Trans. Amer. Soc. Civ. Eng.*, **109**, pp. 1085.

Brune, G. M. 1953. Trap efficiency of reservoirs. *Trans. Amer. Geophys Union.* **34**, pp. 407–18.

Brush, L. M. 1961. Drainage basins, channels, and flow characteristics of selected streams in central Pennsylvania. *U.S. Geol. Survey Prof. Paper* 282F.

Carlston, C. W. 1963. Drainage density and streamflow. *U.S. Geol. Survey Prof. Paper* 442C.

Chorley, R. J. and Dale, R. F. 1972. Cartographic problems in stream channel delineation. *Cartography*, **7,** pp. 150–163.

Costa, J. E. 1975. Effects of argiculture on erosion and sedimentation in the Piedmont Province, Maryland. *Geol. Soc. Amer. Bull.*, **86,** pp. 1281–1286.

Dolan, R., Howard, A. and Gallenson, A. 1974. Man's impact on the Colorado River in the Grand Canyon. *American Scientist*, **62**(4), pp. 392–401.

Drummond, R. R. 1972. When is a stream a stream? *Prof. Geographer*, **XXVI,** pp. 34–37.

Emmett, W. W. 1974. Channel changes. *Geology* 2(6); pp. 271–272.

Eustis, A. B. and Hillen, R. H. 1954. Stream sediment removal by controlled reservoir releases. *Progress. Fish. Cult.*, **16,** pp. 30–35.

Fearnsides, W. G. 1932. The geology of the eastern part of the Peak District. The valley of the Dergyshire Derwent. *Proc. Geol. Assoc. London*, **43,** pp. 153–178.

Fedorev, B. G. 1969. Erosion below hydroelectric dams. *Tr. Tsniievt*, No. 58.

Fraser, J. C. 1972. Regulated discharge and the stream environment. In Oglesby, R. T., Carlson, C. A., and McCann, J. A. (Eds.) *River Ecology and Man*, Academic Press, New York, pp. 263–285.

Frickel, D. G. 1972. Hydrology and effects of conservation structures, Willow Creek Basin, Valley Country, Montana 1954–68, *U.S. Geol. Survey Water Supply Paper 1532-G.*

Gilbert, C. R. and Sauer, S. P. 1970. Hydrologic effects of floodwater retarding structures, Ganza-Little Elm Reservoirs, Texas, *U.S. Geol. Survey Water Supply Paper 1984.*

Goede, A. 1973. Flood-plain stratigraphy of the Tee Tree Rivulet, Buckland, Eastern Tasmania. *Aust. Geog. Studies,* **II**(1), pp. 28–39.

Gottschalk, L. C. 1964. Reservoir sedimentation. In Chow, Y. T. (Ed.) *Handbook of Applied Hydrology,* McGraw-Hill, New York, section 17-1.

Gregory, K. J. 1976a. Drainage basin adjustments and man. *Geographica Polonica,* **34,** pp. 155–173.

Gregory, K. J. 1976b. The determination of river channel capacity. *New England, Research Series in Applied Geography* No: 42.

Gregory, K. J. 1976c. Bankfull determination and lichenometry. *Search,* **7,** pp. 99–100.

Gregory, K. J. and Gardiner, V. 1975. Drainage density and climate. *Geomorph.* **19,** pp. 287–298.

Gregory, K. J. and Park, C. C. 1974. Adjustment of river channel capacity downstream from a reservoir. *Water Resour. Res.* **10,** 4, pp. 870–873.

Gregory, K. J. and Walling, D. E. 1973. *Drainage Basin Form and Process,* Edward Arnold, London.

Gvelesiani. L. G. and Shmalkmzel, N. P. 1971. Studies of storage work silting of H.E.P. plants on mountain rivers and silt deposition fighting. *Int. Assoc. Hyd. Res. 14th Congress,* Vol. 5, pp. 105, 17–20.

Hales, Z. L. Shindala, A. and Denson, K. H. 1970. Riverbed degradation prediction. *Water Resour. Res.* **6,** pp. 549–556.

Hammad, H. Y. 1972. Riverbed degradation after closure of dams. *J. Hydraul. Div. Amer. Soc. Civ. Engrs.,* **98,** pp. 591–607.

Helley, E. J. and La Marche, V. C., Jr., 1973. Historic flood information for northern California streams from geological and botanical evidence. *U.S. Geol. Survey, Prof. Paper 485E,* El–E16.

Hey, R. D. 1972. Analysis of some of the factors influencing the hydraulic geometry of river channels. *Unpublished PhD thesis,* Cambridge.

Hey, R. D. 1975. Design discharges for natural channels. In Hey, R. D. and Davies, J. D. (Eds.) *Science and Technology in Environmental Management,* Saxon House, D. C. Heath Ltd., U.K. pp. 73–88.

Hill, H. P. 1949. The Ladybower dam. *J. Inst. Water Eng.,* **3,** pp. 414–433.

I.C.E. 1933. *Floods in Relation to Reservoir Practice.* The Institute of Civil Engineers, London.

Kennon, R. W. 1966. Effects of Small Reservoirs on Sandstone Creek, Oklahoma. *U.S. Geol. Survey Water Supply Paper 1839C.*

Kilpatrick, F. A. and Barnes, H. H. 1964. Channel geometry of Piedmont Streams as related to frequency of floods. *U.S. Geol. Survey. Prof. Paper 422E.*

King, N. J. 1961. An example of channel aggradation induced by flood control. *U.S. Geol. Survey Prof. Paper 424B,* **15,** pp. 29–32.

Komura, S. and Simons, D. B. 1967. River bed degradation below dams. *J. Hydraul. Div. Amer. Soc. Civ. Engrs.* **93,** pp. 1–14.

La Marche, V. C. Jr. 1968. Rates of slope degradation as determined from botanical evidence. White Mountains, California. *U.S. Geol. Survey Prof. Paper 352-1,* pp. 341–377.

Lane, E. W. 1955. The importance of fluvial morphology in hydraulic engineering, *Proc. Amer. Soc. Civ. Engrs,* **81,** pp. 1–17, 795.

Lawson, C. M. 1925. Does desilting affect cutting power of streams? *Eng. News Rec.,* **95,** 24, pp. 969.

Leopold, L. B., Wolman, M. G. and Miller, J. P. 1964. *Fluvial Processes in Geomorphology,* Freeman, San Francisco.

Leopold, L. B. and Maddock, T., Jr. 1953. The hydraulic geometry of stream channels and some physiographic implications *U.S. Geol. Survey Prof. Paper 252.*

Leopold, L. B. and Maddock, T., Jr. 1954. *The Flood Control Controversy*, Ronald, New York.

Leopold, L. B. and Skibitzke, H. E. 1967. Observations on unmeasured rivers. *Geografiska Annaler*, **494**, pp. 247–55.

Leopold, L. B. and Snyder, C. T. 1951. Alluvial fills near Gallup, New Mexico. *U.S. Geol. Survey Water Supply Paper 1110-A.*

Livesey, R. H. 1963. Channel armoring below Fort Randall Dam. *U.S. Dept. Agric. Misc. Public No. 970*, **54**, pp. 461–470.

Makkaveyev, N. I. 1970. The impact of large water engineering projects on geomorphic processes in stream valleys, *Geomorfologiya*, **2**, pp. 28–34.

Miller, J. P. and Wendorf, F. 1958. Alluvial chronology of the Tesngne Valley, New Mexico. *Journ. Geol.*, **66**, No. 2, pp. 177–194.

Moore, C. M. 1969. Effects of small structures on peak flow. In Moore, C. M. and Morgan, C. W. (Eds.), *Effects of Watershed Changes on Streamflow*, Univ. Texas Press, Austin, pp. 101–117.

Nixon, M. 1959. A study of the bankfull discharges of England and Wales. *Proc. Inst. Civ. Eng.*, **6322**, pp. 157–74.

Nixon, M. 1962. Flood regulation and river training in England and Wales. *Proc. Symp. Inst. Civ. Eng.*, London.

Northup, W. L. 1965. Republican river channel deterioration. *U.S. Dept. Agric. Misc. Pub. 970*, Paper 47, pp. 409–424.

Nunally, N. R. 1967. Definition and identification of channel and riverbank deposits and their respective roles in flood plain formation.*Prof. Geog.*, **19**, 1–4.

Park, C. C. 1975. Stream channel morphology in Mid Devon. *Trans. Devon Assoc.*, **107**, pp. 25–41.

Park, C. C. 1976. The relationship of slope and stream channel form in the River Dart, Devon. *J. Hydrol.*, **29**, pp. 139–147.

Riley, S. J. 1972. A comparison of morphometric measures of bankfull. *J. Hydrol.*, **17**, pp. 23–31.

Rutter, E. J. and Engstrom L. R. 1964. Reservoir regulation. Section 25–111. In Chow, V. T. (Ed.) *Handbook of Applied Hydrology*. McGraw-Hill, New York.

Schumm, S. A. 1960. The shape of alluvial channels in relation to sediment type. *U.S. geol. Survey. Prof. Paper 424B*, pp. 26–7.

Schumm, S. A. 1969. River metamorphosis. *J. Hydraul. Div. Amer. Soc. Civ. Engrs.*, 95, pp. 255–275.

Schumm, S. A. 1973. Geomorphic thresholds and complex response of drainage systems. In Morisawa, M. E. (Ed.) *Fluvial Geomorphology*, State University of New York, Binghamton, pp. 299–310.

Shulits, S. 1934. Experience with bed degradation below dams on European rivers *Eng. News Rec.*, June, pp. 838–839.

Smith, N. 1971, *A History of Dams.* London, Peter Davies.

Stabler, H. 1925. Does desilting affect cutting power of streams? *Eng. News Record*, Dec. **95**, 24, pp. 960.

Straw, A. and Lewis, G. M. 1962. Glacial drift in the area around Bakewell, Derbyshire. *E. Midland Geogr.*, **3**, 2, pp. 72–80.

Tinney, E. R. 1962. The process of Channel degradation. *J. Geophys. Res.*, **67**, 4, pp. 1475–80.

Thornes, J. B. 1970. The hydraulic geometry of stream channels in the Xingu–Araguaia headwaters. *Geog. J.* **136**, 3, pp. 376–382.

UNESCO, 1972. Status and trends of research in hydrology 1961–74. *UNESCO Studies and Reports in Hydrology*, No. 10.

Wolman, M. G. 1955. The natural channel of Brandywine Creek, Pennsylvania, *U.S. Geol. Survey. Prof. Paper 271.*

Wolman, M. G. 1967. Two problems involving river channel changes and background observations. *Quant. Geog. Pt. II N.W. Studies in Geog. 14,* pp. 67–107.

Wolman, M. G. and Miller, J. P. 1960. Magnitude and frequency of forces in geomorphic processes. *J. Geol.,* **68,** pp. 54–74.

Section III

RIVER CHANNEL PATTERN

J. LEWIN
Lecturer in Geography
University College of Wales
Aberystwyth

10

Channel Pattern Changes

Any assessment of geomorphological change requires both initial and subsequent specifications of morphology; the times at which these are made define the timespan over which change is integrated. If alterations to the *rate* of morphological change are being considered then more than two specifications of state will be needed.

This section is concerned with such matters in the context of river channel dynamics (rather than mechanics or kinetics) and planforms (rather than profile characteristics, and the hydraulic geometry and cross-sectional characteristics of the previous section). Such a focusing of interest cannot be exclusively adhered to because planform and planform changes are not independent of other aspects of river geometry, and together with these other aspects they deserve to be considered in relation to the hydraulics of channels with loose boundaries. Nonetheless it is useful initially to consider definitions of form and pattern change, for which morphological interrelationships may be sought and mechanisms may subsequently be provided, and much research over the past decade has been devoted to this end. Earlier work, in which the authors themselves played a notable role, is summarized in the book by Leopold, Wolman, and Miller (1964). Here we shall be considering firstly the ways in which patterns have come to be defined, and therefore the terms in which change may be identified; secondly, the available evidence and the available timespans over which change may be perceived; and thirdly the pattern changes that have been found in field situations.

Defining Geometry

In the recent past, at least six methods have been used to specify channel geometry. These are summarized in Table 10.1. As we shall see, each

TABLE 10.1 METHODS FOR DEFINING CHANNEL PATTERN AND PATTERN CHANGE

Method		Examples
1. Qualitative terminology		Bluck, 1971; Smith, 1974
2. Graphical representation:	(a) Cartographic	Johnson and Painter, 1967; Smith, 1971
	(b) Photographic	Burkham, 1972; Lewin, 1976
	(c) Diagrammatic	Bluck, 1974; Keller, 1972; Kondrat'yev, 1968
3. Curve fitting:	(a) Circular arcs	Chitale, 1973; Brice, 1974a
	(b) Sine-generated curves	Langbein and Leopold, 1966; Daniel, 1971
	(c) Others	Ferguson, 1973
4. Bend statistics:	(a) Representative erosion rates	Handy, 1972; Leopold, 1973
	(b) Wave length, amplitude	Dury, 1964a
	(c) Sinuosity	Schumm, 1963
5. Spatial series:		Speight, 1965a; Chang and Toebes, 1970; Ferguson, 1975
6. Braiding measures:	(a) Brice index	Brice, 1964
	(b) Topology	Howard and coworkers, 1970; Krumbein and Orme, 1972

approach allows an appreciation of pattern change in some more or less satisfactory manner.

Qualitative Terms

A ready distinction can be drawn between rivers with single or multiple channels; however, even single channels have the occasional developing island (e.g., Knighton, 1972), or undulating bedforms which may split the channel when discharges are low. Multiple channels have been further subcategorized using various criteria which may have genetic implications—for example ones which have islands which are small relative to channel width (probably resulting from channel bar development) or large (produced by crevassing during overbank flows). Some recent research has used the term *braiding* as but one class of multiple channel (Schumm, 1968; Smith, 1974) and distinct from *anastomosing*. Dury (1969) has also illustrated use of the terms *anabranching, reticulate,* and *deltaic-distributary.*

Single channels have also been variously described, *meandering* in particular being used in a general sense to describe streams with a sinuous but often irregular path (sometimes called wandering), as well as for those which have a regular and repeated bend geometry (see, for example, Ferguson, 1973, 1975). In the Soviet literature, up to eight meander processes are distinguished (Makkavyeyev and coworkers, 1969), including *free, restricted,*

and *incomplete* meandering (Kondrat'yev, 1968, Kulem‑
dering and *pseudo-meandering* are described as distinct t‑
Hickin, 1969), whilst in the sedimentology literature, ch‑
viewed as complex associations of a series of unit bedfor‑

In these circumstances, with a limited but expandir
unambiguous terminology, it has naturally been appreciateu t‑.
methods of pattern specification are desirable, and that only a cruu‑
distinction of pattern change—involving perhaps dramatic metamorphoses
from single to multiple channels (Schumm, 1969)—is possible in purely
qualitative terms.

Graphical Representation

A graphical plot of channel features on several occasions allows channel
change to be visually appreciated; in numerous recent studies serial air
photographs and serial or superimposed maps have been used by way of
illustration (for example, Alexander and Nunnally, 1972; Blench, 1969;
Brice, 1974a and b; Burkham, 1972; Everitt, 1968; Fahnestock and Brad‑
ley, 1974). Ground photographs may also be very helpful (Burkham, 1972;
Lewin, 1976; Schumm and Lichty, 1963). Often these plots and photo‑
graphs form a prelude to the derivation of metric information, but they do
frequently give details beyond those which are specifically discussed in the
accompanying texts, and may constitute useful information in their own
right. The study of channel planforms has benefitted considerably from the
much wider availability of survey, and particularly remote sensing, informa‑
tion. Much valuable material of a similar kind—often expressed in the form
of idealized diagrams—is to be found in recent sedimentology literature
(Allen, 1965; Bluck, 1971, 1974; McGowen and Garner, 1970; Rust, 1972;
Shelton and Noble, 1974; Williams and Rust, 1969).

At the same time, many graphical plots are not in themselves entirely
satisfying: maps are selective, and in the case of published maps they may be
more than a little arbitrary in the detail they show. Only two-dimensional
information may be provided, with some addition by way of contours or spot
height, and information on exactly what is plotted and at what river stage,
being additionally required. This sort of information can be provided for
maps derived from field survey or from photogrammetric plots, but it may
not be available to the users of published maps from whom the complexities
of channel pattern may remain concealed. It also has to be appreciated that
a number of diagrams modelling channel change represent a putting to‑
gether of spatially coexisting forms, or are based on interpretations of
sedimentary structures, and thus may be of debateable validity.

Curve-fitting

Channel loop planforms frequently, but not invariably, have an appealing
simplicity of outline which, at least in the case of individual bends, has been

roximated by various curve-fitting procedures, notably the sine-
enerated curves of Langbein and Leopold (1966) and by circular arcs
(Brice, 1974a). Deciding which curve to fit may be a problem, and one to be
solved through theoretical justification (Langbein and Leopold, 1966), by
appeal to simplicity (Brice, 1974a), or by following empirical comparison of
respective loop properties (Ferguson, 1973).

Loop movement may then be expressed in terms of the trajectories
followed by, or distances between, reference points or axes as the loops
develop, rather than by reference to fixed points (Daniel, 1971; Brice,
1974a). In this way it has proved possible to quantify the way in which
meanders increase in path length (expand), rotate, or translate downvalley—
in any combination. Loops may equally decrease in radius, or develop
asymmetrical and multiloop forms.

Two reservations may, perhaps, be expressed with reference to this
approach. One is that curves are at times fitted and justified rather subjec-
tively. A second lies in the existence of meander bend asymmetry, either
with right hand bends differing from left, or with a lack of symmetry about
the axis of curvature, as well as the multilooping forms already mentioned
(Ferguson, 1973; Surkan and Van Kan, 1969). Studies of such planforms
have not yet developed to any considerable degree.

Bend Statistics

From available maps, from planimetric surveys, or indeed from the survey
of certain dimensions only, it is possible to extract data representative of
channel pattern or change. Thus one method is to resurvey a monumented
cross-section or series of points (Leopold, 1973; Wolman, 1959), though
some care must be taken over the implications of taking measures at fixed
points in space and therefore not at strictly comparable points on a dynamic
channel pattern, as mentioned earlier (Daniel, 1971). More usually, a series
of channel pattern parameters are derived.

For meanders, the parameters commonly used are: wavelength, which
may be measured as a straight-line distance between equivalent successive
points on a meander train, or as talweg distance (Ferguson, 1975; Speight,
1965a); amplitude; channel width, and radius. Populations of meander
bends may be sampled in this way, and conclusions can be drawn as to the
empirical relationships amongst meander dimensions and other channel-
form or process parameters—such as channel slope, discharge, or sediment
type (Ackers and Charlton, 1970a and b; Carlston, 1965; Dury, 1964a,
1965; Schumm, 1967). Varying regression relationships have been found by
different workers (compare, for example, Ackers and Charlton, 1970b;
Schumm, 1968; Ferguson, 1965) which suggests either that sampling procedures are misleadingly biased, or that parameter derivation is not standardized, or that in effect more than one population is being sampled with real
differences between meander geometries in different process environments.

All of these may be true to an extent, and this may indirectly affect channel-change assessment. It is, for example, notoriously difficult to achieve a measure of objectivity in assessing meander wavelength properties on tortuous and irregular channels by sampling individual meanders. Derivation may not be standardized, especially in the early literature. Thus Jefferson (1902) in studying the width of meander belts selected the *largest* meander in a given river reach and took cut-off loops into account. It is not surprising therefore that the use of his data taken to represent meander amplitude shows a very high amplitude/width ratio compared with other sources. Finally we do not yet have a quantitative theory adequately expressing the multivariate relationships of geometry, hydraulics, and materials, of which meander planform is but *one* variable.

Against these reservations must be set the fact that planform changes have usefully been specified in the terms described above. Dury (1964a and b, 1965) has used wavelength ratios between present streams and the larger valley meanders within which they occur to define degrees of underfittedness. Dury's studies must rank as by far the most comprehensive approach to a particular channel-change pattern on a long-term basis, and although some of the ideas he presents have recently been criticized, they have been vigorously defended and restated.

Other measures of meander scale have also been preferred: Hickin (1974) has argued that this is best measured by his 'critical meander radius', whilst Leeder (1973), faced with the practical problem that indications of channel width are less readily preserved in alluvial sediment than are channel depth, has developed procedures for reconstructing wavelengths and discharges from channel depth estimates.

In its way, therefore, each of several parameters may provide a measure of scale which can be useful in comparing channel patterns at different times. An additional consideration is that of sinuosity or 'wiggliness' (Ferguson, 1975). This may be taken as the ratio of talweg length to air-line distance (Speight, 1965), or to valley length (Schumm, 1963), and it may be assessed over distances ranging from individual loops to extended valley reaches—each of these criteria affecting the results obtained (for example, Speight, 1965a). Schumm (1963) suggested five *types* of sinuosity (involving, for instance, regular and irregular meandering) occurring on a *continuous* sinuosity scale, and related to channel geometry and bank materials.

Channel Patterns as Spatial Series

In recent years there have been some moves away from the sampling and analysis of populations of meander bends, to the use of data for whole reaches treated as a one-dimensional spatial series of direction or curvature as a function of distance along the path. Spectral analyses of the direction, or perhaps preferably direction-change, series have then been performed

J. Lewin

which may (Speight, 1965a; Chang and Toebes, 1970) or may not (Ferguson, 1975) reveal that spectra are polymodal, with more than one 'characteristic' wavelength for any given reach. Such patterns may readily arise in
the case of manifestly underfit streams, a quite normal occurrence in
midlatitudes (Dury, 1970), but whether 'channel in channel' forms, responding perhaps to flows of differing magnitude within the present process
regime, exist in plan requires some verification.

Pattern stability can be analysed using path characterizations of this type,
just as in previous methods (Speight, 1965b; Ferguson, 1975), but equally
the exercise is greatly affected by sampling procedures, operational definitions, and methods, each of which must be satisfactorily standardized to
make comparisons meaningful.

Braiding Measures

Analyses of braided patterns have often used the Brice braiding index—
twice the sum of the length of islands and/or bars in a reach divided by the
length of the reach measured midway between banks (Brice, 1964). Topological measures have also recently been adopted, and some interesting
results obtained (Howard, Keetch, and Vincent, 1970; Krumbein and Orme,
1972).

Available Evidence and Available Timespans

The assessments of channel geometry so far described, and the reassess-
ments necessary to ascertain channel changes, have been undertaken using a
variety of information sources (Table 10.2) These may allow differing
timescales for change to be appreciated, with particular sources being
appropriate for particular change-rates.

Direct Observation

Direct field or laboratory observation, survey or photography may be used
to record changes that occur within hours (Fahnestock, 1963; Krumbein and
Orme, 1972; Mosley 1975a; Schumm and Khan, 1972), days (Smith, 1971),
months (Knighton, 1972; Lewin, 1976), or years (Leopold, 1973). At the
same time, other observations may be usefully undertaken such that it
proves possible to associate change with, for example, stream power, patterns of secondary flow, sediment discharge, and so on. In the field, this may
be with reference to common or to extreme pattern-forming events (compare Knighton, 1972 and Costa, 1974).

Experimental flume studies may be very helpful in interpreting channel
pattern changes providing they are reasonable analogues for field situations,
assuming that the latter are the basic objects of study. Such matching may

TABLE 10.2 AVAILABLE DATA

1.	Direct observation	(a) Qualitative
		(b) Ground survey
		(c) Photography
2.	Historical sources	(a) Diaries, surveyors notes, etc.
		(b) Planimetric surveys
		(c) Photography
3.	Sedimentary evidence	(a) Surface vegetation dating
		(b) Conventional dating techniques
		(c) Sedimentary structures

not always occur: straight channels with meandering talwegs within them are not the same as truly sinuous bankfull channels (Schumm and Khan, 1972). Schumm and Khan in fact were only able to develop truly meandering channels after introducing fine 'suspended sediment' (kaolinite) into their sand-bed flume, and they believed such a laboratory channel to be a pioneering achievement.

Historical Sources

Because the timescale of channel changes in field situations often appears to be longer than that for which individual observers can bear to persist with their observations, it is frequently necessary to make use of documentary evidence. This may be in the form of traveller's jottings, photographs, maps or other surveys which were not generally obtained with channel pattern studies in mind. There are some exceptions, notably the fine series of surveys of the Mississippi (Mississippi River Commission 1939, 1941), but generally research workers must make use of what they can get. For instance, Burkham (1972) in studying channel change on the Gila river from 1846 to 1970, was able to make use of cadastral and soil surveys, topographic maps, ground and aerial photographs, and river cross-sections first taken in 1937. Ruhe recorded the availability of 15 maps of the Otoe bend on the Missouri covering the period 1852–1970 and 11 air photographs from 1925–1966 (Ruhe, 1975).

The availability of historical sources in such proliferation may be exceptional, but in many areas with an extended cultural history surveys have repeatedly been undertaken, and this allows the nature of channel changes to be appreciated in a range of environments (for example Handy, 1972; Johnson and Painter, 1967; Nordseth, 1973; Schattner, 1962; Wilhelmy, 1966). In a British context a minimum of five useful surveys covering the past 130 years is generally available which allows estimates of the incidence of channel change to be made (Lewin and Hughes, 1976).

At the same time, a move from observations to documentary sources involves a loss of information, a probable decrease in attainable levels of precision, and the possibility of misleading interpretation. The general direction and rate of channel change, integrated over the time period between surveys, may be obtained (though even this may not be possible where frequent complex changes occur between infrequent surveys), but these can seldom be related to the events which, cumulatively, may have produced them. Only two-dimensional information is generally available, and this must prove a disadvantage. However, given the timescale of pattern change, it is commonly necessary to use such sources of information whatever the problems may be. Provided these problems are recognized, results which are both useful, and for which there is no alternative, can and have been obtained.

Evidence from Sediments and Sedimentary Surfaces

Most river channel pattern changes are accomplished by bank erosion and sedimentation in alluvial materials: forms on sedimentary surfaces and the internal structure of alluvial deposits may thus indicate former channel patterns and processes, whilst the dating of such features may allow a timescale to be put to the changes that have occurred. Morphological forms, and specifically meander scrolls on ridged and swaled point bars, have been used to trace patterns of evolving channels (for example Kondrat'yev, 1968; Hickin, 1974). These patterns can be complex and difficult to decipher, but Hickin has suggested ways of characterizing meander development using this approach: he called sets of orthogonals through arcuate scroll patterns *erosion pathlines*, with the pathline along which the channel migration rate is greatest being the *erosional axis* of the meander.

A minimum age determination for sedimentation occurring within the past few centuries has been attempted through the dating of floodplain trees (Eardley, 1938; Everitt, 1968; Hickin and Nansen, 1975). Everitt, for example, dated cottonwoods on the Little Missouri River in North Dakota by coring them and counting growth layers, and he was able to relate tree age to floodplain elevation and to plot 25-year isochrones delimiting dates of floodplain reworking. The reliability of such methods depends, of course, on an understanding of the behaviour of the tree species in question—there may, for instance, be a delay of variable length between sedimentation and effective tree colonization—but the method does seem to achieve useful results over a time period of centuries.

Elsewhere, more conventional dating techniques have been used (Dury, 1964b; Alexander and Prior, 1971): these may show, for example, the stability of a present-day channel, or the latest date by which underfittedness commenced.

The stratification, textural, and orientation characteristics of alluvial sediments are relatable to flow environment (for example Allen, 1965; Harms and Fahnestock, 1965), one aspect of which is channel pattern. Study of the sedimentary structures associated with contemporary river activity has been intensively undertaken, especially to help in the identification and interpretation of ancient sediments (Allen, 1965, 1970, 1974; Bluck, 1971, 1974; McGowen and Garner, 1970; Ore, 1964; Rust, 1972; Williams and Rust, 1969; Shelton and Noble, 1974). However, even recent channel-pattern changes may be approached from this viewpoint, and the sedimentology of Quaternary alluvial deposits deserves closer attention in the light of recent advances in sedimentological research. However, once again it should be stressed that knowledge of sedimentation processes is not so complete as to make interpretations foolproof: it is for instance possible for the unwary to misinterpret the results of Holocene slopewash following deforestation as part of a contemporaneous fining-upward sequence. As far as channel patterns are concerned, interpretations based on exposures of very limited extent may be difficult.

Pattern Changes

Essentially two kinds of pattern change may be distinguished:

(1) *Autogenic*, (compare Allen, 1974) which are ones inherent in the river regime and involve channel migration, cut-offs, crevassing, avulsion, etc.
(2) *Allogenic*, which occur in response to systems changes involving for instance climatic fluctuation or altered sediment load or discharges, perhaps as a result of human activity.

To these may be added that condition in which planforms are essentially stationary over a period, with no channel changes at all. Achievement of this state is often the object of river engineering works, and many channel stabilization projects have proved admirably successful. However, alluvial valley floors are usually *created* by river migration, and such mobile streams can also be regarded as in regimen (Bondurant, 1972) or in some form of equilibrium state. Channelization in such circumstances may be costly and inadvisable, with unfortunate ecological and other side-effects (Emerson, 1971; Gillett, 1972).

When changes do occur, they may be described in the terms and using the evidence previously reviewed (Table 10.3)

Autogenic Changes: Single Channels

A number of developmental models, involving progressive transformation from an initial quasi-straight channel through one of increasing sinuosity to some form of stability or end-stage, have recently been elaborated (Hickin, 1969; Keller, 1972, 1975; Lewin, 1976; Tinkler, 1970) using various

TABLE 10.3 SELECTED CHANNEL CHANGE EXAMPLES

	Authors	Data sources (Table 10.2)	Pattern characterization (Table 10.1)	Change[a]
Autogenic, single				
Lower Ohio	Alexander and Nunnally, 1972	2(b); 3(b)	2(a)	V. little
Lower Missouri	Schmudde, 1963	1(a), 1(c); 2(b)	2(a)	$\frac{1}{3}$ alluvial surface 1879–1954
Lower Missouri	Everitt, 1968	2(c); 3(a), 3(c)	2(a), 2(b); 4(a)	1.8×10^6 ft^3 yr^{-1} mile^{-1}
Beaton River	Hickin and Nansen, 1975	1(b), 1(c); 3(a), 3(c)	2(a); 4(a), 4(b)	0.475 m yr^{-1} (related to r_m/w_m and ridge spacing)
Des Moines	Handy, 1972	2(b), 2(c)	2(a); 4(a), 4(c)	6.6 m yr^{-1} down-valley
Watts Branch	Leopold 1973	1(a), 1(b), 1(c)	2(a), 2(b); 4(a)	—
Indiana rivers	Daniel 1971	1(b); 2(b), 2(c)	2(a); 3(b)	—
Indiana rivers	Brice, 1974a, 1974b	2(b), 2(c)	2(a), 2(b), 2(c); 3(a)	—
Numerous Russian rivers	Kondrat'yev and Popov, 1967	2(b), 2(c)	1; 4(a)	Various, 10–15 m yr^{-1} unexceptional
Endrick	Bluck, 1971	2(b); 3(c)	1; 2(a), 2(b), 2(c)	—
Irk	Johnson and Painter, 1967	1(a), 1(b), 1(c); 2(b) 3(c)	2(a), 2(b)	Cut-off
Bollin-Dean	Knighton, 1972	1(a), 1(b), 1(c)	2(a), 2(b); 4(a)	48 cm in 18 months
Brahmaputra	Coleman, 1969	1(a), 1(b), 1(c); 2(b), 2(c); 3(c)	2(a), 2(b), 2(c); 4(a)	Bank erosion 2,600 ft yr^{-1} maximum. Bar movement 5000 ft in one flood
Platte	Smith, 1971	1(a), 1(b), 1(c)	2(a), 2(b)	—
Donjek	Williams and Rust, 1969	1(a), 1(c); 3(c)	1; 2(a), 2(b), 2(c)	—
Allogenic				
Gila	Burkham, 1972	1(a), 1(b), 1(c); 2(a), 2(b), 2(c)	2(a), 2(b); 4(a), 4(c)	—
Cimarron	Schumm and Lichty, 1963	1(a), 1(b); 2(a), 2(b), 2(c); 3(c)	2(a), 2(b), 2(c); 4(a)	—
Murrumbidgee	Schumm, 1968	1(a), 1(b), 1(c); 2(c); 3(b), 3(c)	2(a), 2(b); 4(b), 4(c)	—

[a] Rates are given in the form used by the authors where a brief statement of them is possible: reference to the original source is advisable.

combinations of reasoning, and field and laboratory observation. It appears that straight channels are hydrodynamically unstable once minimum gradients are exceeded (Schumm and Khan, 1972) and stream power exceeds bed material resistance. Even straight channels develop a sinuous talweg with alternate deeps and shallows.

Initiation of meandering in natural streams may be rapidly accomplished following bar formation in secondary flow patterns (Lewin, 1976); this phase may be by-passed where looping channels never achieve quasi-straight reaches even following cut-offs (Kondrat'yev, 1968). A second *developmental* phase is marked by increasing sinuosity and the development of pool, riffle, and point bar forms (Bluck, 1971; Keller, 1972; Noble and Palmquist,

1968). Rates and patterns of change have been very variously reported. Wave length, or channel pattern generally, appears to vary with Froude number (Anderson, 1967; Hayashi, 1973; Hickin, 1969; Kondrat'yev, 1968). Some rivers are characterized by meander loop expansion and cut-offs rather than down-valley translation (Vogt, 1963; Speight, 1965b; Neill, 1970; Schäfer, 1973; Mosley, 1975a). Translation rather than expansion may be found where channels are laterally confined (Schmudde, 1963) and, according to Chitale (1973), on streams with high width-depth ratio which have scour pools *downstream* of the meander apex.

Quite commonly, though, both translation and expansion occur simultaneously. Makkavyeyev and coworkers, (1969) reported from their studies that 'longitudinal displacement' exceeded 'lateral displacement' on average by a multiple of 5.2. Handy (1972), in a dextrous study of the Des Moines River, found that downstream migration remained constant at about 6.6 m yr^{-1}, but that a negative exponential relationship existed between the distance of the river from the edge of the meander belt and the time since migration began following cut-off in about 1880. Thus average rates had slowed from approximately 19.0 m yr^{-1} in the 1880's to 3.0 m yr^{-1} in the 1960's. By contrast, Hickin and Nansen (1975) related rates of migration to the ratio of meander radius to stream width, finding rates were most rapid when this was about 3.0, decreasing both above and below this value.

It seems to be possible for a variety of meander planforms to achieve some stability (Keller, 1972), for instance as in the downvalley migration of loops essentially without changes in form, and perhaps in relation to minimum curvature ratio, but equally *dismemberment* of simple meander forms may occur by various means. Hickin's (1969) flume studies suggested that his pseudo-meanders were terminated by a phase of braiding, and field forms which may be comparable have been described by McGowan and Garner (1970). Elsewhere chute development across broad point-bars has been seen as important, as on the Mississippi where under natural conditions chute cut-offs appear more significant than neck cut-offs (Matthes, 1948), and notably on the River Ob' (Kulemina, 1973). *Incomplete meandering* of the type suggested here is said to be characteristic of severely flooded and easily eroded floodplains.

The significance of cut-offs has recently been reemphasised in a study of the River Bollin in Chesire by Mosley (1975b). He showed an increase in the rates of lateral shift on the river since 1875, with seven cut-offs causing a decrease in sinuosity. Cut-offs are also known to cause rapid and otherwise unexpected developments on adjacent loops (Kondrat'yev, 1968).

Another alternative end-phase involves multi-looping (Brice, 1974a and b; Hickin, 1974); in fact Brice has suggested the existence of about 16 meander form types, involving combinations of simple and compound, and symmetric and asymmetric curves. Hickin discussed six types on the basis of growth patterns deduced from point bar forms.

Autogenic Changes: Multiple Channels

Several authors have recently described channel changes occurring in multiple-channel streams which, as we have seen, are of various types. Amongst braided streams, defined in the restricted sense, distinctions have been drawn between *proximal* and *distal* types, the former characterized by coarse deposits and longitudinal bars, the latter by sandy sediments and transverse bars (Smith, 1971). Channel changes in fact may be linked in part to the migration and development of these mobile bedforms essentially *within* channels, and in part to the shifting of the channel itself (Coleman, 1969).

In his study of the Lower Platte, Smith (1971) showed braiding to be a low-stage phenomenon brought about by the dissection of transverse bars formed during high annual spring discharges: exposure and braiding occurred within days, with bars that escaped complete destruction remaining until they were destroyed during the next spring flows.

Proximal streams may be illustrated by the River Donjeck, in Yukon territory, Canada (Rust, 1972; Williams and Rust, 1969). Here a nested arrangement of channels and bars is present with varying levels of activity: these include relatively permanent islands, inactive and partly vegetated channels and bars, and the present active channel where longitudinal bars in coarse material migrate during peak seasonal flows. Except on steep slopes, the coarse sediment on proximal streams may only be mobile for a very limited time (Fahnestock and Bradley, 1974) when it is extremely difficult to observe; distal streams may fluctuate in their course much more often (Coleman, 1969).

A number of studies of multiple channel systems are now available—including ones involving crevassing (Knight, 1975) and anastomosing (Smith, 1974)—but patterns and rates of change are still not well known. In many cases, rather than having direct observations of change and change rates, we have, for the present, to be content with enlightened inferences as to the pattern of movement derived from observations of the sediment involved (for example, Bluck, 1974).

Allogenic Changes

In the longer term, channel patterns may change in response to climatic fluctuation (Hjulström, 1949), as exemplified by Dury in the case of underfit streams (Dury, 1964a and b, 1965). They may also respond to modifications in both water and sediment discharges, as on the Murrumbidgee (Langford-Smith, 1960; Schumm, 1968). Historically, patterns have changed as a result of mining activity (Gilbert, 1917), changes in vegetation or the level of grazing activity (Grant, 1950; Orme and Batley, 1971), following floods (Burkham, 1972; Schumm and Lichty, 1963), or after river regulation has been undertaken (Schumm, 1969).

Such changes have become known as *metamorphoses*, defined by Schumm (1969, 1971) as involving a 'complete transformation' of river morphology. Metamorphoses need not necessarily be allogenic. River patterns relate to a spectrum of discharges ranging from low-flow and minor adjustments to pool and riffle forms, up through what are sometimes described as 'channel-forming' discharges at or approaching bankfull, to the major effects of extreme floods (for example, Costa, 1974). It is possible to regard extreme events, which can produce pattern changes, as part of the present process system (Tricart and Vogt, 1967), or to treat the channel changes and the slow (Burkham, 1972), or rapid (Costa, 1974), recovery from them essentially as allogenic, and in relaxation path terms. Burkham's study (Burkham, 1972) is extremely instructive concerning the changes that can occur following floods and floodplain land use and vegetation changes.

A second complication lies in the apparent existence of abrupt thresholds between pattern states, with sudden changes occurring when critical limits are exceeded (Schumm, 1974). Hence, a small change in slope may lead to a large change in channel pattern, provided slopes are suitably close to a critical value. This may be achieved by alteration to external controls, or by internal ones—as in the channel shortening or lengthening involved in meander loop expansion or cut-offs.

Finally, not all allogenic changes involve 'complete' transformations of morphology. Dury (1970) has described the Osage type of underfit stream in which channel widths and bedforms have become adjusted to reduced discharge, but meander wavelengths have not.

It seems likely that work on allogenic changes will develop considerably in the future, particularly as the consequences of river regulation become better appreciated. However some such changes may be less than dramatic, and only apparent perhaps following statistical analysis of pattern properties.

Some Conclusions

It should be clear from this brief review that a considerable volume of work on channel pattern change has been accomplished in the past decade: we now have much information, couched in a variety of terms, about the incidence and nature of the changes that have taken place. At the same time no general model—perhaps linking patterns and rates of change to channel shape, discharges of water and sediment, and bank materials—is yet possible, and it often remains very difficult to compare or even reconcile results from different research projects one with another. The purposes, methods, and measures used are diverse and diffuse to a confusing extent.

At the same time, such variety of interest and approach is one of the strengths of river channel study, demonstrating that there are compelling practical reasons for studying pattern change in a number of disciplines.

These include, firstly, an awareness that human activity in the vicinity of river channels has unfortunately proceeded in ignorance of the pattern changes that may be expected (for example, Kondrat'yev, 1968), and that this ought to be corrected. Secondly, channel pattern change is one of the more rapid forms of geomorphological change, with developing forms and patterns of erosion and sedimentation, that should be incorporated more fully into a general understanding of fluvial geomorphic systems. Thirdly, pattern changes involve the reworking of floodplain environments, and the soils, sedimentation, and morphological patterns that result are of very broad concern (Allen, 1971; Burkham, 1972; Davies and Lewin, 1974; Lewin and Manton, 1975; Ruhe, 1975).

Given such multidisciplinary stimuli, and the increasing availability of information and the methodological development that the last decade has witnessed, one can only anticipate a further increase in the volume and quality of channel pattern studies.

References

Ackers, P. and Charlton, F. G. 1970a. Meander geometry arising from varying flows. *J. Hydrol.*, **11**, pp. 230–252.

Ackers, P. and Charlton, F. G. 1970b. The geometry of small meandering streams. *Proc. Inst. Civil Engineers*, Paper 73285, pp. 289–317.

Alexander, C. S. and Nunnally, N. R. 1972. Channel stability on the lower Ohio River. *Annals. Assoc. Amer. Geogr.*, **62**, pp. 411–417.

Alexander, C. S. and Prior, J. C. 1971. Holocene sedimentation rates in overbank deposits in the Black Bottom of the lower Ohio River, Southern Illinois. *Amer. J. Sci.*, **270**, pp. 361–372.

Allen, J. R. L. 1965. A review of the origin and characteristics of recent alluvial sediments. *Sedimentology*, **5**, pp. 89–191.

Allen, J. R. L. 1970. Studies in fluviatile sedimentation: a comparison of fining-upwards cyclothems, with special reference to coarse-member composition and interpretation. *J. Sedim. Petrol.*, **40**, pp. 298–323.

Allen, J. R. L. 1971. Rivers and their deposits. *Sci. Prog.*, **59**, pp. 109–122.

Allen, J. R. L. 1974. Studies in fluviatile sedimentation: implications of pedogenic carbonate units, Lower Old Red Sandstone, Anglo-Welsh outcrop. *Geol. Jour.*, **9**, pp. 181–208.

Anderson, A. G. 1967. On the development of stream meanders. pp. 370–378 *Proc. Twelfth Congress of IAHR*, Fort Collins.

Blench, T. 1969. *Mobile-bed fluviology*. University of Alberta Press, Edmonton.

Bluck, B. J. 1971. Sedimentation in the meandering river Endrick. *Scott. J. Geol.*, **7**, pp. 93–138.

Bluck, B. J. 1974. Structure and directional properties of some valley sandur deposits in southern Iceland. *Sedimentology*, **21**, pp. 533–554.

Bondurant, D. C. 1972. Sediment control methods: B. Stream channels. *J. Hyd. Div. Amer. Soc. Civ. Engrs.*, **98**, pp. 1295–1326.

Brice, J. C. 1964. Channel patterns and terraces of the Loup River in Nebraska. *U.S. Geol. Survey Prof. Paper 422-D.*

Brice, James 1974a. Meandering pattern of the White River in Indiana—an analysis. In Morisawa, M. E. (Ed.) *Fluvial Geomorphology*. State University of New York, Binghamton, pp. 179–200.

Brice, J. C. 1974b. Evolution of meander loops. *Geol. Soc. Amer. Bull.*, **85,** pp. 581–586.

Burkham, D. E. 1972. Channel changes of the Gila River in Safford Valley, Arizona 1846–1970. *U.S. Geol. Survey Prof. Paper 655-G.*

Carlston, C. W. 1965. The relation of free meander geometry to stream discharge and its geomorphic implications. *Amer. J. Sci.*, **263,** pp. 864–885.

Chang, T. P. and Toebes, G. H. 1970. A statistical comparison of meander planforms in the Wabash Basin. *Water Resour. Res.*, **6,** pp. 557–578.

Chitale, S. V. 1973. Theories and relationships of river channel patterns. *J. Hydrol.*, **19,** pp. 285–308.

Coleman, J. M. 1969. Brahmaputra River: channel process and sedimentation. *Sediment Geol.*, **3,** pp. 129–239.

Costa, J. E. 1974. Response and recovery of a Piedmont Watershed from Tropical storm Agnes, June 1972. *Wat. Resour. Res.*, **10,** pp. 106–111.

Daniel, J. F. 1971. Channel movement of meandering Indiana streams. *U.S. Geol. Survey Prof. Paper 732-A.*

Davies, B. E. and Lewin, J. 1974. Chronosequences in alluvial soils with special reference to historic lead pollution in Cardiganshire, Wales. *Environ. Pollut.*, **6,** pp. 49–57.

Dury, G. H. 1964a. Principles of underfit streams. *U.S. Geol. Survey Prof. Paper 452-A.*

Dury, G. H. 1946b. Subsurface exploration and chronology of underfit streams. *U.S. Geol. Survey Prof. Paper 452-B.*

Dury, G. H. 1965. Theoretical implications of underfit streams. *U.S. Geol. Survey Prof. Paper 452-C.*

Dury G. H. 1969. Relation of morphometry to run-off frequency, In Chorley, R. J. (Ed.) *Water, Earth and Man.* Methuen, London, pp. 419–430.

Dury, G. H. 1970. General theory of meandering valleys and underfit streams. In G. H. Dury, G. H. (Ed.) *Rivers and river terraces.* Macmillan; pp. 264–275.

Eardley, A. J. 1938. Yukon channel shifting. *Geol. Soc. Amer. Bull.*, **49,** pp. 343–358.

Emerson, J. W. 1971. Channelization: a case study. *Science*, **173,** pp. 325–326.

Everitt, B. L. 1968. Use of the cottonwood in an investigation of the recent history of a floodplain. *Amer. J. Sci.*, **266,** 417–439.

Fahnestock, R. K. 1963. Morphology and hydrology of a glacial stream—White River, Mount Rainier, Washington. *U.S. Geol. Survey Prof. Paper 422-A.*

Fahnestock, R. K. and Bradley, W. C. 1974. Knik and Matanuska Rivers, Alaska: a contrast in braiding. In Morisawa, M. (Ed.) *Fluvial Geomorphology.* Publications in Geomorphology, State Univ. of New York, Binghamton, pp. 221–250.

Ferguson, R. I. 1973. Regular meander path models. *Water Resour. Res.*, **9,** pp. 1079–1086.

Ferguson, R. I. 1975. Meander irregularity and wavelength estimation. *J. Hydrol.*, **26,** pp. 315–333.

Gilbert, G. K. 1917. Hydraulic mining debris in the Sierra Nevada. *U.S. Geol. Survey Prof. Paper 105.*

Gillett, R. 1972. Crow Creek: case history of an 'Ecological Disaster' (in) Stream channelization; Conflict between Ditchers, Conservationists. *Science*, **176,** pp. 890–894.

Grant, A. P. 1950. Soil conservation in New Zealand. *New Zealand Institute English Proctor*, **36,** 269–301.

Handy, R. L. 1972. Alluvial cut-off dating from subsequent growth of a meander. *Geol. Soc. Amer. Bull.*, **83,** pp. 475–480.

Harms, J. C. and Fahnestock, R. K. 1965. Stratification, bed forms, and flow phenomena (with an example from the Rio Grande) pp. 84–110. In G. V.

Middleton (Ed.) *Primary sedimentary structures and their hydrodynamic inter-*
pretation. Soc. of Economic Palaeontologists and Mineralogists spec. publ. 12,
Tulsa, Okla. 265 pp.

Hayashi, T. 1973. On the cause of meandering of rivers. *International Symposium on*
River Mechanics Proceedings, Vol. I, pp. 667–678, Bangkok.

Hickin. E. J. 1969. A newly-identified process of point bar formation in natural
streams. *Amer. J. Sci.,* **267,** pp. 999–1010.

Hickin, E. J. 1974. The development of meanders in natural river-channels. *Amer. J.*
Sci., **274,** pp. 414–442.

Hickin, E. J. and Nansen, G. C. 1975. The character of channel migration on the
Beatton River, Northeast British Columbia, Canada. *Geol. Soc. Amer. Bull.,* **86,**
pp. 487–494.

Hjulström, F. 1949. Climatic changes and river patterns. *Geog. Annlr.,* **31,** pp.
83–89.

Howard, A. D., Keetch, M. E., and Vincent, C. L. 1970. Topological and geometrical properties of braided streams. *Water Resour. Res.,* **6,** pp. 1674–1688.

Jefferson, M. S. W. 1902. Limiting width of meander belts. *Natl. Geog. Mag.,* **13,** pp.
373–384.

Johnson, R. H. and Painter, J. 1967. The development of a cut-off on the River Irk
at Chadderton, Lancashire. *Geography,* **52,** pp. 41–49.

Keller, E. A. 1972. Development of alluvial stream channels: a five-stage model.
Geol. Soc. Amer. Bull., **83,** pp. 1531–1536.

Keller, E. A. 1974. Development of alluvial stream channels: a five-stage model:
reply. *Geol. Soc. Amer. Bull.,* **85,** pp. 150–152.

Knight, M. J. 1975. Recent crevassing of the Erap River, Papua, New Guinea. *Aust.*
Geogr. Stud., **13,** pp. 77–84.

Knighton, A. D. 1972. Changes in a braided reach. *Geol. Soc. Amer. Bull.* **83,**
3813–3822.

Kondrat'yev, N. Ye. 1968. Hydromorphological principles of computations of free
meandering. 1. Signs and indexes of free meandering. *Soviet Hydrol.,* **4,** pp.
309–335.

Kondrat'yev, N. Ye and Popov, I. V. 1967. Methodological prerequisites for
conducting network observations on the channel process. *Soviet Hydrol.,* **3,**
273–297.

Kulemina, N. M. 1973. Some characteristics of the process of incomplete meandering of the channel of the upper Ob' River. *Soviet Hydrol.,* **6,** pp. 518–534.

Krumbein, W. C. and Orme, A. R. 1972. Field mapping and computer simulation of
braided stream networks. *Geol. Soc. Amer. Bull.,* **83,** pp. 3369–3380.

Langbein, W. B. and Leopold, L. B. 1966. River meanders—theory of minimum
variance. *U.S. Geol. Surv. Prof. Paper,* 422*H.*

Langford-Smith, T. 1960. The dead river systems of the Murrumbidgee. *Geogr. Rev.,*
50, pp. 368–389.

Leeder, M. R. 1973. Fluviatile fining-upwards cycles and the magnitude of
palaeochannels. *Geol. Mag.,* **110,** pp. 265–276.

Leopold, L. B., 1973. River channel change with time: an example. *Geol. Soc. Amer.*
Bull., **84,** pp. 1845–1860.

Leopold, L. B., Wolman, M. G., and Miller, J. P. 1964. *Fluvial Processes in*
Geomorphology. W. Freeman, San Francisco and London.

Lewin, J. 1976. Initiation of bedforms and meanders in coarse-grained sediment.
Geol. Soc. Amer. Bull., **87,** pp. 281–285.

Lewin, J. and Hughes, D. 1976. Assessing channel change on Welsh rivers. *Cambria,*
3, pp. 1–10.

Lewin, J. and Manton, M. 1975. Welsh floodplain studies: the nature of floodplain geometry. *J. Hydrol.*, **25**, pp. 37–50.

McGowen, J. H. and Garner, L. E. 1970. Physiographic features and stratification types of coarse-grained point bars: modern and ancient examples. *Sedimentology*, **14**, pp. 77–111.

Makkavyeyev, N. I., Khmelyeva, N. V., and Gun Go-Yuan. 1969. Formirovaniye meandr. *Eksperm. Geomorfologiya*, **2**, pp. 7–87.

Matthes, G. H. 1948. Mississippi river cut-offs. *Trans. Amer. Soc. Civ. Engrs.*, **113**, pp. 1–39.

Mississippi River Commission. 1939. *Lower Mississippi River, Early stream channels, Cairo, Ill. to Baton Rouge, La* (in 12 sheets) Vicksburg, Miss.

Mississippi River Commission. 1941. *Lower Mississippi River. Stream channels 1930–1932 and 1940–1941, Cairo Ill. to Baton Rouge, La.* (in 12 sheets). Vicksburg, Miss.

Mosley, M. P. 1975a. Meander cut-offs on the River Bollin, Chesire in July, 1973. *Revue de Géom. Dyn.*, **24**, pp. 21–31.

Mosley, M. P. 1975b. Channel changes on the River Bollin, Chesire, 1872–1973. *E. Mid. Geogr.*, **6**, pp. 185–199.

Neill. C. R. 1970. Formation of floodplain lands-discussion. *J. Hydraul. Div. Amer. Soc. Civ. Engrs.* **96**, pp. 297–298.

Noble, C. A. and Palmquist, 1968. Meander growth in artificially straightened streams. *Iowa Acad. Sci.*, **75**, pp. 234–242.

Nordseth, K. 1973. Fluvial processes and adjustments in a braided channel system. The islands of Koppangsöyene on the River Glomma. *Norsk. geogr. Tidsskr.*, **27**, pp. 77–108.

Ore, H. T. 1964. Some criteria for recognition of braided stream deposits. *Univ. Wyo. Contr. Geol.*, **3**, pp. 1–14.

Orme, A. R. and Batley, R. G. 1971. Vegetation conversion and channel geometry in Monroe Canyon, South California. *Yearbook Assoc. of Pacific Coast Geographers*, **33**, pp. 65–82.

Ruhe, R. V. 1975. *Geomorphology* Houghton Miffin, Boston.

Rust, B. R. 1972. Structure and process in a braided river. *Sedimentology*, **18**, pp. 221–245.

Schäfer, W. 1973. Der Oberrhein, sterbende Landschaft? *Natur und Mus.*, **103**, pp. 1–29.

Schnattner, I. 1962. *The Lower Jordan Valley*. Magness Press, Jerusalem.

Schmudde, T. H. 1963. Some aspects of the landforms of the lower Missouri river floodplain. *Ann. Assoc. Amer. Geogr.*, **53**, pp. 60–73.

Schumm, S. A. 1963. Sinuosity of alluvial rivers on the Great Plains. *Geol. Soc. Amer. Bull.*, **74**, pp. 1089–1100.

Schumm, S. A. 1967. Meander wavelength of alluvial rivers. *Science*, **157**, pp. 1549–1550.

Schumm, S. A. 1968. River adjustment to altered hydrologic regimen Murrumbidgee River and Palaeochannels Australia. *U.S. Geol. Survey Prof. Paper 598.*

Schumm, S. A. 1969. River Metamorphosis. *J. Hydraul. Div. Amer. Soc. Civ. Engrs.*, **95**, pp. 255–273.

Schumm, S. A. 1971. Fluvial geomorphology: channel adjustments and river metamorphosis. In H. W. Shen (Ed.) *River Mechanics*, Vol. I, Fort Collins, Colorado, pp. 5–1 to 5-22.

Schumm, S. A. 1974. Geomorphic thresholds and complex response of drainage systems. In Morisawa, M. (Ed.) *Fluvial Geomorphology* State Univ. of New York, Binghamton, New York, pp. 299–310.

Schumm, S. A. and Khan, H. R. 1972. Experimental study of channel pattern. *Geol. Soc. Amer. Bull.*, **83**, pp. 1755–1770.

Schumm, S. A. and Lichty, R. W. 1963. Channel widening and floodplain construction along Cimarron River in southwestern Kansas. *U.S. Geol. Survey. Prof. Paper 352-D.*

Shelton, J. W. and Noble, R. L. 1974. Depositional features of braided-meandering stream. *Amer. Assoc. Petroleum Geologists Bull.*, **58**, pp. 742–752.

Smith, D. G. 1974. Aggradation of the Alexandra-North Saskatchewan River, Banff Park, Alberta. In Morisawa, M. (Ed.) *Fluvial Geomorphology.* Publications in Geomorphology, State Univ. of New York, Binghamton, pp. 201–209.

Smith, N. D. 1971. Transverse bars and braiding in the lower Platte River, Nebraska. *Geol. Soc. Amer. Bull.*, **82**, pp. 3407–3420.

Speight, J. G. 1965a. Meander spectra of the Angabunga River. *J. Hydrol.*, **3**, pp. 1–5.

Speight, J. G. 1965b. Flow and channel characteristics of the Angabunga River, Papua. *J. Hydrol.*, **3**, pp. 16–36.

Surkan, A. J. and Van Kan, J. 1969. Constrained random walk meander generation. *Wat. Resour. Res.*, **5**, pp. 1343–1352.

Tinkler, K. J. 1970. Pools, riffles and meanders. *Geol. Soc. Amer. Bull.*, **81**, pp. 547–552 (see also discussion by Keller and Tinkler, *ibid.*, **82**, pp. 279–282).

Tricart, J. and Vogt, H. 1967. Quelques aspects du transport des alluvions grossières et du façonnement des lits fluviaux. *Geogr. Annlr.*, **49A**, pp. 351–366.

Wolman, M. G. 1959. Factors influencing erosion of a cohesive river bank. *Amer. J. Sci.*, **257**, 204–216.

Vogt, H. 1963. Aspecte der Morphodynamik des mittleren·Adour (SW-Frankreich). *Petermanns Geogr. Mitteilungen*, **107**, pp. 1–13.

Wilhelmy, H. 1966. Der 'wandernde' Ström. *Erdkunde*, **20**, pp. 265–276.

Williams, P. F. and Rust, B. R. 1969. The sedimentology of a braided river. *J. Sedim. Petrol.*, **39**, pp. 649–679.

I. A. LACZAY

Research Associate
Research Institute for Water Resources
Development, Budapest

11

Channel Pattern Changes of Hungarian Rivers: The Example of the Hernád River

The proper understanding of meander development and channel pattern changes of alluvial rivers is very important for all river engineering projects. Meandering is one of the means through which rivers tend towards the so-called dynamic or quasi-equilibrium state. River training and the construction of training works are also designed to achieve a proper equilibrium. The more the planned channel pattern, channel geometry, slope conditions, etc. correspond to the natural conditions of the river in question, the better will the river accept the new semi-artificial state.

The quasi-equilibrium state of a meandering river, or a certain longer reach of it also, includes the dynamic equilibrium of the river-length. The slow successive long-term increase in length caused by meander development is compensated by the short-term, self-shortening, process effected by local cut-offs of overdeveloped meanders.

One method of training meandering rivers is by the construction of artificial cut-offs. From the viewpoint of the equilibrium length, the natural and artificial cut-offs provide similar results.

Investigations into the channel pattern changes of the Hernád river in northern Hungary have been extended to include the effects of artificial cut-offs and the corresponding river-length changes. The practical aim of the research work was to collect data and to draw conclusions for river training design.

The Hernád River

The Hernád river has its source on the southern slopes of the Tatra Mountains in Czechoslovakia from where it flows eastwards and then southwards. The total length of the river is 286 km and the catchment area is about 5400 km^2. Before the confluence with the Sajó river, which belongs to the Tisza–Danube system, the Hernád flows for 108 km on Hungarian territory, where the average valley slope is about 0.5 m/km declining in the downstream direction. The river meanders extensively. Because the sediment transported from the upper reaches, especially the bed load during high floods, is deposited in various parts of the channel, the flow is deflected and heavily erodes the loose banks. The mean particle size of both the bed load and bed material is about 5–15 mm, decreasing in the downstream direction, and that of both the suspended load and bank material is between 0.04–0.08 mm on the Hungarian river-reach.

The river flows in its valley which is between 3 and 5 km in width. According to the maps available since 1789, the meanders have shifted covering a 600- to 800-m wide belt mainly on the left-hand side of the valley. The average normal or well-developed meanders have an average meander length of 800–1200 m, and an average bend radius of 200–250 m. The channel is generally 50–60 m wide and 3–4 m deep in the crossings, and 5 m deep in the apexes of the bends. The monthly mean flow in the March–May period is about 30 m^3/sec and during September–October is 10 m^3/sec. The maximum observed flow is about 700 m^3/sec and the minimum is 3 m^3/sec.

In order to protect bridges, settlements, etc., the river along some 24 km of its length is partially trained by bank revetments and groynes. On a 24-km long stretch, flood protecting levees were also built. Three smaller barrages constructed at 14, 54, and 66 km, respectively, up-stream of the confluence (Figure 11.3) store the low flows in the channel. The barrages, fixing three points of the river, cause definite change in the meander patterns.

Patterns of Meander Development

Using the maps of river surveys in 1937, 1957, and 1972, various patterns and phases of meander development including natural and artificial cut-offs can be investigated (Laczay, 1973).

A special case of meander development can be seen in Figure 11.1A. The meander of about 2 km in length is fully overdeveloped yet unchanged for decades since, for some local reasons, the neck has been protected. This situation is a counter example to the case of river training with artificial cut-offs. Without this bank protection, the neck (Figure 11.1A) should have been cut off long ago causing 2 km local shortening of the river.

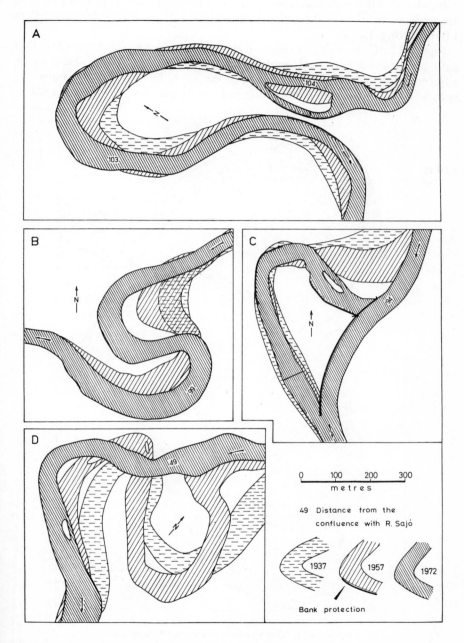

FIGURE 11.1 MEANDER DEVELOPMENT AND CUT-OFFS ON THE RIVER HERNÁD
BETWEEN 1937 AND 1972

Figure 11.1B is an example of accelerated meander development. The apex of the bend moved southwestwards some 80 m during 20 years, then 110 m during 15 years.

Figure 11.1C shows an example of natural meander development together with an artificial cut-off. It can be seen that due to the presence of an adjacent railway line, further shifting of the meander was already prevented before 1937. However, maintaining the *status quo* has consumed a lot of

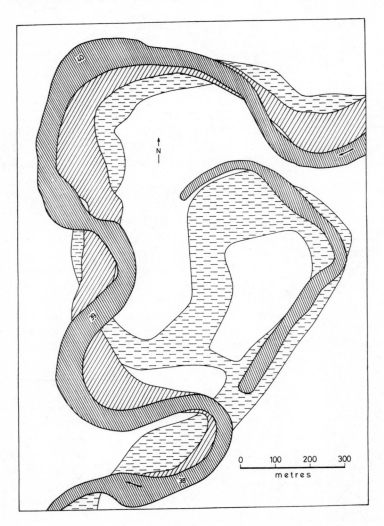

FIGURE 11.2 DEVELOPMENT OF AN OXBOW LAKE

Shading as in Figure 11.1

work and cost. Finally, in 1971, a well-traced artificial cut-off was constructed and the meander abandoned. The situation is a good example of river training employing cut-offs.

Figure 11.1D shows an example of a natural cut-off. The overdevelopment of the bend can be traced between 1937 and 1957. The cut-off occurred in 1965 during a high flood in the neck of the meander. Subsequently, the shortening caused secondary bank erosion some 100 metres in the downstream direction.

In Figure 11.2 five developing apexes can be seen. Most interesting is the development of the oxbow lake in the centre. In 1937 the neck of the former big overdeveloped meander had just been cut off. The 1957 field survey was of the active channel and did not include the abandoned meander. The 1972 status clearly shows not only the remainder after 35 years, but also that the apex had shifted further after the 1937 cut-off in the northwest direction. Figure 11.2 also shows a special case of the confinement of meander development (see Chapter 14). The southeast shore-line of the oxbow lake practically coincides with the edge of the valley. The adjacent hills have restricted the normal meandering and caused the right-forward shift of the apexes.

Concerning meander development on the Hernád river a number of practical conclusions can be drawn. Firstly, the river-reaches not affected by river training are in a fully meandering state. The movement of the apexes, i.e. the erosion rate of the banks, is about 5–10 m/year. Secondly, meander development finally leads to natural cut-offs. The cut-offs occur usually during floods at neck width not much narrower than the channel width. After the primary cut-off, the meander to be abandoned can develop further until the new channel achieves a sufficient width. Thirdly, due to human activity and geological confinements, the meander pattern and development are rather irregular on most reaches. Some good examples of river training with artificial cut-offs can be seen and this method corresponds to the natural tendency of the river.

Effects of Natural and Artificial Cut-offs

Using the river maps from 1937, 1957, and 1972, the length changes caused by cut-offs were determined and are illustrated in Figure 11.3. Between 1937 and 1957 two big natural and six artificial cut-offs caused altogether 5840 m shortening. In the period 1957–72 six cut-offs occurred with 3680 m shortening.

In the absence of river training interferences, the development of further cut-offs can be predicted. Because the neck of the meander shown in Figure 11.1B is as narrow as the channel width, the formation of a natural cut-off can be expected by 1979 causing 640 m shortening of the present channel length. In Figure 11.2 the upper meander has approached the oxbow lake

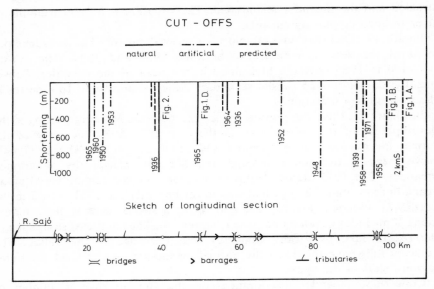

FIGURE 11.3 LONGITUDINAL SKETCH OF THE HERNÁD RIVER AND ITS SHORTENING
BY CUT-OFFS

for some 40 metres. The cut-off should occur in or around 1977 with 3 km shortening obviously likely to influence channel development on the downstream reach. Such a substantial change can only be allowed to occur if it does not endanger valuable agricultural land or other interests.

As a result of the cut-offs, the water surface slope and the flow velocity increase locally causing additional erosion of the banks on the downstream reaches. This effect can be well seen in Figure 11.3, especially in the reach between 20–24 km, where, after an artificial cut-off in 1950, another one had to be performed in 1960 which caused a natural cut-off in 1965 still farther downstream. At about 60 km a cut-off was made in 1936; it was followed by a natural cut-off in 1964, and another one can be envisaged not far from it in the downstream direction. However between 90 and 100 km the same tendency does not occur so clearly, because the banks are more protected.

All the facts mentioned above emphasize the simple and well-known practical rule that after artificial cut-offs there will be increased meandering and bank erosion on the downstream reaches, and this must be prevented by proper bank protection and training works. The over-exaggerated straightening is contrary to the nature of the river, and too much shortening requires very comprehensive and expensive bank protection and maintenance intervention afterwards.

Length-changes and Equilibrium Length

Using the river maps from 1957 and 1972 the length-changes of the channel were determined between successive fixed points such as bridges and barrages. The data are summarized in Table 11.1. It can be seen that reasonable shortening occurred only on subreaches where the above-mentioned cut-offs took place with 3680 m total shortening. Comparing this value to the total 3220 m shortening (Table 11.1), the difference means that on all subreaches the length-increase caused by meander development is generally the dominating factor.

The overall length-increase on the 108 km long Hungarian reach was 1950 m during 15 years. This increase is equivalent to 18 m/km, 130 m/year or $1.2 \text{ m km}^{-1} \text{ year}^{-1}$. Therefore the changes in general are only about 0.1% per year. Taking into consideration the fact that the neck of the big meander in Figure 11.1A without protection would have been cut off causing 2 km shortening, the result of length-changes would be practically zero. Therefore the length of the river seems to be in a dynamic equilibrium as a result of the long-term general increase and the local shortenings.

TABLE 11.1 LENGTH-CHANGES ON THE HUNGARIAN REACH OF THE RIVER HERNÁD IN THE PERIOD 1957–1972 (see also Figure 11.3)

Fixpoints upstream from the confluence (km)		Length-changes of subreaches (m)	
1957	1972	Increase	Shortening
0.000	0.000		
10.250	10.975	725	
11.370	12.080		15
12.795	13.555	50	
13.970	14.730	0	0
23.235	22.995		1000
24.580	24.355	15	
29.140	29.400	485	
42.130	43.940	1550	
48.600	50.005		405
49.720	51.100		25
52.790	54.320	150	
57.790	59.045		275
64.315	65.510		60
78.620	80.385	570	
84.000	86.350	585	
91.940	93.230		1060
95.720	96.630		380
105.990	107.940	1040	
	Total:	5170	3220

The major length increases in Table 11.1 occurred naturally on those subreaches where the river is less trained. The length increase on these reaches accounts for 10% during 15 years which is six times greater than the average value. Natural cut-offs are also expected on these reaches as a consequence of the trend towards the equilibrium length. These results are in close agreement with those mentioned above, and show the importance of planning a proper horizontal plan and length for the particular river-reach to be trained.

Conclusions

The channel pattern changes and problems of river training are very similar on several Hungarian rivers amongst which the river Hernád is only one example. Other Hungarian rivers differ in their hydromorphological characteristics, and they are generally more comprehensively trained including flood protecting levees along the banks.

Employing both laboratory and field experiments and investigations, various authors have derived several relations between morphologic and hydrologic parameters of rivers. Although not denying the necessity and usefulness of theoretical and laboratory results, it has to be remembered that practical river training design requires a great deal of experience and knowledge of the behaviour of the river in question. Rivers are perhaps the most difficult systems of nature within the natural environment. General relationships which have been derived have to be checked against every river. Therefore in this field much work remains to be done by morphologists, geographers, and by engineers.

Reference

Laczay, I. A. 1973. Meander development on the river Hernád. (in Hungarian). In *Atlases of Hungarian Rivers, No. 16 'The river Hernád'*. Research Institute for Water Resources Development, Budapest.

D. J. HUGHES

Formerly Department of Geography
University of Liverpool

12

Rates of Erosion on Meander Arcs

Alluvial river channels are controlled by the processes of erosion and deposition of sediment in response to variations in amount of discharge. The morphology of an alluvial channel, therefore, is an expression of fluctuating discharge and sediment balance within a reach. Changes in channel morphology as a result of erosional and depositional processes have been well documented in map and diagram form, but only a few studies have measured the actual rate of channel migration and morphological change. Wolman (1959) reported rates of erosion in cohesive river banks based upon measurements collected from erosion pins secured in the bank and recognized the importance of seasonality in rates of erosion. Daniel (1971) monitored the effect of erosional change on the form of the outer bank of meander bends along streams in Indiana. There is still a need, however, to measure actual rates of erosion at channel margins in relation to the magnitude and frequency of discharge events.

Wolman (1959) commented that bank erosion occurs under a variety of discharges but few conclusions have been made concerning the amount and rate of erosion that occurs in relation to different discharge magnitudes. Most studies of channel adjustment have stressed the importance of the annual flood (Dury, 1961), but Harvey (1975) remarked on the effectiveness of intermediate discharges. Wolman and Brush (1961) recognized the importance of the unstabilizing effects of fluctuating discharges experienced in nature.

Field Observations

This paper reports an investigation of rates of erosion around meander arcs in relation to peak discharges, based upon the analysis of data recorded at meander locations in the river Cound catchment, Shropshire. The catch-ment drains an area of $100 \, km^2$ with an altitudinal range of 75 to 550 m. Although the headwaters rise in the Caer Caradoc and Longmynd hills composed of Pre-Cambrian and Lower Palaeozoic rocks, most of the catchment lies on Triassic sandstones and marls mantled by Pleistocene sands and gravels. These sands and gravels are the main type of sediment supplied to the channel.

Measurements of bank erosion and variations in discharge were moni-tored from January 1972 to September 1974 for a reach of the Cound river near Condover village. This reach (Figure 12.1) is approximately 1 km in channel length and cuts into friable, sandy alluvium with gravel lenses, which forms the present flood plain and two low terraces. The Cound channel is wide and shallow with a width to depth ratio of about 15, and it is characterized by irregularly spaced pools and riffles and a series of meander bends of variable amplitude. The three meander arcs 1, 2, and 3 in Figure 12.1 represent areas of major erosion and channel migration, contrasting with the remaining stable sections of the reach.

Rates of erosion were monitored around the eroding arcs using 92 bank pegs surveyed along the channel margin and spaced 3 m apart. Measure-ments of distance from peg to channel margin were recorded at monthly intervals and more frequently during periods of high discharge. Twenty periods of field observation were made between January 1972 and Sep-tember 1974. Profiles of the channel margin were measured every 6 months at 15 sites to show the nature of bank retreat in relation to the rates of erosion recorded at the pegs. A series of cross-sections taken at profile sites indicates the total morphological changes of both bed and bank of the river in response to the discharges experienced. Discharge was monitored at the

RIVER COUND AT CONDOVER

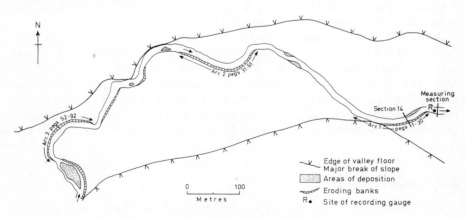

FIGURE 12.1 THE STUDY REACH

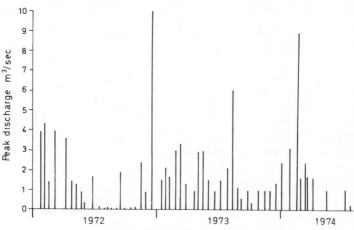

FIGURE 12.2 PERIOD PEAK DISCHARGES

downstream end of the study reach. A peak stage recorder was installed in a stable section one stream's width upstream of a riffle crest. This gauge yielded instantaneous stage heights and a peak stage height since the previous reading. Stage heights were rated by current meter and a rating curve was established for discharges up to $3\,m^3/s$. Higher discharges were estimated from the rating curve.

Figure 12.2 shows a plot of period peak discharges over the 32-month period of observations. Three major floods with peak discharges in excess of $6\,m^3/s$ occurred in December 1972, July 1973, and February 1974. The pattern of minor flood peaks includes higher winter discharges reaching $4\,m^3/s$ and rather more variable summer peaks. The summer of 1972 was fairly dry with discharges never exceeding $0.1\,m^3/s$, but the summers of 1973 and 1974 were wetter with peaks of around $1\,m^3/s$.

Erosion Rates

An examination of the erosion data for the whole reach at Condover reveals a mean erosion rate of 1.71 m per peg site for the period January 1972 to September 1974. Within each of the eroding arcs there is some variation from this mean rate (Table 12.1).

A closer analysis of mean rates of erosion for each period of observation is illustrated in Figure 12.3 which shows that the pattern of period rates of erosion is similar for each of the three individual arcs. The values of mean loss per site for each arc are also similar, especially in the case of the two major periods of erosion in excess of 0.3 m mean loss per peg site, resulting from the two peak discharges recorded in Figure 12.2 in December 1972

D. J. Hughes

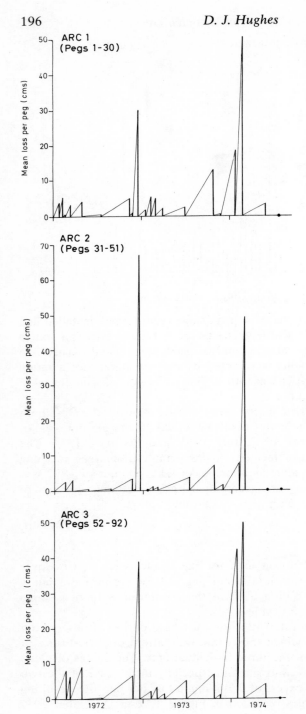

FIGURE 12.3 MEAN LOSS PER
PEG SITE

TABLE 12.1 MEAN LOSS PER SITE FOR STUDY PERIOD

Location	Number of peg sites	Mean loss per site
Arc 1	30	1.61 m
Arc 2	21	1.51 m
Arc 3	41	1.90 m
Whole reach	92	1.71 m

and February 1974. There was little response, however, to the flood peak of July 1973. The low rates of erosion in all three arcs at this time were probably the result of drying out of the bank sediment which produced a more cohesive bank, capable of withstanding the single peak discharge (Wolman, 1959). The importance of the December 1972 and February 1974 flood peaks is further shown because these two floods together were responsible, not only for the greatest amounts of erosion recorded in the study period, but also contributed some 52% of total erosion in arc 1, 78% in arc 2, and 47% in arc 3.

Although the pattern of mean rates of erosion is broadly similar for each arc, individual peg sites do shown considerable variation. (Table 12.3). There was no single occasion when all peg sites recorded erosion, and only 71 out of 92 sites responded at the time of maximum erosion associated with the February 1974 flood. Even the two major flood peaks produced variations in rates of erosion at individual sites (Table 12.2).

Erosion Rates Around a Meander Arc

A study of individual site responses to flood events is a rather crude description of rates of erosion around a meander arc. Analysis of losses at a single location may obscure any trend in the rate of erosion for a series of neighbouring sites and so hide the delayed effects of marginal retreat caused by slumping and collapse of the bank. Thus, the variable pattern of erosion within an arc may be examined using cumulative erosion details for each peg site. Figure 12.4 illustrates the accumulated totals for peg sites 1 to 30 around meander arc 1. The effects of the two major floods in December 1972 and February 1974 are again evident, in that nearly all sites suffered considerable erosion (Figure 12.4). However, the pattern of total erosion for the period 1972 to 1974 does vary for different sections of the arc. This

TABLE 12.2 PEG LOSSES IN RELATION TO MAJOR FLOODS

Flood	Total pegs	Number of pegs with loss over 1 m	0–1 m	Loss from single site	Mean loss per site
Dec. 72	92	16	42	2.00 m	0.43 m
Feb. 74	92	18	53	1.65 m	0.51 m

RIVER COUND AT CONDOVER, ARC 1
Bank erosion 1972-74

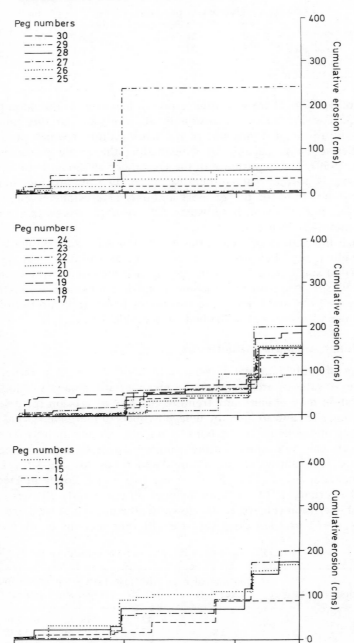

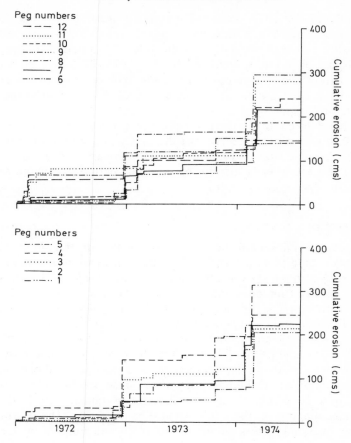

FIGURE 12.4 ARC 1: CUMULATIVE BANK EROSION 1972–74

suggests that responses to lesser floods are not evenly spread throughout the arc but are concentrated in particular sections of the meander. At the upstream end of the arc at peg sites 25–30, virtually no erosion occurred until December 1972 flood. On this occasion, site 27 suffered the greatest loss of more than 1.5 m, whilst all sites 25–28 responded to some extent. Subsequently, no erosion occurred at sites 29 and 30 and only minor changes at sites 25–28.

Further downstream, the pattern of erosion is rather more complex. A few sites in the middle area (sites 13–24) suffered erosion in early 1972, followed by a total response to the December 1972 flood. In addition, some sites indicate small changes in relation to minor floods in 1973 with a second major response from most sites to the February 1974 flood. At the lower end of the arc, there was little change until December 1972. Subsequently, the lesser floods of 1973 produced marginally greater response from the lower sections (sites 1–12) than the middle areas (13–24). Even so, there was little direct change in relation to the July 1973 flood, but considerable loss in respect of a 1 m^3/s flood in October 1973. This may be a delayed effect of erosion due to collapse of the bank. The reaction to the February 1974 flood was even greater at the lower end of the arc than at the sites in the middle and upper areas.

The general trend that emerges from the cumulative erosion diagrams (Figure 12.4) is that of a distortion of the meander arc pattern at the lower end in the area of sites 1–12. This downstream area of the arc suffered between 2 and 3 m of bank loss during the study period, compared to 2 m or less in the upstream areas of arc 1.

Profiles and Cross Sections

Peg measurements record the surface loss at the upper part of the bank, and therefore only illustrate the end product of erosion and bank retreat over a period of time. At the same sites considerable loss of sediment can occur as a result of undermining of the bank without an immediate change in peg measurements at the bank top. In order to account for such changes, bank profiles were taken at several sites throughout the reach.

A typical effect of undermining of the bank is illustrated at site 14 in Figure 12.5. The first profile taken in April 1972 displays an almost vertical bank, with slight overhang. By September 1972 the bank had been undercut with erosion of 70 cm at the base, but was represented in the peg measurements by only 20–30 cm change. At the end of January 1973, the upper portion had collapsed resulting in a vertical bank section. The peg measurement now recorded a loss of 60 cm although the bulk of material was removed during two previous periods of undermining. Through August 1973 the bank surface receded a further 60 cm with little development of an overhang. Instead, slumped blocks appeared at the base of the bank. By July 1974 another major period of erosion had resulted in a bank recession of 1 m at the surface and greater than 1 m below, thus creating a slight overhang. The mass of material at the base was being removed and the channel showed considerable signs of scour immediately beyond the remains of the slumped material.

Minor variations in sediment type and strength within the banks produce different changes in profile form and appear to affect the method of bank retreat. The profiles throughout the reach suggest four possible mechanisms operating in marginal recession, namely vertical retreat, basal scouring and undermining, collapse of overhanging blocks, and rotational slipping. Following the major flood of February 1974, most profiles had returned to an almost vertical section. Thus during the study period as a whole, a similar rate of loss was maintained for most profiles in eroding sections, despite

RIVER COUND AT CONDOVER
PROFILES AND CROSS SECTION AT PEG 14

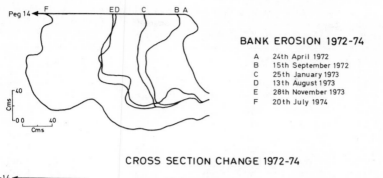

BANK EROSION 1972-74

A	24th April 1972
B	15th September 1972
C	25th January 1973
D	13th August 1973
E	28th November 1973
F	20th July 1974

CROSS SECTION CHANGE 1972-74

——— 15th September 1972
········· 13th August 1973
– – – 20th July 1974
——— Water level

FIGURE 12.5 PROFILES AND CROSS-SECTION AT PEG 14

different mechanisms of bank retreat in response to individual flood events.

This removal of bank material around the outer margins of eroding arcs is related to progressive changes in channel morphology. Such tendencies to progressive change in channel shape are illustrated in the cross-sections at site 14 (Figure 12.5). All cross-sections within the reach indicate a net removal of material with very little compensating aggradation. Accumulation of sediment along the bed of the channel is only evident in two of the 15 sites examined. In all eroding sections of the reach, there is a progressive widening of the channel accompanied by a lateral shift of the channel margin and scour of the bed in the close proximity of the bank. Such scouring action at some cross-sections is producing a transitional situation from a riffle to pool bed form.

Interpretation

Analysis of the data collected at the peg sites reveals periods of substantial erosion around the meander arcs associated with two major floods of greater than $8 \, m^3/s$. This discharge magnitude appears to represent a threshold for major channel changes along the Condover reach of the River Cound. Of equal importance, is the discharge required for a threshold of minimum activity below which little channel change takes place. In Table 12.3, the amounts of erosion during the 20 periods of observation are listed

TABLE 12.3 NUMBER OF PEGS WITH LOSS IN ERODING ARCS, MEAN LOSS PER
PEG SITE, AND PEAK DISCHARGES

Measurement period	Number of peg sites with loss			Mean loss per site (cm)	Erosion Class	Peak discharge (m³/s)
	Arc 1	Arc 2	Arc 3			
26.1.72	7	6	—	3.19	B	3.90
9.2.72	7	0	17	5.39	B	4.30
14.2.72	5	0	0	0.21	C	1.39
6.3.72	11	3	18	4.61	B	3.90
24.4.72	8	0	22	5.18	B	3.60
13.7.72	5	0	2	0.35	C	1.68
18.11.72	15	8	17	5.47	B	2.30
28.11.72	5	4	0	0.36	C	1.75
18.12.72	22	12	21	43.12	A	10.00
15.1.73	2	0	0	0.60	C	1.49
7.2.73	10	3	15	3.29	B	2.10
26.2.73	9	2	10	3.56	B	2.90
28.3.73	5	2	8	1.91	B	3.30
2.7.73	8	5	14	4.25	B	2.90
23.10.73	11	7	12	9.34	B	6.00
19.11.73	1	4	4	1.16	C	0.87
31.1.74	17	9	32	26.95	B	3.10
21.2.74	22	13	36	50.69	A	8.90
21.5.74	8	0	7	3.19	B	2.30
20.7.74	0	0	0	0.00	C	1.00

for the three eroding arcs together with the peak discharge for each erosion period.

From an examination of mean loss per site and the number of sites within each arc indicating a loss, three classes of erosional response have been identified. The criteria for distinguishing the three classes are presented in Table 12.4.

Without exception, all events involving considerable erosion (Class B) had peak discharges of at least 2.1 m³/s, whilst all periods of minimal erosion (Class C) involved peak discharges of less than 1.75 m³/s. Hence, the discharge representing the erosion threshold for the reach as a whole lies between 1.75 m³/s and 2.1 m³/s. It is possible, therefore, to recognize a range of discharges associated with each of the three erosional classes as indicated in Table 12.4. It is also important to establish the frequencies of these three classes of event. Unfortunately, no long-term stream gauging station exists on the Cound, and the nearest station operated by the Severn–Trent Water Authority is on the Rea Brook, to the west. Although the Rea Brook has a larger catchment than the Cound, both streams drain areas of similar rock type and relief. Therefore observed instantaneous discharges and period peak discharges from the Cound were correlated with

TABLE 12.4 CLASSES OF EROSIONAL RESPONSE

Class	Type of erosional response	Criterion of class	Discharge range
A	Major erosional changes affecting many areas of the channel.	Mean loss per site over 30 cm and some 50% of peg sites recording a loss.	8 m³/s and over
B	Moderate to considerable erosion affecting certain areas of the channel but not all sections.	Mean loss of at least 1.9 cm and more than 10 peg sites showing a loss.	2 to 8 m³/s
C	Minimum erosion with response only at isolated sites. Very little bank change.	Mean loss per site of no more than 1.2 cm and no more than nine pegs in the reach and five pegs in any one arc showing a loss.	Less than 2 m³/s

the equivalent daily mean and period peak discharges from the Rea Brook record. This indicated that a crude but significant relationship could be established between the discharges of the Cound and Rea Brook. Flood frequency for the Cound could then be estimated from the flood frequency curve (Figure 12.6) for the Rea Brook, which was derived for a 13-year period 1962–1975 based upon the partial duration series for instantaneous peak discharges (Langbein, 1949).

For the two erosion thresholds identified on the Cound, the lower threshold representing minimum erosion conditions at approximately 2 m³/s on the Cound is thought to be equivalent to 4 to 5 m³/s on the Rea Brook. These discharges have a long term frequency of 10 to 12 times per year (Figure 12.6). The higher threshold associated with widespread channel erosion and discharges of 8 m³/s on the Cound is equivalent to about 20 m³/s on the Rea, which gives discharges with a return period of about 1.5 years.

A long term return period of approximately 1.5 years has also been recognized by other workers (Wolman and Miller, 1960; Dury, 1961) as representing discharges associated with major, channel-changing events. In addition, Harvey (1975) commented upon the range of moderate flows which possess the competence to carry out some marginal adjustment and erosion. These moderate flows are equivalent to the discharges between 2 and 8 m³/s on the Cound.

Along the river Cound it is possible to recognize substantial rates of erosion around meander arcs, which reflect two stages of channel adjustment. Firstly, a slower rate of bank retreat and sediment removal at

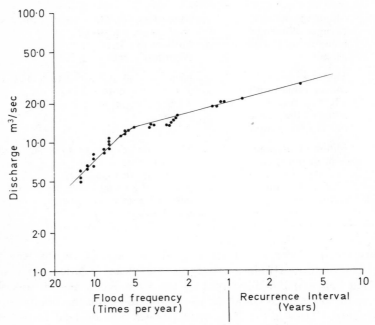

REA BROOK – FLOOD FREQUENCY CURVE, 1962-75
Based on the partial duration series for peak discharges

FIGURE 12.6 REA BROOK FLOOD FREQUENCY CURVE 1962–1975

individual sites associated with moderate flows of 2 to 8 m^3/s and occurring 10 to 12 times per year. Secondly, major adjustments linked to flood peaks of over 8 m^3/s which have a return period of 1.5 years and result in widespread channel pattern change.

Acknowledgement

The author wishes to thank the Natural Environment Research Council, for a research studentship to carry out the field study in the Cound catchment; the Severn–Trent Water Authority, for their assistance in providing data for the Hookagage gauging station on the Rea Brook; his supervisor, Dr A. M. Harvey, for his assistance in the preparation of this paper, and the staff of the drawing office and photographic department of the Department of Geography in the University of Liverpool for the preparation of maps and diagrams.

References

Daniel, J. F. 1971. Channel movement of meandering Indiana streams. *U.S. Geol. Survey. Prof. Paper 732A*, pp. 1–18.

Dury, G. H. 1961. Bankfull discharge; an example of its statistical relationships. *Int. Assoc. Sci. Hydrol. Bull.*, **5,** pp. 48–55.

Harvey, A. M. 1975. Some aspects of the relations between channel characteristics and riffle spacing in meandering streams. *Amer. J. Sci.*, **275,** pp. 470–478.

Langbein, W. B. 1949. Annual floods and the partial duration series. *Trans. Amer. Geophys. Union*, **30,** pp. 879–881.

Wolman, M. G. 1959. Factors influencing erosion of a cohesive river bank. *Amer. J. Sci.*, **257,** 1 *pp.* 204–216.

Wolman, M. G. and Brush, L. M., Jr. 1961. Factors controlling the size and shape of stream channels in coarse noncohesive sands. *U.S. Geol. Survey Prof. Paper 282G*, pp. 183–210.

Wolman, M. G. and Miller, J. P. 1960. Magnitude and frequency of forces in geomorphic processes. *J. Geol.*, **68,** pp. 64–74.

D. HITCHCOCK

Lecturer in Geography
Polytechnic of Central London

13

Channel Pattern Changes in Divided Reaches: An Example in the Coarse Bed Material of the Forest of Bowland

This paper deals with channel pattern changes and sediment movement of divided reaches in coarse bed material. The study was undertaken in a particularly mobile channel, characterized by a flashy regime where frequent changes could be expected to occur. These changes were investigated as a response to individual storm events. The watershed on which this study is focused, Langden Brook in the Forest of Bowland, is underlain by early Namurian gritstones and these provide the exceptionally coarse flood gravels of the flat valley floor. On the valley floor, which is up to 100 m wide with an overall downvalley slope of 0.0303, are numerous former channels. Some of these are occupied as overflow channels during high flows or carry small streams from the valley side to the main stream. The main anabranches and active bars are composed of coarse reworked sediments, incorporating few fines on the surface, and are devoid of a contiguous vegetation cover. Away from the active channel the gravels are overlain by sands and interbedded reworked peats. The 15 km^2 watershed ranges in altitude from 150 m to 450 m and receives a mean annual precipitation of 1825 mm. The steep valley sides, which approach angles of 25° in places, are mantled by coarse superficial deposits eventually giving way to flatter, peat-covered, interfluves.

On several separate occasions between 1972 and 1975, major changes in channel pattern occurred in response to major flood events. In addition, there were numerous occasions when moderate floods caused less widespread channel change. However, the magnitude of channel change in

D. Hitchcock

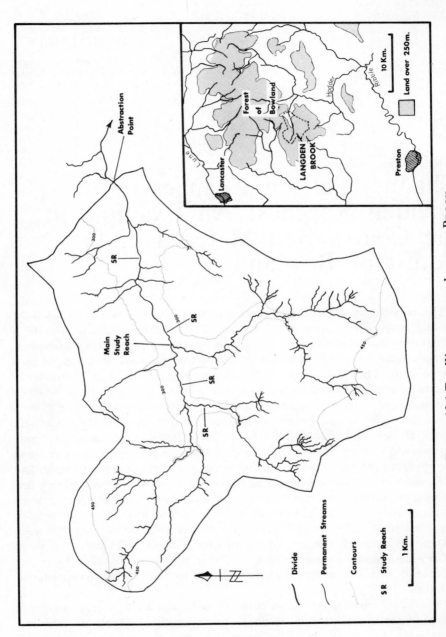

FIGURE 13.1 THE WATERSHED OF LANGDEN BROOK

response to an individual flood appears to be only partly related to the flood magnitude itself and partly to the localized short-term sequence of channel instability.

Measurement of Change

Several reaches characterized by braided channels were selected for detailed study. Examination of air photograph and map evidence indicated that these reaches experience extensive and repeated changes in channel pattern. A detailed record of channel change from 1972 to 1975 was made using repeated bank offset measurements, channel cross-section surveys, and oblique photography. On the basis of these observations within the main study reach, a classification of channel changes was devised based on a hierarchial sixfold interval scale. The criteria for the differing magnitudes of change are shown in Table 13.1. Class 0 indicates no physical change; Class 1 very minor changes, up to Class 5 includes instances of the formation of a completely new channel. Any channel change which had taken place since a previous visit was classified according to this hierarchy, which is based upon amount of channel change rather than the hydraulic conditions prevailing at any time. Changes were recorded at a total of 20 stations over the total 800 m of channel of the main study reach. The majority of the 20 stations were selected at the beginning of the study period at points considered likely to undergo significant changes. Several stations were subsequently discarded because they proved to be more stable than originally considered, while others were included later after they exhibited significant changes. Frequent and regular observations between December 1972 and April 1975 provided a record of 17 occasions when noticeable change, i.e. Class 2 or above (Table 13.1), could be detected at least at some of the stations. A summary of these changes is given in Table 13.2 and the number of occasions on

TABLE 13.1 CRITERIA USED FOR DETERMINING HIERARCHY OF CHANGE

0: No physical change observed. Accommodation change associated with small increase or decrease in stage height, or hydraulic accommodation changes associated with long period of base flow recession and gradually falling stages.
1: Sand grades moved, algal growth removed. Not usually detectable except during prolonged periods of low flows.
2: Pebble and gravel grades moved (up to T.N.D. 80 mm). Limited bank collapse of previously unstable banks, but collapse not necessarily due to hydraulic effect, more often due to sheep or human activity or overhanging, drying out bank. Imbricate structure more apparent.
3: Larger stones T.N.D. 80 mm transported. Considerable activity. Bars modified. Stream competent to move all but coarser material. Bank scour widespread.
4: Material with T.N.D. 200 mm moved. Bars completely changed. Extensive scour. New channel forms appear, may take on form of shift to formerly abandoned channel.
5: Complete new channel formed. Old channel need not necessarily be abandoned.

TABLE 13.2 SUMMARY OF CHANGES IN MAIN STUDY REACH

Date	Duration of period (days) —since last observation—	Maximum daily precipitation (mm)	Maximum Peak flow at abstraction point (m³/s)	Total number of stations available	Number of stations where magnitude of change exceeded class shown ≥2	≥3	≥4	Maximum magnitude of change at any station
1.12.72	Beginning of observations							
26. 4.73	27	42.5	18.64	15	12	8	1	4
15. 5.73	11	25.6	3.68	15	4	0	0	2
6. 8.73	1	36.8	12.08	17	9	2	0	3
31. 8.73	1	10.2	4.46	17	1	0	0	2
20.10.73	7	41.1	9.46	17	8	2	0	3
14.11.73	25	65.4	28.92	20	19	13	8	5
24. 1.74	14	29.1	11.29	20	8	0	0	2
12. 2.74	15	40.8	7.88	20	5	0	0	2
11. 6.74	21	22.6	3.40	20	1	0	0	2
3. 7.74	1	36.3	11.56	20	2	0	0	2
22. 7.74	19	27.0	11.03	20	6	0	0	2
30. 7.74	3	20.7	13.66	20	7	0	0	2
23. 9.74	14	52.0	12.08	19	9	2	0	3
31.10.74	26	38.0	9.45	19	3	0	0	2
20.11.74	10	33.9	7.35	19	2	0	0	2
16. 1.75	26	30.5	10.50	19	3	0	0	2
22. 1.75	5	44.1	20.22	19	16	8	1	4
23. 4.75	30	29.3	15.76	19	8	1	0	3

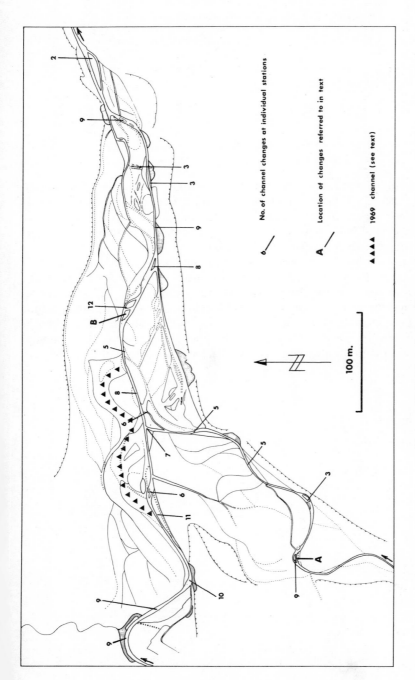

6 —— No. of channel changes at individual stations.

A —— Location of changes referred to in text

▲▲▲▲ 1969 channel (see text)

100 m.

FIGURE 13.2 MAIN STUDY REACH

Number of occasions on which individual stations underwent changes of Class 2 or above

TABLE 13.3 THRESHOLDS OF CLASSES OF CHANGE IN
TERMS OF DISCHARGE AT ABSTRACTION POINT

Class of change	Range of abstraction point discharges resulting in change (m³/s)
5	22
4	17–20
3	10–16
2	3–14

which individual stations underwent such changes is illustrated in Figure 13.2. This record of channel change was supplemented by a more general reconnaisance survey of the channel outside the detailed study reaches and was designed to identify gross changes in the channel morphology and to put the study reaches into an overall basin context.

The results (Table 13.2) indicate that similar flood discharges can affect various parts of the channel to different extents. Although the larger flood events are responsible for the major channel changes, for any given flood the relative amount of change at different locations along the channel cannot easily be predicted from past events. It is apparent from Table 13.2 that a number of possible thresholds can be identified. The only Class 5 change was associated with an abstraction point discharge in excess of 25 m³/s. The

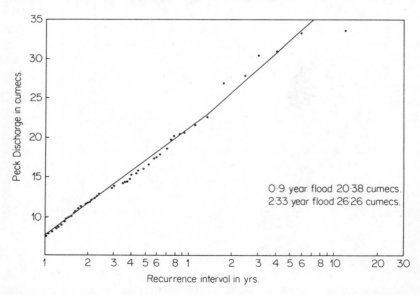

FIGURE 13.3 PARTIAL DURATION SERIES, LANGDEN BROOK 1962–1975

other Classes are associated with the ranges of discharges (Table 13.3) deduced from the data in Table 13.2. These thresholds may be identified (Table 13.3) despite the paucity of more extreme events and the overlap of values for Classes 2 and 3, which is attributed to the nature of localized instability in the channel.

A study of storm events over the period 1962–74 enabled a partial duration series to be used (Figure 13.3) tentatively to relate the classes of channel change to hydrological events of a particular magnitude.

Analysis of Change

Major changes in pattern were observed on three occasions between December 1972 and April 1975 as a response to particularly large floods. Such events resulted in some change at the majority of stations in the main study reach and changes of Class 4 and 5 were evident. The other study reaches (Figure 13.1) all showed marked changes as a result of these events, and in two cases new channels were cut and the entire valley floor underwent considerable revision. These events resulted in pattern changes at numerous points within the watershed, but it was within the study reaches that changes were most numerous.

The stream was competent to move most particles, except those firmly embedded in the streambed, as a result of storms expected to occur approximately four times a year. Such events resulted in numerous Class 3 events with the potential to bring about Class 4 events in cases of extreme channel instability. Following such flows, particles weighing between 10 and 15 kg were found with all their edges abraded by other particles, indicative of the transport of all such particles. Field experiments have shown that in Langden Brook particularly large particles, which often project above the bed and so may be more susceptible to entrainment, do not usually move more than a few metres with the passage of such storm events, although such particles have been found to move distances in excess of 20 m.

When discharges in the main study reach exceeded approximately 2.5 m^3/s, which represents a return period of 10 times per year, channel changes begin to occur. Not all stations were expected to undergo changes as a result of such flows but most were subject to Class 2 changes although individual sites experienced Class 3 changes. As a result of such flows, bars may be trimmed and well-defined turf-capped banks regularly retreated. Mean vertical profile velocity in midstream may exceed 1.5 m/s and cobbles of TND of the order of 80 mm were in motion over the bed.

The stream became competent to move some sediment whenever discharge showed a marked increase following prolonged or heavy rain. The catchment characteristics of Langden Brook result in a particularly flashy regime and events competent to move sediment occur more than 20 times per year. At several points within the study reaches, the large storm events

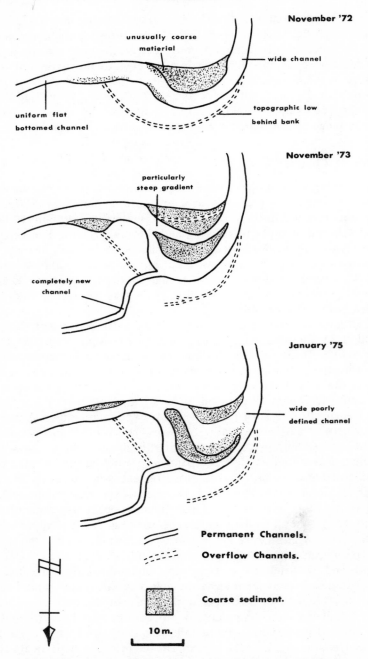

November '72

unusually coarse
matierial

wide channel

uniform flat
bottomed channel

topographic low
behind bank

November '73

particularly
steep gradient

completely new
channel

January '75

wide poorly
defined channel

Permanent Channels.

Overflow Channels.

Coarse sediment.

10 m.

FIGURE 13.4 SEQUENCE OF CHANNEL CHANGES AT LOCATION A.

Location shown on Figure 13.2

brought about important changes which were subsequently followed by a period of instability, during which changes were apparently triggered off by lower flow conditions than would normally be expected. Hence, by use of the classification it has been possible to identify the characteristics of channel pattern change as a response to flood flows.

The way in which the sequences of changes may be initiated and the way in which they develop, are illustrated by examination of two locations A and B in the main study reach, Figure 13.2. The examples show how a completely new channel may form on the floodplain or how, more usually, channels already in existence may be substantially modified. At point A, Figure 13.4, no change had been apparent for some years. Following significant changes and the creation of a completely new channel during a storm in November 1973, several changes occurred (Figure 13.4).

The notion that some channel segments are chronically unstable may be developed further in that several cases of cyclical change were recognized, illustrated by location B (Figure 13.5). In October 1972 two separate channels were in use, the main channel being to the north, characteristically wide and shallow, carrying over half the normal total flow. During the winter 1972–73, a number of changes ensued. The old south channel was blocked and scour along the north bank resulted in the total flow being carried through what had been the northern anabranch. By April 1973, a well-marked central bar had developed towards the downstream end of this channel resulting in a division of flow into two anabranches, the southern anabranch being both larger and more regular. The sediment on the bar was of cobble grade material with occasional larger particles, but with very few fines present for most of the time.

By December 1973 further changes had occurred with scour continuing along the north bank. The flow pattern around the bar had changed as a result of the main bar migrating downstream, and a second bar appearing downstream of the first. Secondary anastomosis (Church 1972), a process whereby former channels are reopened and become active again, also appears to have been important. An overflow channel, to the north of the main channel, was opened up in this way and subsequently occasionally occupied. There had been considerable scour in the anabranch to the south of the two bars which had enlarged the anabranch, locally increased the surface water gradient, and trimmed the bars.

Throughout 1974, although there were seven occasions when discharges were sufficient to move sediment, no major changes occurred. This appears to be associated with a stability of the north bank. The third and final major change within the study period came in January 1975, after which the channel took on an appearance much like that of 1972. The channel to the north of the two bars took the total flow, leaving the stream undivided and the two former braid bars as lateral bars. In addition, the original south channel and transverse channels, some 50 m upstream, were again occupied during intermediate and high flows. These changes resulted from a high

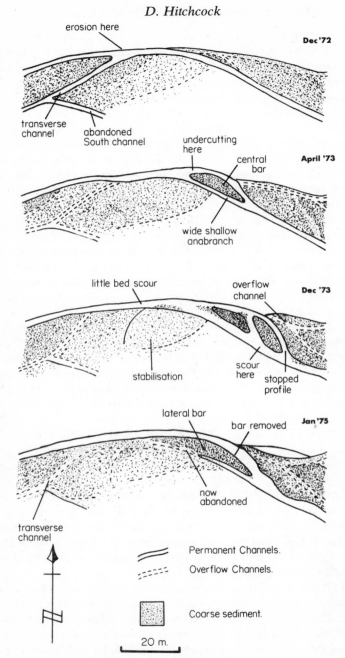

FIGURE 13.5 SEQUENCE OF CHANNEL CHANGES AT LOCATION B.

Location shown in Figure 13.2

throughput of sediment from upstream. After January 1975, the morphology of the channel was in many ways similar to that prevailing before the sequence of changes began, and the channel had again attained a degree of stability whereby frequent changes no longer occurred.

Discussion

Cartographic and air photographic evidence has been used to determine when former channels were occupied. In 1969 the main channel is known to have followed a course indicated by the triangles in Figure 13.2, some 40 m north of its contemporary course. By late 1972 this former channel was abandoned in favour of the present channel. Its lower part had become vegetated, whilst the upper part, although unoccupied by normal flows, remained more active, being occupied by occasional high flows. These major changes resulted from hydrological conditions with no flood with a return period in excess of 3 years.

Cartographic and air photographic evidence indicate that changes of the order of magnitude recognized in the short term of the study period are sustained over longer time periods. The development of the channel pattern may thus be explained in terms of an adjustment to relatively frequent events without recourse to more infrequent extreme discharges. While an attempt to define a consistent bankfull stage for such a channel would be impractical, flood events with a return period of the order of 0.9 years, as described by Nixon (1959), appear to be important.

It appears in Langden Brook that channel pattern change results most commonly from braiding, where localized channel widening, associated with high sediment transport rates, results in channel modification. Secondary anastomosis appears to be the most important cause of major channel changes. Such a situation is only to be expected on a flood plain of coarse friable material on whose surface there is a myriad of abandoned channels which have resulted from frequent channel shifts in the past.

In addition, however, two important trigger mechanisms may be recognized in Langden Brook which influence the nature of channel changes. These are both associated with the coarse nature of the bed material and the flashy flow regime. Firstly, local distortion of the flow direction due to random erosion and deposition within the channel may initiate an entirely new sequence of channel instability. Secondly, the plugging of the upstream end of a branch channel by the deposition of coarse material, or scour in the main channel, may trigger off a sequence of channel instability downstream.

Both mechanisms were observed to operate within the study reaches on several occasions. Local distortion was often associated with the deposition of a group of particularly coarse particles which directed flow towards the channel banks. This resulted in bank erosion and subsequent changes in channel hydraulics, resulting in further channel pattern changes. Localized concentration of flow in several cases resulted in scour of the main channel stream-bed, which in turn induced further change. The plugging of anabranches at their upstream end had a considerable influence on the development of channel pattern in Langden Brook. The immediate result was a reduction in the number of channels occupied by running water, the abandoned channels soon beginning to show signs of stabilization. Such old abandoned anabranches, marked by a topographic depression, provided a line which could be opened up by secondary anastomosis at some later date. The increased water and sediment in the main channel resulted in changes in hydraulic conditions which resulted in further change in the main channel, as has been shown at location B (Figure 13.5).

Other studies have shown that braided channels may exist in very different environments ranging from proglacial sandurs (Krigstrom, 1962) to semi-arid environments, and in a wide range of bed and bank materials from sand grades (Brice, 1964) to coarse cobbles (Fahnestock, 1963). Various factors have been identified as being of importance in the development of a braided pattern. The availability of an abundant, although not necessarily an excessive, supply of sediment (Fahnestock, 1963), has been recognized by numerous workers as a prerequisite for braiding. In Langden Brook large quantities of coarse sediment are derived from the reworking of channel bars, which could be described as normally loose using Shield's terminology, lacking the interstitial matrix on the bar surface as described by Church (1972) on Baffin Island. Further sources of sediment are incohesive channel banks, whose association with wide shallow channels and braiding was emphasized by Lane (1957). The banks of Langden Brook are flood gravels overlain by sand grades. Since they are almost devoid of clay and silt grades, the banks lack cohesion and so permit the existence of a wide shallow channel (Schumm, 1963). Braided streams, although locally intrinsically unstable, may develop in a state of dynamic equilibrium (Leopold and Wolman, 1957), this equilibrium being maintained through the interdependence between the wide shallow channel, high stream gradient (Leopold and Wolman, 1957), and a rapid decrease in channel resistance with increasing flow (Church, 1972). The importance attributed to increased width: depth associated with channel division (Leopold and Wolman, 1957) has been shown (Fahnestock, 1963) to be, at least in part, more apparent than real due to the method of computation. The importance of fluctuating discharge (Doeglas, 1962) may reflect an increase in braiding index (Brice, 1964) brought about by hydraulic accommodation adjustment as stage heights increase rather than by any physical change. However discharge fluctuations may in some circumstances aid the development of a braided pattern.

Fahnestock (1963) observed that braided channels on the White River, Washington, were better developed during the summer months when flows were high, and when sediment loads were also high, while Ore (1964) also maintained that seasonal and other fluctuations in regime may be important, being associated with the supply of sediment, a feature noted by Church (1972).

No single process appears to account for the development of braided streams in all environments. Brice (1964) recognized the importance of the build up of lateral dune-like forms in sand bed rivers, which were subsequently broken through to form braid bars. Leopold and Wolman (1957) stressed the importance of excessive width leading to the deposition of coarse particles in the centre of the stream, which constitute the nucleus of a bar. Once initiated, further accretion occurs and eventually the bar becomes stabilized and vegetated if development is unimpeded. Church (1972) describes secondary anastomosis as an important process whereby flood discharges reoccupy and enlarge former abandoned channels. The latter two mechanisms are important in Langden Brook.

In an environment such as Langden Brook, adjustment to a wide range of events takes place. Exceptionally rare floods undoubtedly do occur in the watershed, although the development of channel pattern takes place by a sequence of changes which occur as a response to more normal floods. The higher magnitude events with return periods of about a year appear to bring about major changes, while more frequent events bring about minor changes. Such minor events can have a considerable impact upon the channel, depending upon the local situation, due to the inherent variability of conditions within the channel.

Acknowledgements

The author wishes to thank Dr A. M. Harvey, Department of Geography, University of Liverpool, for valuable assistance given both during the period of field research and throughout the preparation of this paper. Thanks are also due to the staff of the Preston and District Water Supply Unit for making hydrological data available and for background information about the watershed. The study was undertaken while the author was in receipt of an N.E.R.C. Research Studentship.

References

Brice, J. C. 1964. Channel patterns and terraces of the Loup River in Nebraska. *U.S. Geol. Survey Prof. Paper 422-D*, pp. 1–41.
Church, M. 1972. Baffin Island sandurs: A study of arctic fluvial processes. *Geol. Survey of Canada Bulletin 216.*
Doeglas, D. J. 1962. The structure of sedimentary deposits of braided streams. *Sedimentology*, **1**, pp. 167–190.
Fahnestock, R. K. 1963. Morphology and hydrology of a glacial stream—White River, Mount Rainier, Washington. *U.S. Geol. Survey Prof. Paper 422-A.*
Krigstrom, A. 1962. Geomorphological studies of sandur plains and their braided rivers in Iceland. *Geog. Annaler*, **44**, pp. 328–346.

D. Hitchcock

Lane, E. W. 1957. A study of the shape of channels formed by natural streams flowing in erodible material. *U.S. Army Corps of Engrs., Missouri River Div., Omaha, Nebraska; Sediment Ser. 9.*

Leopold, L. B. and Wolman, M. G. 1957. River channel patterns—braided, meandering and straight. *U.S. Geol. Survey Prof. Paper 282B*, pp. 39–85.

Nixon, M. 1959. A study of the bankfull discharges of rivers in England and Wales. *Inst. Civil Engineers.* Paper *6322*, pp. 157–174

Ore, H. T. 1964. Some criteria for recognition of braided stream deposits. *Univ. of Wyoming. Contrib. Geol., 3*, pp. 1–14.

Schumm, S. A. 1963. A tentative classification of alluvial river channels. *U.S. Geol. Surv. Circular 477*, 10pp.

Wolman, M. G. and Miller, J. P. 1960. Magnitude and frequency of forces in geomorphic processes. *J. Geol., **68**,* pp. 54–74.

J. LEWIN

Lecturer in Geography

and

B. J. BRINDLE

Department of Geography
University College of Wales
Aberystwyth

14

Confined Meanders

Most discussions of meanders deal explicitly or implicitly with forms in homogenous media, particularly those in alluvial deposits, solid rock or glacier ice. However, in Britain contemporary river channels commonly pass to and fro across alluvial valley floors to impinge with some frequency against bluffs of valley infill or valley sides of strictly limited erodibility, so that considerable distortion of meander planforms can result. These *confined meanders* nonetheless have a pattern to their distortion, with repeating forms and modes, and rates of development. In a British context this phenomenon appears of considerably more importance to the development of meander forms than the inhomogeneity of alluvial deposits, and such a 'normal' feature of actual channel patterns seems worthy of closer examination, particularly as the nature and rate at which channel adjustment occurs are such as to be of practical concern.

Meander confinement appears occasionally in the literature, though often only in passing, and with a varied terminology. The term as used here appears to have been coined by Lane (1957), but many other writers have used similar concepts. Hence, Jefferson (1902) wrote delightfully of meanders that were 'hindered, embarrassed or bluff-bound'; Matthes (1941), followed also by Lane (1957), used the term 'deformed' meanders, whilst Fisk (1944) in particular appreciated the significance of floodplain inhomogeneity, a theme recently reemphasized by Ferguson (1975). Soviet literature commonly distinguishes between free, limited, and incomplete meandering (e.g. Kopeliana and Romashin, 1970, Kulemina, 1973). Mak-kavyeyev and coworkers (1969) discuss types, literally translated as 'heaped' and 'boxed', developed as a response to confinement. This last distinction seems similar to that drawn by Chitale (1973) between 'hairpin bends or acute meanders as against flat or broad meanders'.

It appears from this short literature review that the phenomenon of confinement must be widespread, even if discussion of it is neither extensive nor terminologically consistent. Indeed as river migration plains are finite in extent some marginal confinement must occur, but confinement is particularly marked in mid-latitude environments at present because of Quaternary valley developments and the entrenchment of streams into valley infills. Hence the phenomenon is apparent in both the recent arroyo trenching of the southwestern United States (see Haynes, 1968) and in the glacially, proglacially, and periglacially modified valley floors of Britain (see, for example, Dury, 1964; Jones, 1965; Cross and Hodgson, 1975).

This paper draws on studies undertaken in mid-Wales to demonstrate the nature and extent of confined meanders, the dynamics of the confined channels, and processes observed in confinement as illustrated by a case study on the River Elan.

Definition and Extent

Meanders can be described as confined when they impinge against, or are partly developed in, media which alter the forms or rates of development from those found in local channel materials where such confinement is absent. The degree of confinement can vary from minor bank irregularities to a near-complete distortion of loop form; it may also be viewed as an interesting transitional stage between primitive rock-cut channels and alluvial meanders freely developing across extensive floodplains.

In Wales, free development of meanders may be confined by three main agencies. First, the width of alluvial valley floor between solid-rock valley walls is limited but very variable and not simply to be related to catchment size. This is well illustrated by the River Teifi (Jones 1965) where the river flows alternately through gorge and floodplain sections, with parts of a prior system of valley meanders plugged with glacial deposits. Quite commonly streams are manifestly underfit (Dury 1964), with small-scale meanders within a larger meandering valley, as in the case of the River Monnow at Grosmont (Figure 14.1C). Such spatial constraints are highly significant in determining the *active* development of channel patterns in the present environment.

Secondly, there exists within rock-walled valleys a variety of unconsolidated valley deposits. Rivers may impinge against materials of glacial, periglacial or fluvial origin, and marginal bluffs may be of considerable height, so that for a given increment of lateral migration, a much greater

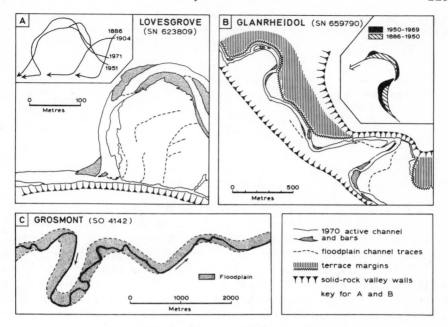

FIGURE 14.1 EXAMPLES OF CONFINED PATTERNS

A: a developing confined loop on the River Rheidol; the inset shows channel location on four occasions in the last century. B: two further confined loops on the Rheidol, the inset showing areas eroded 1886–1950 and 1950–1969. C: the channel and floodplain on a reach of the manifestly underfit River Monnow

volume of material has to be removed than would be the case on a low-relief floodplain. For example, a survey of an 8-km reach of the upper Elan showed that the channel was confined against left and right banks for a total of 3.68 km, predominantly by solifluction terraces or by solid rock (Table 14.1), and that such bluffs may be over 4 m high. A survey of the distribution and extent of confinement in upland Wales is currently being completed, and this suggests that the type and extent of confinement varies downvalley. Thus tills and solifluction deposits dominate in headwaters, with terraces of fluvially deposited gravels in the lower sections of valleys, here confining rivers to a lesser extent.

River migration is finally inhibited by human structures, manifestly by those specifically designed to stabilize river courses, but less obviously by such features as roads and railways which for convenience (and yet at some hazard) occupy floodplain locations, and which over the past few centuries have increasingly constrained and confined river development. This may be illustrated by the case of the Rivers Rheidol and Ystwyth in mid-Wales (Table 14.2). The migration plains of the rivers have been reduced in area

TABLE 14.1 CONFINEMENT MATERIALS ON AN 8-KM REACH OF THE UPPER ELAN[a]

	Right bank (km)	Left bank (km)	Average (%)
Rock	0.53	0.53	29.0
Solifluction deposits	0.75	0.74	40.1
Fluvial gravels	—	0.25	6.8
Sections of complex superficial deposits[b]	0.52	0.34	24.1
Total	1.82	1.86	100.0

[a] Meander belt axis length is 5.3 km, giving a sinuosity of 1.51.

[b] These include combinations of peat, fluvial gravel, and solifluction deposits.

from 4.9 km^2 and 4.2 km^2, respectively, at the onset of the nineteenth century to 3.3 km^2 and 2.4 km^2 today. Protection measures have frequently to be undertaken where rivers impinge on road or railway which create artificial confinement situations. This is well shown by the River Tywi at Golden Grove (Figure 14.2). Here the railway was constructed close to the 1841 channel; subsequent meander loop development has involved the formation of a deep scour pool and an unstable channel reach adjacent to the rail track. Information on the 'natural' pattern of channel migration and the possible effects of confinement would clearly be helpful in this and similar situations.

Degrees of Confinement

The major factors so far discussed have confined channel development to differing extents and it may be useful to distinguish three degrees of

TABLE 14.2 DECREASING EFFECTIVE AREA OF THE RHEIDOL AND YSTWYTH MIGRATION PLAINS

	Rheidol (m^2)	(%)	Ystwyth (m^2)	(%)
Area removed by:				
Aberystwyth-Ponterwyd turnpike, 1812	335,625	6.8	—	—
Aberystwyth + Welsh coast railway, 1864	325,625	6.6	—	—
Manchester and Milford railway, 1867	—	—	1,815,000	43.1
Devil's Bridge light railway, 1904	760,000	15.5	—	—
Waste dumping	186,875	3.8	—	—
Active migration plain remaining:	3,308,750	67.3	2,396,250	56.9
Total:	4,916,875	100.0	4,211,250	100.0

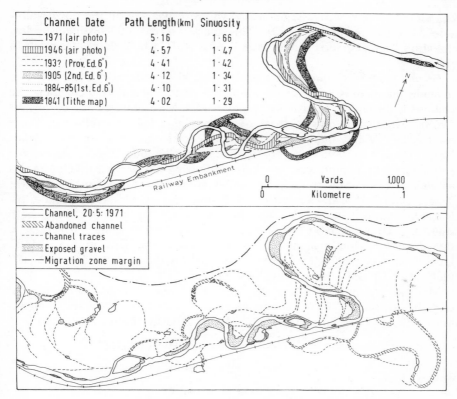

Channel Date	Path Length(km)	Sinuosity
1971 (air photo)	5·16	1·66
1946 (air photo)	4·57	1·47
193? (Prov. Ed. 6")	4·41	1·42
1905 (2nd. Ed. 6")	4·12	1·34
1884-85 (1st. Ed. 6")	4·10	1·31
1841 (Tithe map)	4·02	1·29

Channel, 20·5·1971
Abandoned channel
Channel traces
Exposed gravel
Migration zone margin

Railway Embankment

Yards 1,000
Kilometre 1

FIGURE 14.2 THE RIVER TYWI AT GOLDEN GROVE (SN 5821)

Upper diagram shows channel location on six occasions since 1841; sinuosity has increased and the confined channel has moved up-valley against the rail embankment. Lower diagram shows lineations and abandoned channels distinguished on 1971 vertical air photography

confinement. *First degree confinement* occurs in wide-floored valleys, where the channel may impinge irregularly against confining agents, creating local pattern distortions, or perhaps abandoned or active arcuate scars along valley sides. This type of meander has been called 'restrained' by Lane (1957). In frequency, confinement occurs at most with a spacing equal to one meander wavelength, but commonly channels impinge only occasionally against confining media. However, in length terms they may hug the confining bank for considerable distances, as against the concave walls of valley meanders (Figure 14.1C) and this is one way in which underfit patterns of the Osage type (Dury, 1966) may be actively created. In places, the long-wave pattern of spatial constraints appears to override the short-wave pattern of current process by critically preventing outward-turns of the channel for long concave segments of valley, especially on the down-valley side of valley meander bends. This makes it easy to distinguish flow direction from geometry alone, as in Figure 14.1C where flow is from left to right.

Second degree confinement takes place where the width of the free meander medium is less than the amplitude of the meandering channel. The channel is 'boxed' or 'sinusoidal' in pattern (cf. Makkavyeyev and coworkers 1969) such that it crosses from one confining edge to another in every wavelength. Valley-floor alluvial areas are discontinuous, forming 'headlands' (Sundborg 1956) or 'bottomlands' (Schmudde 1963). Again the picture may be complicated by channels flowing along the confining media as well as frequently crossing the valley from side to side (e.g. the River Tawe, National Grid Reference SN 7406), and by Osage-type underfits. Meander patterns may also not be dominated by a single-wavelength sinuosity that can reliably be determined by eye (Speight 1965). Nonetheless, this degree of confinement can be recognized especially in the deeply-incised, narrow valleys of South Wales, such as those of the Rhondda (SS 9895, Taff ST 0893) and Neath (SN 8403).

Meander amplitude, which together with valley width determines the existence of second degree confinement, is believed to be a variable meander property. Leopold and Wolman (1960) gave the following relations:

$$A = 18.6 \; w^{0.99} \; \text{(Jefferson data, relation by Inglis)}$$

$$A = 10.9 \; w^{1.04} \; \text{(Bates data, relation by Inglis)}$$

$$A = 2.7 \; w^{1.1} \; \text{(their data and relation)}$$

where A is meander amplitude and w is channel width. However, Jefferson's data refers to the largest meander in a reach and incorporates cut-offs; Leopold and Wolman's own equation, in which the constant is remarkably low, is here recalculated using their own data giving:

$$A = 4.4 \; w^{1.02}$$

Thus free meander amplitude may be a little less variable than has appeared, though further sampling would seem desirable before confinement (amplitude/valley width) relationships with respect to discharge or channel width are established.

Third degree confinement occurs where meander geometry appropriate to discharges in the free-meander medium is not allowed to develop. There may be no meandering pattern in any meaningful sense, as is commonly the case where channels are closely confined by artificial structures.

The three degrees of confinement distinguished here relate to the erosional environment of currently active streams; no necessary assumptions have been made about valley-floor history and development, or the reasons why dichotomous media should have become juxtaposed.

Patterns of Confinement

It is generally recognized that single-bank confinement may locally produce deepening and narrowing of the river channel, and sharp changes in
channel direction (or smaller radius of curvature) that would not otherwise
occur (e.g. Sundborg, 1956; Konditerova and Popov, 1966; Carey, 1969).
Pool depth and radius of curvature have been empirically related (Konditerova and Popov, 1966), while bar formation may occur downstream of
the deepened scour pool. Similar effects have been documented in laboratory flume studies in which bank stabilization experiments provide natural
confinement analogues (Friedkin, 1945). In fact, different patterns may also
be formed depending on the geometry of the confinement and on the
erodibility of the confining medium, as may be illustrated by two developing
meanders on the River Rheidol.

The first is a meander loop, developed essentially over a period of 90
years, that impinges against a solid-rock valley side which is in effect
non-erodible. The channel takes a sharp change of direction with an
abnormally deep scour pool. As the meander has migrated down-valley,
remains of such a pool have been left as a deep dead slough against the
confining bank (Figure 14.1A). Headward erosion from this channel by
water percolating through the floodplain gravels threatens eventually to cut
off the main meander upstream. Such channels commonly occur in confined
meanders of this kind. The active scour pool may migrate down-valley, or,
depending on the geometry of the freely developing loop of the meander, it
may reoccupy former positions and move up-valley, as on the Tywi (Figure
14.2). The point bar zone within the sharpened meander loop can be an area
of rapid *erosion* (Carey, 1969) or of accretion.

A second mode of development is illustrated by an impeded meander cut
into a gravel terrace (left-hand loop in Figures 14.1B and 14.3). Here a
smoothly curving loop has been created, with dimensions comparable to
those of adjacent free meanders, but lately this development has been at a
lesser rate than channel development in the adjacent meander loop upstream which is cut into low-relief floodplain deposits. The result is that the
locus of maximum bank erosion had moved upstream by 1946 (Figure 14.3);
deposition of bluff material occurs in mid-channel downstream (but still
within the meander loop) with its own associated area of bank erosion (1969
and 1973); eventually the confined loop will be overtaken and abandoned.
A distorted rock-confined loop which exists upstream (Figure 14.1B) forms
a 'heaped' pattern (Makkavyeyev and coworkers, 1969) and is being overtaken in the same way.

Thus two modes of development can occur. Firstly a locally-boxed or
sinusoidal pattern, with deepened scour pools, may creep down-valley with
the pattern maintained intact, or, where the channel to some degree
becomes ensconced in the confining medium, mid-channel sedimentation

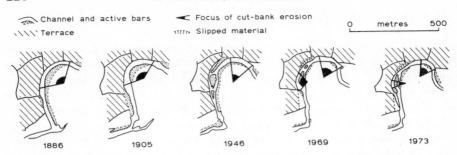

FIGURE 14.3 DEVELOPMENT OF A CONFINED LOOP ON THE RIVER RHEIDOL AT GLAN RHEIDOL

Maps of 1886 and 1905 show a smoothly curving outer bank without gravel accumulation, together with inner point-bar gravels. In 1946 and subsequent air photography, the outer bank is undercut only at the upstream end, with mid-channel sedimentation downstream and the development of a secondary erosion focus

occurs and meander loops may become cut-off. In fact without a measure of confinement neck cut-offs rarely occur in the area studied although point-bar chutes can be found. Creeping appears more likely with second degree confinement where free loop development does not lead to up-valley orientation of the channel against the confinement; cut-offs appear more common with first degree or impeded meander situations.

Processes in Confinement

Processes in this second type of confinement, with loops developed at least in part within confining media, appear especially interesting since they are both more common and they constitute a more significant erosion problem. Accordingly a programme of field observations was undertaken on the Upper Elan (SN 8575). The catchment area is 4.24 km^2 and the valley is developed in Upper Ordovician and Silurian sedimentary rocks in which shales, mudstones, and siltstones predominate; much of the landscape is of glacial and periglacial origin. In the study area a series of meander loops are confined on the northeast by the valley wall and on the southwest dominantly by soliflucted till and gravel. A reach was selected for study (Figure 14.4) in which two loops were confined against the northern valley wall (A and C in Figure 14.4), three against the southern terraces (B, D, and F), and one loop was relatively unconfined. Examination of historic maps and air photographs show that loops C, D, and E have been developing slowly and have remained very approximately in the same position at least since 1885, but that loop F has become confined within the last century. Other channel changes can be seen to have occurred in the vicinity, but channel traces on the floodplain suggest that *lateral* growth, loop expansion

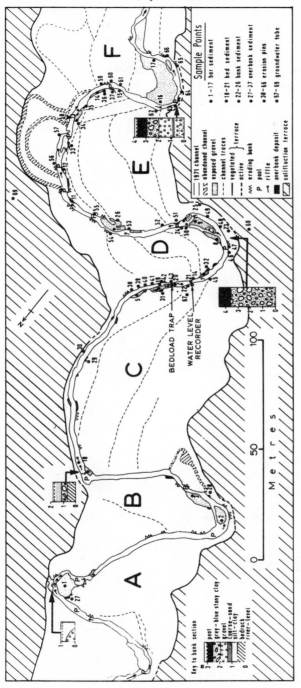

FIGURE 14.4 THE STUDY REACH ON THE RIVER ELAN (SN 864748)

Channel characteristics, bank sections, and instrument and sampling sites are indicated

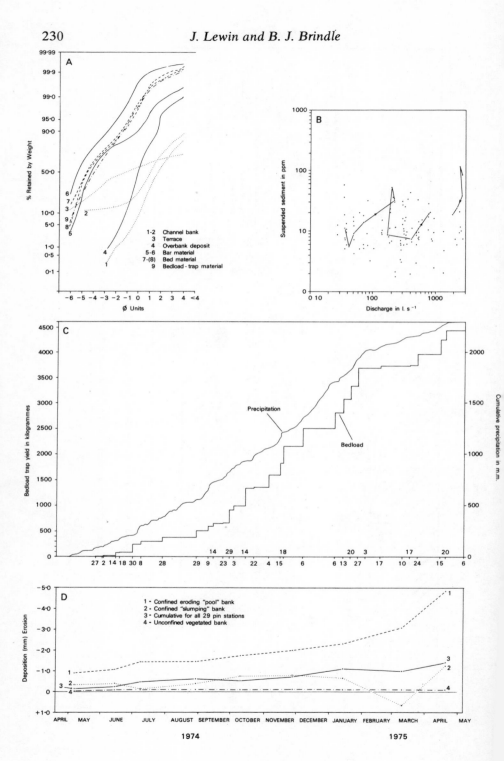

and relative stabilization against the confining media are the dominant channel change characteristics.

A number of channel bed, bar, overbank, and terrace deposits were sampled and analysed (Figures 14.4 and 14.5A), and a regular programme of field observations was undertaken from April 1974 to May 1975. This involved periodic sampling of bed material movement using a removable box trap of full-channel width and 30 cm breadth and depth within a container permanently buried in the bed; sampling of suspended sediment using a rod-mounted hand sampler; continuous water level recording using a Munro IH 125 float recorder; velocity observations using an Ott C1 current meter, and the observation of bank erosion rates using inserted wire pins at the sites shown in Figure 14.4.

Results of the programme are summarized in Figure 14.5. A total of 4400 kg of trap material was extracted during the study period, 61% in the 4-month period from October to January preceding an unusually dry spring (Figure 14.5C). This material appears very similar in size distribution to that of the channel bed elsewhere, suggesting that trap material is reasonably representative of transported bed material. Only on one occasion did the trap fill between observations, and despite the well-known difficulties in using traps of this kind, it is believed that a reasonable estimate of the transport of coarse material was obtained.

Yields of suspended sediment proved more problematical: only low concentrations were observed, with a wide scatter on the plot of data against discharge, and little consistency even in the trace of concentrations during individual discharge days (Figure 14.5B). The picture derived from the erosion pin data is similarly far from straightforward. An average loss of 1.4 mm for all 29 stations during the study period was achieved, but this conceals much variety in the timing and rates of erosion to be found at individual sites (Figure 14.5D).

Some of this variability may be explained in terms of the technique itself: erosion pins may remain apparently unaltered in exposure while fine matrix sediment is removed from around the larger fragments, while the sloughing and removal of discrete blocks of bank material may go unrecorded. Nonetheless, the variability in pin observations and the scatter on the suspended sediment/discharge plot may be both meaningful and not unrelated. Individual cut-bank sites appear to undergo localized slumping and removal of sediment by wash derived from the base of the peat top and flowing down the bluff faces; in this way slugs of sediment are injected into the stream irregularly in space and time and with respect to river discharge.

FIGURE 14.5 SUMMARY OBSERVATIONS ON THE UPPER ELAN 1974–5

A: size analysis of selected sediment samples. B: suspended sediment concentration and discharge; the lines show the relationship between data points for individual days. C: cumulative daily precipitation at Bodtalog (SN 869791) and bed load trap yield. D: selected erosion/deposition rates from erosion pin data

Hence at any particular time and discharge, the sediment yield can be highly variable when measured downstream of but a few active sediment sites.

The individual confined cut banks do form a most significant sediment source in the catchment: 4 m high bluffs can provide $0.02\,m^3$ of sediment per metre of bank per annum. This quantity is of course greatly in excess of the volume per unit length on the low-relief floodplain, and channel forms are characterized by local sediment accumulation in, and adjacent to, the stream at and below the high undercut bluffs (Figure 14.4). Furthermore, the increase in pool depth and decrease in channel width, found in cases of confinement against relatively inerodible materials, is not generally characteristic of the Elan meanders, although local planform inflexions and scour pools are in evidence.

Development of the confined loops appears to occur as follows. Fine and coarse material is fed into streams, particularly following the slumping of discrete slugs of sediment during winter months and the washing of material across bluff faces. Fines may be removed in a range of discharges, though only low concentrations were observed, with a small proportion of fine sediment redeposited locally on the floodplain (Figures 14.4 and 14.5A). Coarser materials may remain at the base of the undercut slopes to be removed by the occasional high flow. The rate of bank retreat is not lower, and the volume of sediment contribution is considerably higher, at confined bends than at other channel bank erosion sites. Thus it is possible for confined meander planforms to be preserved which are not always greatly distorted. Confinement provides the major local source of sediment at the present time, with individual confined bends apparently remaining active for at least a century.

Conclusions

We have shown that confinement is commonly to be found in a mid-latitude environment at the present time as a result of both Quaternary and subsequent valley floor developments, and of human activity. The degree of confinement varies considerably, and may affect the patterns achieved by currently active streams. Two modes of development are suggested: one in which a meandering pattern creeps down-valley intact though with a modified geometry, and a second in which loops become ensconced within confining media and cut-offs occur. In a field study of the second type, it has been shown that erosion of confined banks forms a major source of sediment in upland valleys, but that these erosion sources local to the river appear to behave in a complex manner with respect to river discharge. It is believed that in the present process environment, the phenomenon of confinement is of considerable significance not only for the development of channel patterns but also for the sediment dynamics of active river systems.

References

Carey, W. C. 1969. Formation of floodplain lands. *J. Hydraul. Div. Amer. Soc. Civ. Engrs.*, **95,** pp. 981–994.

Chitale, S. V. 1973. Theories and relationships of river channel patterns. *J. Hydrol.*, **19,** pp. 285–308.

Cross, P. and Hodgson, J. M. 1975. New evidence for the glacial diversion of the River Teme near Ludlow, Salop. *Proc. Geol. Ass.*, **86,** pp. 313–331.

Dury, G. H. 1964. Principles of underfit streams. *U.S. Geol. Survey Prof. Paper 452-A.*

Dury, G. H. 1966. Incised valley meanders on the lower Colo River, New South Wales. *Aust. Geogr.*, **10,** pp. 17–25.

Fisk, H. N. 1944. *Geological investigations of the alluvial valley of the lower Mississippi River.* Mississippi River Commission, Vicksburg, Miss.

Friedkin, J. F. 1945. *A laboratory study of the meandering of alluvial streams.* Mississippi River Commission, Vicksburg, Miss.

Ferguson, R. I. 1975. Meander irregularity and wavelength estimation. *J. Hydrol.*, **26,** pp. 315–333.

Haynes, C. V. 1968. Geochronology of Late-Quaternary alluvium. In Morrison, R. B. and Wright, H. E., *Means of correlation of Quaternary successions.* University of Utah Press, Salt Lake City, pp. 591–631.

Jefferson, M. S. W. 1902. Limiting width of meander belts. *Natl. Geog. Mag.*, **13,** pp. 373–384.

Jones, O. T. 1965. The glacial and post-glacial history of the lower Teifi valley. *Quart. J. Geol. Soc. Lond.*, **121,** pp. 247–281.

Konditerova, E. A. and Popov, I. V. 1966. Relation between changes in the horizontal and vertical characteristics of river channels. *Soviet Hydrol.*, **5,** pp. 515–527.

Kopeliana, Z. D. and Romashin, V. V. 1970. Channel dynamics of mountain rivers. *Soviet Hydrol.*, **5,** pp. 441–452.

Kulemina, N. M. 1973. Some characteristics of the process of incomplete meandering of the channel of the Upper Ob' river. *Soviet Hydrol.*, **6,** pp. 518–534.

Lane, E. W. 1957. *A study of the shape of channels formed by natural streams flowing in erodible material.* U.S. Army Engineer Divn., Mississippi River Corps of Engineers, Omaha, Nebr. (M.R.D. Sediment Series No. 9).

Leopold, L. B. and Wolman, M. G. 1960. River meanders. *Geol. Soc. Amer. Bull.*, **71,** pp. 769–794.

Makkavyeyev, N. I., Khemelyeva, N. V., and Gun Go-Yuan 1969. Formirovaniye meander. *Eksperim. Geomorfologiya*, **2,** pp. 7–87.

Matthes, G. H. 1941. Basic aspects of stream meanders. *Trans. Amer. Geophys. Union* **22,** pp. 632–636.

Schmudde, T. H. 1963. Some aspects of landforms of the lower Missouri floodplain. *Ann. Assoc. Amer. Geogr.*, **53,** pp. 60–73.

Speight, J. G. 1965. Meander spectra of the Angabunga River. *J. Hydrol.*, **3,** pp. 1–15.

Sundborg, A. 1956. The River Klarälven, a study of fluvial processes. *Geogr. Annlr.*, **38,** pp. 27–316.

R. I. FERGUSON
Lecturer in Earth and Environmental Science
University of Stirling

15

Meander Migration: Equilibrium and Change

The migration of meander bends is a characteristic feature of alluvial rivers and one of the most conspicuous changes affecting fluvial landscapes. Bank recession rates of 10–15 m yr^{-1} are by no means exceptional in large rivers, and 100 m yr^{-1} is not unknown (Kondratyev and Popov, 1967). However, such rates apply to certain bends only: others on the same channel at the same time shift more slowly, if at all. As every elementary textbook notes, it is this spatial variability which explains the occurrence of neck cut-offs and oxbow lakes. Individual bends become distorted and alter their relative positions, sometimes so much that they meet and thereby shorten the path of the river. The overall meander pattern is irregular and its details change from year to year, decade to decade, or century to century, according to the overall level of channel activity.

While recognizing this irregularity and variability, geomorphologists have come to look on channel geometry in general, and meander patterns in particular, as manifestations of a dynamic equilibrium between hydrologic regime and local geologic environment. Local changes may occur wherever and whenever streamflow exceeds the threshold for erosion of the materials composing the banks of the channel, but self-regulatory mechanisms are thought to come into play which tend to restore the overall form of the channel (Langbein and Leopold, 1964). In this approach attention is focused on the assumed equilibrium state and its relationship to prevailing hydrologic and geologic conditions. Generally both channel form and environmental conditions are quantified using numerical parameters, bankfull discharge, meander wavelength, and the like, which summarize mean or aggregate conditions. Apart from some hydrologic parameters, though, these indices generally relate to a single time, and we cannot know whether at this time the channel was in its equilibrium state or temporarily away from it.

The implications of this uncertainty depend on our aims. In comparing different rivers in order to infer general relationships between channel form and environmental conditions, the problem can be minimized by studying as large a sample as possible. A difference between two rivers at a particular time may indicate a true difference in equilibrium form, but could also, if it is not too great, be explained by temporary departures in opposite directions from a common equilibrium. If, however, information is available for a large number of rivers, at the same or at different times, their deviations from equilibrium may tend to cancel out. The only situation in which this is not true is when channels are measured at similar times relative to some widespread departure from equilibrium, for example just after a major regional flood. With this proviso, sample relationships between channel form and environmental conditions in this kind of study should provide a blurred but reasonably unbiased guide to the assumed underlying equilibrium relationships.

More serious problems arise in studies of river metamorphosis in response to environmental change. The work of Dury (e.g. 1965) and Schumm (e.g. 1969) has stimulated interest in the response of meander patterns to changes in hydrologic regime, with a view to using meander geometry as an indicator of the occurrence and extent of climatic change. A shift in channel equilibrium may consist of a catastrophic jump as some threshold is crossed, or it may be a more gradual evolutionary tendency, but in neither case is there any reason to suppose that short-term fluctuations cease to occur. The resulting ambiguity is illustrated in Figure 15.1 which represents variations in the sinuosity of a hypothetical meandering reach as bends grow and are cut off over a period of perhaps a century. In the upper sketch the channel is in equilibrium, with no overall trend in sinuosity, yet measurements at times t_1 and t_2 would suggest an increased meandering tendency; two other observations might indicate the opposite. In the lower diagram, on the other hand, a real overall trend would not be apparent from observations at times t_3 and t_4.

The extent of this ambiguity should not be overrated. In the long run, climatic change may produce unmistakeably big differences in meander pattern; for example the underfit meanders studied by Dury are typically only one-fifth to one-tenth the size of their predecessors. In short-term modern investigations there may be additional evidence, documentary or in the form of floodplain scrolls, of channel changes between times of accurate mapping or aerial photography. Nevertheless there is still some danger of missing a real trend or inferring a spurious one. This is especially so when only two points in time are compared, and when the characterization of the meander pattern is based on only a short stretch of river. To take an extreme example, it would clearly be unwise to infer a change in equilibrium

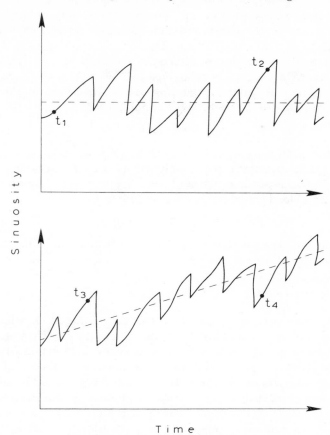

FIGURE 15.1 EQUILIBRIUM OR CHANGE?

The ambiguity of comparing an irregularly migrating meander pattern at two points
in time

from an increase in the size of a single bend over an interval of 10 years or
so, since observations of the next bend downstream might show the opposite
tendency. For this reason studies of meander equilibrium should be carried
out on reaches sufficiently long to encompass a representative selection of
bends, though not so long that external conditions alter appreciably along
the channel.

Steady State and Dynamic Equilibrium

These introductory comments show that the central methodological problem posed by the spatial irregularity and temporal variability of channel
geometry is how to disentangle general tendencies from local or temporary
changes. We have already seen that a statistical approach is the obvious

answer in comparative studies of different rivers, but in developmental studies the possibility of deterministic modelling of channel change must be considered.

One deterministic situation of interest is what will here be termed steady state migration, in which a regular meander waveform shifts down-valley without distortion. Regular meandering can be described by the so-called sine-generated curve (Langbein and Leopold, 1966)

$$\theta = \omega \sin ks \qquad (1)$$

in which the parameters ω, k determine the sinuosity ($P = 1/J_0(\omega)$, where J denotes a Bessel function) and wavelength ($\lambda = 2\pi/k$, measured along the channel) of the repeating pattern (see Figure 15.2). Migration of this waveform at a steady rate c/P can be described by

$$\theta = \omega \sin\{k(s - ct)\} \qquad (2)$$

The explicit equation for channel direction as a function of time and space can also be viewed as a solution of the wave equation

$$\frac{\partial^2\theta}{\partial s^2} = \frac{1}{c^2}\frac{\partial^2\theta}{\partial t^2} \qquad (3)$$

which links time and space rates of change of direction. Characteristic relationships between curvature (the rate of change of direction with distance) and bend growth (which involves changes in direction over time at each locality) have been found in flume and field studies of meander development (e.g. Bagnold, 1960; Hickin and Nansen, 1975), so a partial differential equation such as (3) may have a physical basis as well as allowing complete prediction of the future course of the river.

Steady state migration of this kind seems plausible in a perfectly uniform environment, for example a flume with homogeneous sediments and constant discharge. A regular but evolving pattern can also be envisaged if discharge, and with it wavelength and sinuosity, is systematically increasing rather than remaining constant. In both situations the spatial regularity of the waveform removes all ambiguity about the equilibrium or otherwise of the pattern. It is, however, difficult to see how a regular pattern can survive for long under natural conditions. Quite apart from the effects of streamflow fluctuations, channel and valley-floor sediments are rarely uniform and the lateral redistribution associated with bank cutting and point bar construction introduces size sorting, as the simulation models of Bridge (1975) demonstrate. Continued migration, now with spatially variable boundary conditions, must inevitably lead to distortion of the waveform with some bends, or parts of bends, eroding faster than others and the pattern as a whole becoming irregular. This process is cumulative, since the greater diversity of depositional conditions in an irregular pattern, particularly if backwaters are produced by cut-offs, amplifies the sedimentological and

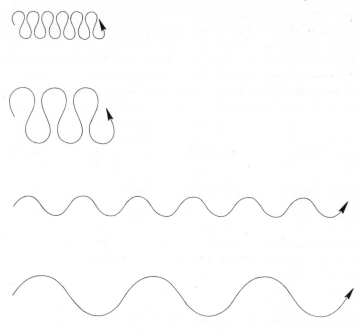

FIGURE 15.2 REGULAR MEANDERS

Examples of sine-generated curves with different combinations of high/low wavelength and high/low sinuosity

topographic variability of the valley floor and produces even more varied conditions for future bank erosion, as documented by Kolb (1963) in the case of the lower Mississippi.

An irregular meander pattern is even less likely to be in a steady state, and a deterministic analysis of its development would be extremely complicated. We can, however, envisage a statistical equilibrium in which the pattern retains its aggregate characteristics despite changes in detail. If some bends grow, but others decline or are eliminated, the scale and degree of meandering, and the overall level of irregularity, may remain more or less constant over the years. In more formal statistical terms, we can treat the process of meandering as deterministic but assume that the local valley-floor conditions which constrain it vary randomly from place to place. Since the river occupies different positions at different times it experiences different spatial sequences of these conditions, or of disturbances about the average condition. The precise course of the channel depends on the detailed pattern of these disturbances, but the overall nature of the waveform need not alter.

Two conditions are necessary for such a statistical equilibrium. First, the deterministic meandering propensity of the river must be constant, which presumably means no change in hydrologic regime or sediment load. Second, the irregularly varying valley floor conditions along the channel at different times must have similar statistical properties (mean, variability, autocorrelation structure), that is to say the disturbance sequences must be realizations of a single stationary stochastic process. This requires an unchanged overall mix of valley floor materials, with the migrating channel not impinging on some part of the valley floor with systematically different conditions. In particular, migration into a less erodible zone will induce a 'stacked' pattern (Tanner, 1955) with a systematic increase in sinuosity and distortion of the overall meander belt, as well as individual bends.

Statistical Characterization of Irregular Meanders

If for any of these reasons the pattern is progressively altering its average properties, it is in neither a steady state nor a dynamic equilibrium, but is instead undergoing metamorphosis or evolution. To prove such a transformation it is not sufficient to show that some bends are altering in a certain way: there must be a definite overall tendency, an accumulation of local changes which do not cancel out. Determining whether this is the case is likely to involve statistical techniques of some kind.

The two main approaches are to treat the meander pattern as a sequence of bends, each characterized by suitable morphometric indices, or to approximate it by a series of direction or curvature values at regular intervals along the channel. If individual bends are considered, their variability can be summarized by statistics of the distributions of their morphometric properties (Brice, 1964, 1974; Ferguson, 1973). Any disequilibrium is expected to show up on comparing bend statistics for different times. Drawbacks to this method are that the number of bends distinguished depends on the resolution at which the meander pattern is mapped and analysed (Ferguson, 1973), and that the spatial order of the bends, and thus the interaction between adjacent bends, is ignored.

The alternative series analysis approach focuses specifically on spatial autocorrelation, and appears to be little affected by the choice of sampling interval in obtaining the direction series (Ferguson, 1975). Attention is transferred from individual bends to the pattern as a whole, but characteristic properties such as wavelength can still be estimated (Speight, 1965; Ferguson, 1975). The most informative technique for studying meander series is spectral analysis, which decomposes the total variance of direction (or curvature) along the channel into contributions from different oscillatory frequencies or wavebands. The spectrum of either direction or curvature for regular meandering is a single spike at the frequency given by the reciprocal of the wavelength. Natural meanders have broader, though not necessarily polymodal, spectra. The curvature spectrum usually has a fairly well-defined peak and appreciable high-frequency variance, whereas the direction series

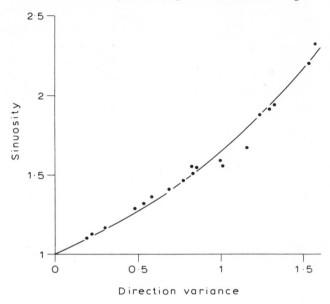

FIGURE 15.3 MEANDER SINUOSITY AND DIRECTION VARIANCE

Data for 19 English and Scottish rivers, and a theoretical curve assuming normal distribution of direction (see Ferguson, 1977)

is often dominated by low-frequency variance associated with valley bends. Peaks define characteristic wavelengths, the scatter about them is some indication of the degree of irregularity, and the total area under the direction spectrum (the series variance) is proportional to the sinuosity of the channel (Figure 15.3). A spectral analysis therefore provides a complete characterization of a meander pattern, and spectra for the same river at different times should reveal any systematic changes in pattern.

As an illustration, the technique is applied here to two reaches of the White River in Indiana, U.S.A. This river is chosen because it appears to be migrating actively, and because Brice (1974) has already investigated its changing pattern using the alternative method based on bend statistics. Channel centrelines of the two reaches according to air photos taken in 1937 and 1966–68 and of the West Fork according to a map published in 1880, are shown in Figures 15.4 and 15.5. Comparison of these plans together with bend statistics led Brice to conclude that after 1880 the pattern became less regular and sinuous, perhaps through accelerated migration following nineteenth century deforestation of the floodplain, but that little systematic change was apparent since 1937 except for the obliteration of loops with a radius of less than 80 m in the East Fork (Brice, 1974, p. 191–3).

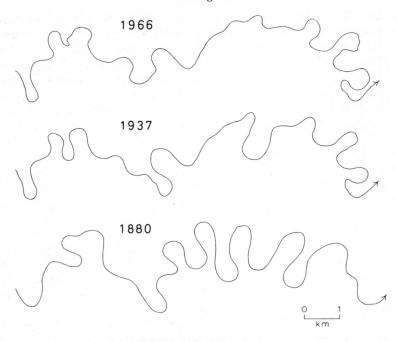

FIGURE 15.4 WEST FORK, WHITE RIVER

Changing centreline of reach upstream of Edwardsport, Indiana. (After Brice, 1974, and reproduced by permission of the State University of New York)

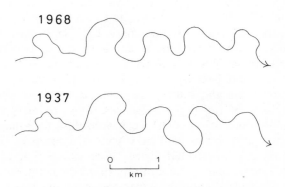

FIGURE 15.5 EAST FORK, WHITE RIVER

(After Brice, 1974, and reproduced by permission of the State University of New York.)

To see whether the series analysis approach confirms these conclusions, and whether it reveals any more, the mapped centrelines have been digitized, converted to direction series with a step length (Δs) of 100 m for the West Fork and 60 m for the East Fork, and differenced to obtain direction-change ($\Delta\theta$) or curvature series as well. The results of spectral analysis of both series for each channel configuration are shown in Figure 15.6. Apart from the second peak in the 1937 curvature spectrum for the East Fork, the plots are typical of meander patterns. The direction spectra, higher because of the greater variance of direction than of curvature, are completely dominated by low frequencies and show no evidence of multiple wavelengths (their rightward extensions, omitted for clarity in Figure 15.6, continue to fall off rapidly). The curvature spectra show negligible variance at very low frequencies, rise sharply to a peak at a somewhat higher frequency (lower wavelength) than do the direction spectra, then tail off fairly steadily with the exception already noted. This anomalous second peak occurs at a frequency of 0.18 cycles per 60 metres, corresponding to a wavelength of 333 m or, using Brice's figure for the mean wavelength to radius ratio in this reach, a radius of about 50 m. The disappearance of this spectral peak between 1937 and 1968 is thus consistent with Brice's observation about elimination of tight bends.

The spectral analysis also reveals certain less dramatic changes in the pattern of the West Fork reach. Its direction and curvature spectra both

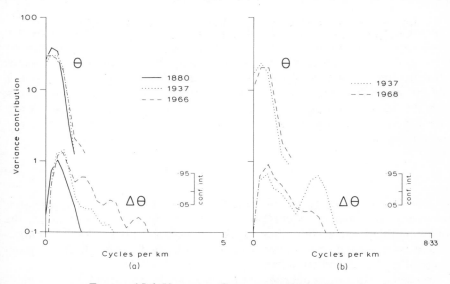

FIGURE 15.6 VARIANCE SPECTRA OF WHITE RIVER

Direction (θ) and curvature ($\Delta\theta$) spectra for West Fork (a) and East Fork (b) in different years, with 90% confidence intervals

tailed off more rapidly in 1880 than subsequently, confirming the increased irregularity of the pattern after that date. The 1880 spectra also peak at a higher wavelength, suggesting some reduction since then in the scale of meandering as measured along the channel. A continuing decline in regularity is also apparent on comparing the 1937 and 1966 curvature spectra, even though the modal wavelength is unchanged. This trend is accompanied by the transformation of the direction spectrum peak into a shoulder, with maximum variance in the zero-frequency band. At a purely descriptive level, then, spectral analysis can reveal changes in meander patterns which are not obvious from maps and which agree with, and perhaps extend, the results of bend analysis.

A Model for Meander Spectra

The similarity in general form of meander spectra from different rivers hints at the possibility of some parametric model applicable to most or all rivers, and to any one river at different times. Such a model can in fact be obtained as a by-product of a simple but seemingly realistic stochastic model for meander direction series (Ferguson, 1976a). The starting point is the differential equation

$$\frac{\partial^2 \theta}{\partial s^2} = -k\theta \qquad (4)$$

obtained from the sine-generated curve of equation (1). This undamped oscillation about the direction $\theta = 0$ can be randomized by replacing θ by $\theta - \varepsilon$ in the right-hand side of (4), where $\{\varepsilon\}$ is a zero-mean random process, representing the spatial sequence of local valley floor disturbances responsible for meander irregularity in the way discussed above. A damping term is also introduced so that the influence of each disturbance dies out downstream to offset the continuing impact of new ones. The resulting disturbed periodic model

$$\theta + \frac{2h}{k}\frac{\partial \theta}{\partial s} + \frac{1}{k^2}\frac{\partial^2 \theta}{\partial s^2} = \varepsilon(s) \qquad (5)$$

is stable for damping factors in the range $0 < h < 1$, and generates waveforms irregular in proportion to h; the limiting case $h = 0$ is simply the sine-generated curve. For discrete direction series, (5) is closely approximated by a second-order autoregression, with coefficients which are functions of the scale and irregularity parameters k and h.

Expected statistical properties of this model are best written for the continuous case but can be assumed to hold also for discrete series if the scale parameter is expressed in terms of Δs. If the disturbance process $\{\varepsilon\}$ is statistically stationary with variance σ^2 and spectrum $S\varepsilon(f)$, the direction process $\{\theta\}$ defined by (5) is also stationary. Its expected variance spectrum is

$$S_\theta(f) = \frac{S_\varepsilon(f)}{(1 - \lambda^2 f^2)^2 + (2h\lambda f)^2} \tag{6}$$

where $\lambda = 2\pi/k$, and the total direction variance is

$$V_\theta = \frac{k\sigma^2}{4h}. \tag{7}$$

Channel curvature, approximated by the direction change $\Delta\theta$ in the discrete case, has expected spectrum

$$S_{\Delta\theta}(f) = (2\pi f)^2 S_\theta(f) \tag{8}$$

with total variance

$$V_{\Delta\theta} = k^2 V_\theta. \tag{9}$$

These equations hold for any stationary disturbance process, but do not define unique spectra until $\{\varepsilon\}$ is specified. The simplest assumption is of white noise (uncorrelated) valley-floor irregularity, with the flat spectrum

$$S_\varepsilon(f) = 2\sigma^2 \tag{10}$$

in the range $0 < f < \frac{1}{2}\Delta s$ relevant to empirical analysis.

As an example of the realism of this model, and of its application in studies of meander changes, Figure 15.7 shows expected spectra of disturbed periodic meandering with constant direction and disturbance variance but different wavelengths, chosen to agree roughly with the 1880 and 1937 patterns of the West Fork of the White River (cf. Figure 15.6a). From equations (7) and (9), the fall in wavelength is balanced by an increase in irregularity, reflected in wider direction and curvature spectra, and in direction-change variance, apparent in the increased area under the curvature spectrum. Considering that the parameters are in no way best-fit estimates, the similarity between observed and theoretical spectra is striking, and provides strong backing for the suggestion of reduced scale and increased irregularity of meandering after 1880 in this river.

It is also possible to utilize the disturbed periodic model in estimating the wavelength of natural meanders. The statistical nature of wavelength estimation for irregular patterns was stressed in Ferguson (1975), in which the curvature spectrum peak was recommended as perhaps the best indicator. This empirical conclusion gains support from the model proposed here, in which the curvature spectrum does indeed have its peak at the fixed wavelength λ, whatever the degree of irregularity, given the assumption of uncorrelated disturbances. Another possibility is the estimator

$$\lambda = 2\pi\Delta s \sqrt{\frac{V_\theta}{V_{\Delta\theta}}} \tag{11}$$

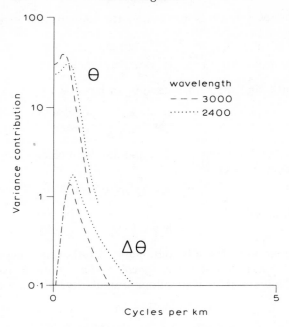

FIGURE 15.7 THEORETICAL VARIANCE SPECTRA

Disturbed periodic model with fixed direction and disturbance variance but different
wavelengths (in metres)

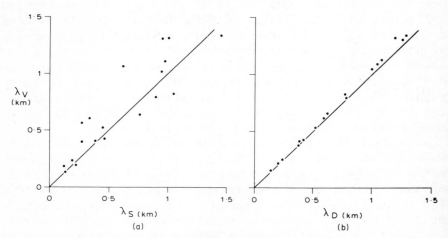

FIGURE 15.8 COMPARISON OF WAVELENGTH ESTIMATES

Wavelengths for 19 English and Scottish rivers, estimated from direction/curvature
variance ratio (λ_V), quadratic interpolation of curvature–spectrum peak (λ_S) in a, and
autoregressive direction model (λ_D) in b. Lines of equality included as a guide

derived from equation (9). This is very easy to use, does not suffer from the lack of resolution inescapable in discrete spectral analysis, and makes no assumptions at all about valley floor irregularity (indeed, it applies also to regular meanders). Wavelengths calculated in this way correlate well with estimates from curvature spectrum peaks (Figure 15.8a), and almost perfectly (Figure 15.8b) with those obtained by fitting the discrete approximation of equation (5) to direction series by least squares. In both cases the variance-ratio estimates tend to be higher, but only by a small percentage. They also increase with Δs, by about 8% for a 50% increase in Δs for the White River, because of pattern smoothing. This should not however be a serious problem if Δs is chosen with due regard to channel scale. Equation (11) thus appears to offer a novel but simple way of estimating meander wavelengths.

Research Needs

Spectral analysis is advocated here as a descriptive technique, but the underlying view of meander migration as a space–time stochastic process raises some inferential issues. If meander patterns are realizations of stochastic processes, properties such as wavelength and direction variance are random variables, and one is forced to ask how big a difference between the values of some pattern property at two different times could occur by chance. Differences between spectra can in fact be tested formally by analysis of variance (Priestley and Rao, 1969), allowing statistical significance testing of meander changes. It should also be possible to attach confidence limits to parameter estimates for the disturbed periodic model. In this context, one topic requiring attention is the estimation of meander irregularity, since different estimators suggested in Ferguson (1976) give rather varied and inconsistent results. Perhaps the best method is the simplest: visual comparison against theoretical curvature spectra with round-number parameters as in Figure 15.7.

A more general problem is just how far to take the statistical analysis of meander waveforms, or indeed other aspects of channel geometry. Visual inspection and comparison of meander spectra is wholly useful, and with the aid of a computer and coordinate digitizer obtaining spectra from a mapped pattern involves only a few minutes' work. The simple model outlined above should also provide a useful basis for assessing empirical spectra and estimating meander parameters. However it seems likely that further progress in understanding channel pattern changes is more likely to result from widespread application of these methods, or other existing ones, in an essentially descriptive way, than from individual investigations using much more sophisticated and time-consuming statistical techniques. More fundamentally, no amount of statistical analysis can compensate for an incomplete understanding of the processes of channel change.

References

Bagnold, R. A. 1960. Some aspects of the shape of river meanders. *U.S. Geol. Surv. Prof. Paper 282-E.*

Brice, J. C. 1964. Channel patterns and terraces of the Loup River in Nebraska. *U.S. Geol. Surv. Prof. Paper 442-D.*

Brice, J. C. 1974. Meander pattern of the White River in Indiana—an analysis. In Morisawa, M., (Ed), *Fluvial geomorphology*, State University of New York, Binghamton, pp. 178–200.

Bridge, J. S. 1975. Computer simulation of sedimentation in meandering streams. *Sedimentology*, **22**, pp. 3–43.

Dury, G. H. 1965. Theoretical implications of underfit streams. *U.S. Geol. Surv. Prof. Paper 452-C.*

Ferguson, R. I. 1973. Regular meander path models. *Water Resour. Res.*, **9**, pp. 1079–86.

Ferguson, R. I. 1975. Meander irregularity and wavelength estimation. *J. Hydrol.*, **26**, pp. 315–333.

Ferguson, R. I. 1976. Disturbed periodic model for river meanders. *Earth Surface Processes*, **1**, pp. 337–347.

Ferguson, R. I. 1977. Meander sinuousity and direction variance. *Geol. Soc. Amer. Bull.*, **88**, pp. 212–214.

Hickin, E. J. and Nansen, G. C. 1975. The character of channel migration on the Beatton River, north eastern British Columbia, Canada. *Geol. Soc. Amer. Bull.*, **86**, pp. 487–494.

Kolb, C. R. 1963. Sediments forming the bed and banks of the lower Mississippi River and their effect on river migration. *Sedimentology*, **2**, pp. 227–234.

Kondratyev, N. Y. and Popov, I. V. 1967. Methodological prerequisites for conducting network observations of the channel process. *Soviet Hydrology*, **3**, pp. 273–297.

Langbein, W. B. and Leopold, L. B. 1964. Quasi-equilibrium states in channel morphology. *Amer. J. Sci.*, **262**, pp. 782–794.

Langbein, W. B. and Leopold, L. B. 1966. River meanders—theory of minimum variance. *U.S. Geol. Surv. Prof. Paper 422-H.*

Priestley, M. B. and Rao, T. S. 1969. A test for non-stationarity of time series. *J. Roy. Stat. Soc.*, B, **31**, pp. 140–9.

Schumm, S. A. 1969. River metamorphosis. *J. Hydraul. Div., Amer. Soc. Civ. Engrs.*, **95**, pp. 255–273.

Speight, J. G. 1965. Meander spectra of the Angabunga River. *J. Hydrol.*, **3**, pp. 1–15.

Tanner, W. F. 1955. Geologic significance of 'stacked' meanders. *Geol. Soc. Amer. Bull.*, **66**, p. 1698.

E. J. HICKIN

Associate Professor of Geography
Simon Fraser University
British Columbia

16

The Analysis of River-Planform Responses to Changes in Discharge

The planform geometry of meandering channels is the subject of a very substantial body of literature. Most detailed analyses of meander planform have been undertaken as a starting point in the search for the underlying physical causes of the meandering phenomenon. Although one of the earliest references to the origin of meanders is in the work of Targioni-Toyzetti in the mid-eighteenth century (from Mather and Mason, 1939), the period of most intensive investigation of the geometry of meander planform dates from the pioneer work of Jefferson (1902), and centres on the post-war studies of Friedkin (1945) and Inglis (1949). More recent studies, including those by Leopold and Wolman (1957, 1960), have refined and firmly established the ideas developed by Inglis.

Although a single theoretical model of river behaviour, sufficient to explain the complexities of meandering, is yet to be developed, several geometric and hydraulic relations (or meander 'laws') have been very widely accepted as characterizing the typical meandering channel. 'Typical' is usually taken to mean the channel trace which would develop in an actively migrating river which is free of the constraints imposed by heterogeneous boundary materials.

These geometric laws of meanders have not only influenced most of the various approaches to the problem of meander initiation by focusing attention on discharge, but also they have found application to other types of problems such as the assessment of hydrologic and climatic changes during the Pleistocene (Dury, 1954, 1964, 1965).

The meander laws are based on two very important assumptions: that a single dominant channel-wavelength can be recognized in a meander array, and that it is associated with a single dominant discharge. It has only been the last decade or so that has seen serious challenge to these assumptions. (Speight, 1967; Carlston, 1965).

Unfortunately, most of the recent analyses of meander planform follow Speight's lead in utilizing the relatively poorly understood statistical technique of power-spectral analysis. Spectral analysis cannot be directly applied to non-functional meander traces and the resulting wavelength spectra of the transformed array are not easy to interpret.

The purpose of 'this paper is twofold: to examine meander planform changes with increasing discharge, and to examine the interaction of meander shape and flow character at the level of the individual channel bend. The first task will involve an assessment of the meander laws on the basis of relatively conventional wavelength frequency and variance analyses on the one hand, and for the purpose of comparison, of power-spectral analysis on the other. The second task will be to offer some explanations of the geometric properties of meander arrays in terms of the statistical methods and of the hydraulics of flow in a developing channel-bend.

The Field Study Area

Data for this study have been obtained from the Beatton River in Western Canada. Although several factors influenced the choice of the Beatton River for planform analysis, the most important is that, for much of

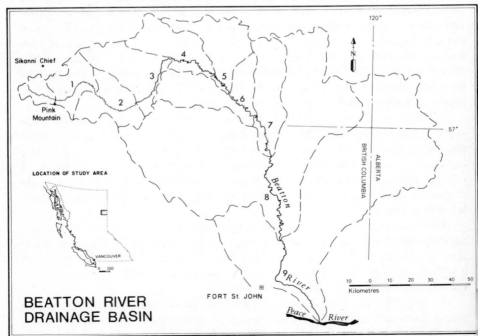

FIGURE 16.1 LOCATION OF THE TEST REACHES ON THE BEATTON RIVER

its length, it is a markedly sinuous and free-meandering channel. There seems to be general agreement that, because such rivers can freely adjust their planform to changing flow conditions, they are most likely to have discharge-sensitive meander geometries. Furthermore, because the river has been the subject of intensive hydraulic, sedimentological, and geomorphic surveys (Hickin, 1974; Hickin and Nanson, 1975; Nanson, in preparation), there exists a considerable amount of field data as useful background to the present study. In addition, large-scale topographic maps (1:50,000) and high-resolution aerial photographs (1:16,000) are available for the complete meander trace.

The location of the Beatton River in northeast British Columbia, and the positions of nine study reaches along the river, are shown in Figure 16.1. Reach lengths range from 7720 m (Reach 2) to 63,440 m (Reach 8). Each one is located between major tributary junctions and can be regarded as being sensibly discharge constant along its length. Bankfull discharges cited in this study were determined by field survey on reaches 5, 6, and 10 (data for Reach 10 was supplied by Rolf Kellerhals, Civil Engineer, Vancouver) and reliable estimates based on channel form were obtained for the remaining reaches.

Methods of Meander Planform Analysis

Planform analyses of meanders can be grouped into one of two basic types depending on the data collection process: those adopting the 'representative reach' concept as the basis for analysis, and those based on the complete meander trace.

The representative reach concept recognizes that many parts of a channel trace include distortions resulting from pockets of unusually strong or weak boundary materials, from the formation of cut-offs, and from other 'disturbing' factors, while others are free of such constraints and are thus truly representative of the pure meander form and scale. While many geomorphologists would recognize the extreme cases of such a distinction, the definition of a representative reach is perhaps a subjective procedure.

Nevertheless, the meander laws are based on this mode of analysis (Leopold and Wolman, 1957, 1960). In the present context the most important of these empirical relationships are as follows:

$$\frac{\lambda}{W} \cong 10 \tag{1}$$

$$\frac{\lambda}{R} \cong 4 \tag{2}$$

$$R/W \cong 2 \tag{3}$$

$$\lambda \cong 10 \, Q^{0.5} \tag{4}$$

$$W \cong Q^{0.5} \tag{5}$$

E. J. Hickin

where λ, W, R, and Q are, respectively, meander wavelength, bankfull channel width, radius of channel curvature (all in metres) and bankfull discharge (m³/s). The planform parameters are summarized in Figure 16.2. In spite of the apparently subjective basis for equations (1) to (4), some subsequent research in open-channel hydraulics suggests that such relations should indeed exist (for example, see Bagnold, 1960).

The alternative approach, primarily developed because of dissatisfaction with the subjectiveness of the representative-reach concept, considers all direction changes in the meander trace whether the result of distortions or not. At an operational level this method usually generates a set of cartesian coordinates for points along the channel thalweg, commonly digitized at an interval of one or two channel widths.

This type of data can be analysed in a number of ways but in practice has been almost exclusively the subject of spectral analysis. There is an almost complete absence of studies which have processed this type of data in terms of frequency distribution or related statistics similar to those on which the meander laws are based. The present study is, in part, an attempt to rectify this situation.

The three types of planform analysis presented here are based on the nine reaches identified in Figure 16.1. Each reach was digitized on the channel centre line at an interval of one channel width; centre lines and bankfull-width series were obtained from topographic maps and aerial photographs respectively.

The first type of analysis considers the distribution of the radius of channel curvature for each reach. Each successive and continuous triad of coordinates was assumed to be common to a given circle, the radius of which was accordingly computed. Radii exceeding 200 channel widths are defined as representing straight reaches, and the remainder were used to construct frequency distributions of channel curvature. Because of the constant meas-

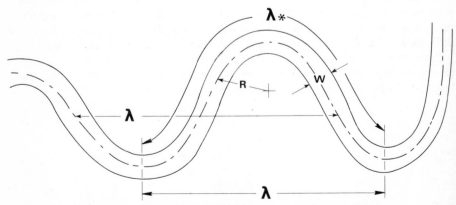

FIGURE 16.2 PLANFORM GEOMETRY OF A MEANDER TRAIN

urement interval used to represent the trace, relative frequency is a close measure of the proportion of the total channel length contributing to various channel curvatures. Channel-curvature arrays described in this way appear to be more meaningful than those based on selective sampling such as that adopted by Brice (1973, 1974).

The second type of planform analysis considers the frequency and variance distributions of channel wavelength, λ. Very high frequency wavelengths present in the initial data set are very likely the product of consistent digitizing errors and are therefore effectively removed from the record using the line-generalizing algorithm of Douglas and Peucker (1973). Figure 16.3 graphically illustrates this procedure. The end points 1 and 2 of the array define a base line; point 3 is subsequently located as that coordinate which terminates the longest normal to that line. Provided the length of the normal exceeds a given tolerance, new base lines 1–3 and 3–2 are formed, and points 4 and 5 respectively are identified as the coordinates of maximum orthogonal departure from them. The process is repeated until maximum departure is less than the specified tolerance limit, at which stage the base line is accepted as the average centre line for the segment in question. Throughout this study the tolerance limit has been set at half the magnitude of the bankfull width.

When all points of non-significant departure are filtered from the total array, each point of sign change in the direction series is designated an inflection point from which linear half wavelengths are computed. The area enclosed by each channel bend and the locus of the inflection points is also computed to form the basis of the variance distributions of wavelength.

Clearly, this mode of analysis also filters low-frequency wavelengths from the record. This type of filtering would appear to be an advantage because the low-frequency components are likely to be related less to the scale of the flow system than they are to regional topographic variability. Nevertheless, the locus of the primary inflection points can be reprocessed as a new series in order to characterize the secondary array, which in turn can similarly yield a tertiary meander array for analysis, and so on. Characteristics of these higher order arrays of the Beatton River will not be considered in this study.

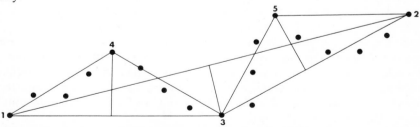

FIGURE 16.3 GRAPHIC REPRESENTATION OF THE DOUGLAS AND PEUCKER LINE-GENERALIZATION ALGORITHM

The third type of planform description considered here is that provided by power-spectral analysis. Spectral analysis has been adopted in a number of investigations of meander planform (Speight, 1965, 1967; Chang, 1969; Thakur and Scheidegger, 1970; Chang and Toebes, 1970; Church, 1970, 1972; Yalin, 1971; Sperare, 1974; Ferguson, 1975) and a number of similar studies are in progress. The technique will not be examined here in detail (instead see Blackman and Tukey, 1958; Jenkins, 1965; Jenkins and Watts, 1968; Rayner, 1969) but it will be useful before examining the results of the present study briefly to consider the general character of spectral analysis and some of its limitations specific to the present context.

The computation of the spectrum of any data series initially involves the formation of the autocorrelation function. In this operation, the series is correlated with itself, using a variety of lags, so that a comparison of the fit of the series to itself with its time origin shifted is obtained. The auto-correlation function is in turn subjected to harmonic analysis to determine the finite sum of sine and cosine terms present in the record, and to assign each term its relative contribution to the series. Usually Fourier transformation technique determines the harmonics by fitting sine and cosine curves by least squares to the observed cycle, although in the present case the more accurate Tukey method has been adopted. The specific contribution to the total variance of the spectrum by a particular frequency is termed the power of that frequency in the series, hence the name power-spectral analysis.

However, it is usually not possible to apply spectral analysis directly to a meandering trace. Because meanders often wind back on themselves, the trace is a multivalued 'function' and is unsuitable for analysis. Transformation to a true single valued function is achieved by converting the meander trace to a new series of angular deviations in which left or right relative deflections are distinguished by plus and minus signs. This transformation results in the same information loss at low frequencies that was earlier associated with the wavelength-distribution analysis.

In the cases where the original meander trace is single valued, direct application of the spectral analysis may still be inappropriate. The analysis assumes that the harmonics are summed in cartesian space when in fact that appears not to be the case. In simple two-component meanders in which a single high-frequency cycle is superimposed on a low-frequency primary cycle, the former appears to be added orthogonally with respect to the latter. There is no reason to suppose that the same process of summation does not apply to the most complex meander traces. Although violation of the assumption of addivity is undoubtedly reflected in the computed wavelength spectra, its precise effect is unknown and is the subject of current investigations (M. C. Church, University of British Columbia, and C. Brown, University of Alberta).

Finally, it should be noted that power spectral analysis yields wavelength spectra in the form of pathlengths (λ_*) and not of meander wavelengths

(Figure 16.2). The two parameters are related through the sinuosity index (l) where $l = \lambda_*/\lambda$ (Leopold and Wolman, 1957).

Results of the Beatton River Planform Analysis

Composite distributions of meander radius of curvature and of bankfull channel width, standardized on the group mean for the nine reaches, are shown in Figure 16.4. Radius of curvature is distributed log-normally and thus conforms to the findings of Brice (1973, 1974). Straight channel segments, defined as those with $R > 200\,\bar{W}$, are excluded from the radius of curvature distributions; they constitute approximately 20% of the total channel length in each test reach. Bankfull width is predictably normal in distribution and is very strongly peaked.

Figure 16.5 shows frequency and variance distributions of wavelength and the corresponding power spectra for each of the nine reaches. Taken as a group, the frequency distribution of meander wavelength is approximately normal. However, there are clearly considerable non-normal elements in specific cases.

The grouped variance distribution of wavelength is strongly bimodal with peaks at the 30th and 70th (λ_{30v} and λ_{70v}) percentiles.

The corresponding power spectra, in general, display a single dominant peak, although some reaches have a pronounced secondary peak at a higher frequency (for example, see reaches 3, 5, and 8). The dominant peak in each case is associated with substantial power distribution over a broad range of wavelengths. The sharp low-frequency peak (broken line in Figure 16.5) corresponds to the digitizing interval and should be treated as background noise.

Selected relations among bankfull discharge (Q), mean bankfull width ($\bar{W}$), mean meander wavelength by frequency (λ_f), modal wavelength by variance (λ_{30v}, λ_{70v}), and the dominant peak in the wavelength spectrum

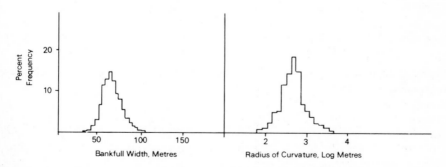

FIGURE 16.4 STANDARDIZED FREQUENCY-DISTRIBUTIONS OF BANKFULL WIDTH AND RADIUS OF CHANNEL CURVATURE FOR THE NINE TEST REACHES

E. J. Hickin

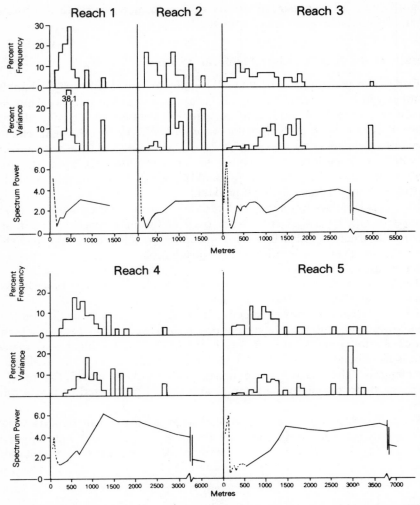

FIGURE 16.5 FREQUENCY AND VARIANCE DISTRIBUTIONS AND SPECTRAL ESTI-
MATES OF WAVELENGTH FOR EACH OF THE NINE TEST REACHES

$(\lambda_s = \lambda_*/l)$, for the nine test reaches are summarized by the following
regression equations:

$$\bar{W} = 4.2Q^{0.48} \qquad (r = 0.99) \tag{6}$$

$$\lambda_f = 54.0Q^{0.54} \qquad (r = 0.97) \tag{7}$$

$$\lambda_{30v} = 43.5Q^{0.59} \qquad (r = 0.95) \tag{8}$$

$$\lambda_{70v} = 121.4Q^{0.49} \qquad (r = 0.88) \tag{9}$$

$$\lambda_s = 81.7Q^{0.51} \qquad (r = 0.71) \tag{10}$$

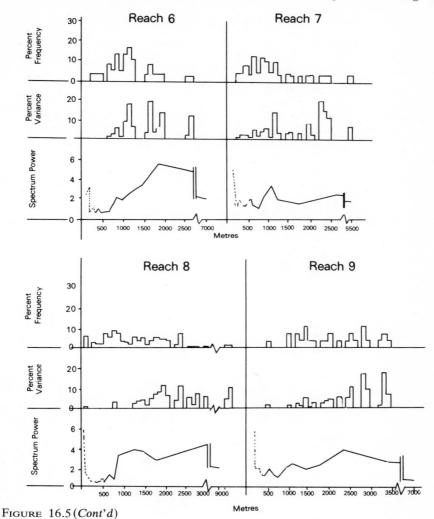

FIGURE 16.5 (*Cont'd*)

The relations among the planform parameters are linear $(0.84 < r < 0.97)$ and ratios of means are listed in Table 16.1.

Equation (6) indicates that there is a close relationship between bankfull discharge and channel width of the type suggested by equation (5). However, the coefficient of 4.2 is considerably greater than the value of unity associated with the traditional relationship.

Equations (7) to (10) sensibly accord with the general meander law $\lambda \propto Q^{0.5}$ but once again the coefficients are all considerably greater than the value of 10 in equation (4); the spectral estimates of wavelength are poorly

TABLE 16.1 DISCHARGE AND SELECTED CHANNEL-GEOMETRY PARAMETERS FOR THE NINE TEST REACHES ON THE BEATTON RIVER

Reach No.	Q	$\bar{W}$	λ_f	λ_{30v}	λ_{70v}	λ_s	R	$R/\bar{W}$	$\lambda_f/\bar{W}$	$\lambda_{30v}/\bar{W}$	$\lambda_{70v}/\bar{W}$	$\lambda_s/\bar{W}$	λ_f/R	λ_{30v}/R	λ_{70v}/R	λ_s/R
1	68	34	462	450	830	442	207	6.1	13.6	13.2	24.4	13.0	2.2	2.2	4.0	2.1
2	115	40	734	874	1240	985	348	8.7	18.4	21.9	31.0	24.6	2.1	2.5	3.6	2.8
3	155	48	930	1100	1670	2218	342	7.1	19.6	23.2	35.2	46.7	2.7	3.2	4.9	6.5
4	200	60	932	850	1420	974	358	6.0	15.5	14.1	23.6	16.2	2.6	2.4	4.0	2.7
5	235	64	1226	1140	2940	2098	372	5.8	19.1	17.7	45.7	32.6	3.3	3.1	7.9	5.6
6	285	71	1152	1170	1840	1100	479	6.7	16.3	16.5	26.0	15.5	2.4	2.4	3.8	2.3
7	335	75	1162	1200	2300	2053	763	10.2	15.6	16.1	30.1	27.6	1.5	1.6	3.0	2.7
8	590	89	1494	1920	2660	2265	1035	11.6	16.8	21.6	30.0	25.5	1.4	1.9	2.6	2.2
9	800	108	2110	2400	2960	2265	1308	12.1	19.6	22.3	27.5	21.0	1.6	1.8	2.3	1.7
Average		65	1134	1234	1984	1600	579	8.3	17.2	18.5	30.5	24.8	2.2	2.3	4.0	3.2

related to bankfull discharge; equations (7) and (8) are clearly very similar to each other, and distinctly different to (9) and (10) respectively.

In summary, and with respect to the meander laws, the frequency, variance, and spectral analyses will describe the pattern of change in meander wavelength with increasing discharge, but not the absolute scale of the system.

The failure of the present analysis to identify the absolute scale of meandering consistent with equations (1) to (4) is well illustrated in Table 16.1. The ratio $R/\bar{W}$ (where R is the modal radius of channel curvature) should be 2.0 but averages 8.28; the ratio $\lambda/\bar{W}$ should be about 10.0 but averages from 17.15 to 30.47 depending on the measure of wavelength. The table also illustrates the similar wavelength to width values yielded by the frequency and lower-mode variance analyses; the spectral estimates of $\lambda/\bar{W}$ are generally intermediate to those yielded at λ_{30v} and λ_{70v}.

Interpretations and Conclusions

The interpretation of these results is made difficult because they are partly an artifact of the statistical techniques used and partly the product of a number of distinctly different fluvial processes.

It should not be surprising that frequency analysis on the one hand, and the variance and spectral analyses on the other, yield rather different results. If, in a given channel, there is present a large number of small-amplitude meanders, frequency analysis will assign considerable relative importance to such features. Variance and spectral analyses, however, may assign little importance to them because their contribution to the total variance will probably be very small. Similarly, a few very large amplitude meanders may represent a significant contribution to total variance and thus be identified by variance and spectral analyses as dominant wavelengths; in frequency analysis they would simply be assigned an importance of a few among many.

Although all three methods are entirely valid means of describing a meander pattern, if the purpose is to identify planform properties which are sensitive to the scale of flow, then variance and spectral analyses are of rather limited use. The frequency with which a given wavelength can be observed in a reach is the best measure of its importance as an indicator of flow scale.

Variance analysis in the present study appears to detect the few large meander loops as primary modes (λ_{70v}), and the most frequently occurring wavelengths as secondary modes (λ_{30v}). The secondary modes in the variance analysis are thus sensibly equivalent to the mean wavelength of the frequency distribution (see equations 7 and 8).

Nevertheless, as already indicated, even frequency analysis of wavelength has not yielded the standard relations described by equations (1) to (4).

There appear to be at least three factors related to the process of bend development that contribute to these results. The first is that the Beatton River planform includes many straightened reaches produced by cut-offs. Although such features can be regarded as transitory, they are very common elements in the planforms of most actively migrating rivers.

Secondly, the oscillatory motion generated by boundary shear fluctuations in open channels may not be completely reflected in the planform. Actively migrating channels usually have easily deformed beds and the oscillatory motion may largely be reflected in bed deformation rather than in planform adjustments.

The formation of channel bends on the Beatton River has been discussed in detail elsewhere (Hickin, 1974) but a brief outline of the process will be useful in the present context. Figure 16.6 illustrates the typical pattern of bend development in which very long and sensibly straight meander limbs (a) are produced which show no tendency to meander. Channel migration is constrained by the development of very tightly-curved channel segments in which $R/\bar{W} = 2.0$ (b). When such a segment develops, channel migration on the limbs (c) initially opens the bend once again, but inevitably $R/\bar{W}$ will approach a value of 2.0 on the new migrating reach (d) and the process repeats itself until cut-off finally occurs. The result is that very large meander loops are produced in which only very small segments of the channel planform directly reflect the scale of flow. Thus the wavelength of such loops reflects not only the constraint of minimum channel curvature,

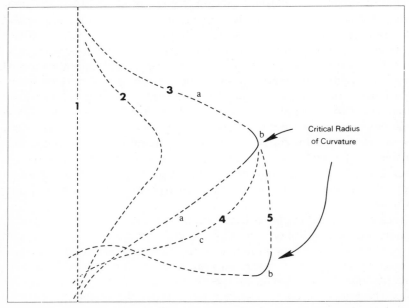

FIGURE 16.6 TYPICAL CHANNEL-MIGRATION PHASES IN A DEVELOPING MEANDER LOOP

but also the sequence of migration phases, a factor probably unrelated to the scale of flow. It is therefore not surprising that the present analysis yields much larger wavelengths per unit channel width than is predicted by the meander laws (reflecting the influence of the migration pattern), but that the rate of change in wavelength with discharge and channel width is sensibly consistent with equations (1) to (4) (reflecting the independence of the migration pattern).

Thirdly, it is now apparent that 'free' meanders produced by relatively rapid migration through easily erodible material are much more likely to develop severe distortions in response to small material heterogeneities than are meanders developed in more resistant materials. The Beatton River contains many such distortions because it is not perfectly free of pockets of unusually weak or resistant material. It has been observed elsewhere (Langbein and Leopold, 1966) that deeply incised meanders are often more sinusoidal and regular than those which are relatively free to migrate laterally.

It is clearly the case that the present methods of 'objective' analysis are not capable of filtering from a meander array the effects of straight-limbed loop development, cut-offs, and material-induced bend distortion, leaving the discharge-dependent elements of the trace as a residual.

It would seem appropriate to reassess the representative-reach concept in which such filtering is claimed to be highly efficient. It does not seem unreasonable that an experienced operator can identify 'distorted' reaches of a meander trace. When the degree of operator error in wavelength estimation has been established, it may be the case that the representative-reach concept is much less subjective than previously has been supposed.

Nevertheless, the fact remains that the Beatton River displays a planform which, although a product of contemporary processes, has the characteristics of a manifestly underfit stream! If it is typical of actively migrating channels there is clearly a need to exercise caution in research where the meander laws form a framework for analysis.

Acknowledgements

This study forms part of a project which is funded by the National Research Council of Canada. The writer is indebted to Dr M. Church (Department of Geography, University of British Columbia) and to Messrs David Mark and Ken Rood (Department of Geography, Simon Fraser University) for their assistance with computer programming and data processing.

References

Bagnold, R. A. 1960. Some aspects of the shape of river meanders. *U.S. Geol. Survey Prof. Paper 282E.*, pp. 135–144.

Blackman, R. B. and Tukey, J. W. 1958. *The Measurement of Power Spectra*, Dover, New York.

Brice, J. C. 1973. Meandering pattern of the White River in Indiana—an analysis. In Morisawa, M(Ed.). *Fluvial Geomorphology*. State University of New York, Binghamton, pp. 180–200.

Brice, J. C. 1974. Evolution of meander loops. *Geol. Soc. Amer. Bull.*, **85,** pp. 581–586.

Carlston, C. W. 1965. The relation of free meander geometry to stream discharge and its geomorphic implications, *Amer J. Sci.*, **263,** pp. 864–885.

Chang, T. P. 1969. Statistical analysis of meandering river geometry, *Ph.D. Thesis*, Purdue University, University Microfilms Ltd., High Wycombe, England.

Chang, T. P. and Toebes, G. H. 1970. A statistical comparison of meander planforms in the Wabash Basin, *Water Resour. Res.*, **6,** pp. 557–578.

Church, M. A. 1970. Baffin Island sandurs: a study of Arctic fluvial environments, *Ph.D. thesis*, University of British Columbia.

Church, M. A. 1972. Baffin Island sandurs: a study of Arctic fluvial processes, *Canada Geol. Survey Bull.*, **216.**

Douglas, D. H. and Peucker, T. K. 1973. Algorithms for the reduction of the number of points required to represent a digitized line or its caricature, *The Canadian Cartographer*, **10** (2), pp. 112–122.

Dury, G. M. 1954. Contribution to a general theory of meandering valleys, *Amer. J. Sci.*, **252,** pp. 193–224.

Dury, G. H. 1964. Principles of underfit streams, *U.S. Geol. Survey Prof. Paper 452-A*.

Dury, G. H. 1965. Theoretical implications of underfit streams, *U.S. Geol. Survey Prof. Paper 452-C*.

Ferguson, R. I. 1975. Meander irregularity and wavelength estimation, *J. Hydrol.*, **26,** pp. 315–333.

Friedkin, J. F. 1945. A laboratory study of the meandering of alluvial rivers, *U.S. Waterways Exp. Station*, Vicksburg, Mississippi.

Hickin, E. J. 1974. The development of meanders in natural river-channels, *Amer. J. Sci.*, **274,** pp. 414–442.

Hickin, E. J. and Nanson, G. C. 1975. The character of channel migration on the Beatton River, Northeast British Columbia, Canada, *Geol. Soc. Amer. Bull.*, **86,** pp. 487–494.

Inglis, C. C. 1949. The behaviour and control of rivers and canals, *Res. Publ. Poona* (India), **13,** 2 Vols.

Jefferson, M. 1902. Limiting width of meander belts, *Nat. Geogr. Mag.*, **13,** pp. 373–384.

Jenkins, G. M. 1965. A survey of spectral analysis. *Applied Statistics*, **14,** pp. 2–32.

Jenkins, G. M. and Watts, D. 1968. *Spectral Analysis and Its Applications*. San Francisco, Holden-Day.

Langbein, W. B. and Leopold, L. B. 1966. River meanders—theory of mimimum variance. *U.S. Geol. Survey, Prof. Paper 422-H*.

Leopold, L. B. and Wolman, M. G. 1957. River channel patterns: braided, meandering and straight. *U.S. Geol. Survey Prof. Paper 282-B*.

Leopold, L. B. and Wolman, M. G. 1960. River meanders. *Geol. Soc. Amer. Bull.*, **71,** pp. 769–794.

Mather, K. F. and Mason, S. L. 1939. *A Source Book in Geology*. McGraw-Hill, New York.

Rayner, J. N. 1969. *An Introduction to Spectral Analysis*. Pion Press, London.

Speight, J. G. 1965. Meander spectra of the Angabunga River. *J. Hydrol.*, **3,** pp. 1–15.

Speight, J. G. 1967. Spectral analysis of meanders of some Australasian rivers. In J. N. Jennings and J. A. Mabbutt (Eds.), *Landform Studies from Australia and New Guinea.* University Press, Cambridge, pp. 48–63.

Sperare, B. 1974. Statistical analysis of plan and depth forms of three rivers. *M.Sc. Thesis,* University of Alberta, Edmonton.

Thakur, T. R. and Scheidegger, A. E. 1970. Chain model of river meanders. *J. Hydrol.*, **12,** pp. 25–47.

Yalin, M. S. 1971. On the formation of dunes and meanders, *14th Congress Internat. Assoc. Hydraulic Res.*, **3,** pp. 101–108.

J. M. HOOKE
Lecturer in Geography
Manchester Polytechnic

17

The Distribution and Nature of Changes in River Channel Patterns: The Example of Devon

The characteristics of channel patterns and the controls upon the type of pattern have long been topics of study in fluvial geomorphology (Jefferson, 1902; Friedkin, 1945; Dury, 1954; Leopold and Wolman, 1957). Until the last decade or so little attention has been paid, except by engineers, to changes in planform. Knowledge of the extent, nature, and distribution of changes in channel pattern in Britain is particularly sparse, and probably the general impression amongst geomorphologists is one of the stability of streams, at least in lowland areas. An initial aim in approaching the problem of changes in channel pattern is to establish whether lateral movement of channels can be discerned from the various forms of evidence available. If changes are taking place this poses questions on their spatial and temporal distribution, particularly the degree of regularity; questions about the controls on pattern and on movement, including the effect of specific disturbances such as alteration of discharge or sediment load, and the effect of structures. Analysis of such questions must depend upon knowledge of the relationship between form and movement, and upon identification of stable forms. Geomorphological theory is at present inadequate to explain or predict planimetric movement because of the number of variables involved and the complexity of their interaction in the natural environment. Engineers have attempted to simulate and model situations, often simplifying conditions, but further empirical knowledge of field situations is still needed.

Two major approaches have been used in the assessment and measurement of the lateral movement of channels. Firstly, the investigation of contemporary erosion by monitoring pins in river banks, (e.g. Wolman, 1959; Twidale, 1964; Hill, 1973). Such studies provide information on

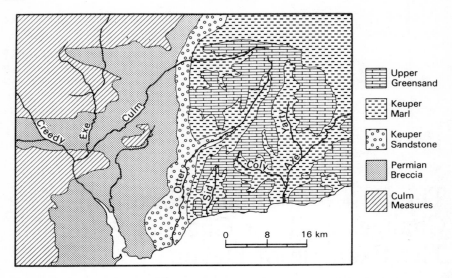

FIGURE 17.1 LOCATION OF STREAMS AND GEOLOGY OF EAST DEVON

processes, rates, and temporal distribution of changes as well as on their
spatial distribution. The second approach employs maps, aerial photographs,
and other historical evidence to investigate spatial changes over longer
periods of time. Such evidence is valuable in providing information for the
100–150 year period which is often beyond the scope of empirical observation. Using such evidence, Duncanson (1909), Fisk (1952), Carey (1969),
and many others in the United States have analysed changes in pattern and
similar case studies have been made by Bluck (1971), Lewin (1972), and
Mosley (1975) in Britain. Recent work by Daniel (1971), Brice (1974), and
Hickin (1974) has examined types of change more generally.

This paper presents the methods and results using the second approach,
an analysis of cartographic evidence of movement of channels. The methods
could be applied in any region but the streams studied are located in East
Devon where preliminary evidence indicated that change of pattern had
occurred. Almost all the streams are meandering in nature and hence the
discussion concerns this type of pattern only. The streams are the Culm,
Creedy, Lower Exe, Yarty, Axe, and Otter (Figure 17.1). They vary in size
from the Yarty, catchment area $110 \, km^2$ and an average width at the
downstream end of 9 m to the Axe with a catchment area of $295 \, km^2$ and
width of 18 m, and the Exe, catchment area $620 \, km^2$ and width in the
lowermost part of 25 m.

Sources of Evidence

Maps of 1:2500 scale proved the most useful for detailed measurement
and analysis of movement. Three classes of maps are available for the east

Devon area and for most other parts of England at this scale. Ordnance Survey National Grid Series maps provide the basic, most recent, coverage; the major survey of the study area was made in the late 1950s and only one edition of most sheets is available. The second group, the Ordnance Survey County Series maps of Devon are of two dates: the major survey of around 1888 and a full revision published as the second edition in 1905. The third and earliest form of map evidence of considerable use are tithe maps. These are detailed maps which were produced for each tithe district, usually a parish, in compliance with the Tithe Commutation Act of 1838. Most of the maps are at a scale of 3 or 6 chains to 1 inch (1:2376 and 1:4752 respectively) and were surveyed between 1838 and 1845.

Measurements of lateral changes in channel courses can be meaningful only when the accuracy and reliability of the river courses plotted on the maps are established and the errors involved in compilation assessed. The Ordnance Survey make their own tests of accuracy on the National Grid Series maps. The bank line is mapped as the 'normal winter level' of water (Harley, 1975) and local surveyors who mapped the rivers of east Devon consider the eroding bank line to be accurate to about 1 m. The County Series are considered to be of the same order of accuracy and no discrepancy in unchanged detail has been found in this study between the two Series. Little is known of the accuracy of the tithe maps and so more detailed tests were necessary. The tithe maps vary considerably in standard of survey, cartography and preservation. Measurements were made of areas and distances on a sample of 23 tithe maps and compared with corresponding measurements from the Ordnance Survey maps (Hooke and Perry, 1976). These produced a mean absolute error in area of 3.94% and in linear measurements of 2.71%. Observation and mapping by plane table of old channel courses and structures, such as bridges, have substantiated the changes indicated by the tithe maps. To compile the evidence a base tracing of the 1:2500 National Grid Series maps was made and the courses from the County Series and tithe maps superimposed upon this. The maps were adjusted to scale by matching detail near the river and any discrepancy was minimized, thus further reducing errors. This gave an estimated overall accuracy of about ±3 mm for the tithe maps, ±7.5 m on the ground, or generally an accuracy of $\frac{1}{4}-\frac{1}{3}$ of the channel width.

Analysis

The problem in analysing any aspect of channel pattern is to find a method which adequately demonstrates the properties of the course planform, especially where meanders are ill-defined or irregular. The aim here was to characterize the channel pattern as objectively as possible whilst also being able to identify and measure change and movement in a meaningful way. Two basic approaches are available. First and most frequently used, to divide the channel planform into individual bends and to make direct measurements of parameters such as wavelength, breadth, sinuosity, and radius of curvature (e.g. Inglis, 1947; Carlston, 1965; Dury, 1958; Chitale, 1973). The second approach, a more recent development, is to analyse the characteristics of the course statistically, most conveniently using discrete data series generated from equally spaced points along the channel. This was used by Speight (1965) on the Angabunga River, by Chang and Toebes (1970) on the Wabash River, and has been discussed by Ferguson (1975). The first method is difficult to apply objectively and to use on complex meander patterns. The second method does not easily permit comparison of the characteristics of individual bends, providing only a generalized summary of the properties of a section. Thus a compromise scheme combining the two approaches has been devised.

Data were produced by digitising the centre-line of the course of each river for each different map date (Figure 17.2). This involved choosing the spacing of the points, usually about $1\frac{1}{2}$ channel widths, and following the centre-line with the cursor. The cursor and digitising table were linked to a computer which was programmed to punch out on cards the coordinates of points along the channel at the spacing specified. Angles and difference in angles (curvature) between adjacent points were computed from the coordinate data and points of inflexion located where the difference in angles passed through zero. Meander parameters were calculated between adjacent points of inflexion; the parameters include wavelength, sinuosity, maximum curvature, radius of curvature, and a shape parameter approximating wavelength/radius of curvature. The difference in angle series were also used as the data input for spectral analysis to elucidate periodic components in the data. If there is a regular oscillation corresponding to the wavelength of meanders, this should be revealed by a peak in the power spectrum at that wavelength or frequency. The power spectrum is a measure of the amount of variance accounted for by various frequency or wavelength components. Cross-spectral analysis compares the two series for correspondence of these frequency components, and measures the amount they vary together or are

Example of digitised river courses

Points of Inflection
+1903 *1958

• Coordinate points at equal spacing

FIGURE 17.2 EXAMPLE OF DIGITISED MAP

lagged relative to one another. This was used to compare the courses of different dates. Some measurements of amount of change on individual bends were made direct from the maps by measuring the conventional meander parameters such as wavelength, meander breadth, and radius of curvature between common points on the courses of different dates, and comparing the values of the parameters obtained.

Spatial distribution of channel changes can be considered at several scales, all of which are part of a continuum; five scales have been identified as suitable for analysis; interregional, regional, basin, section, and bend. The first two do not use the more detailed analyses of the 1:2500 maps but are included to cover the range of scales. At the *interregional* scale, comparison of streams in Devon could be made with those from other parts of Britain from rather different climatic and geological environments. Little such information is available at present, however; of the few studies in Britain, most have been in upland areas concerned with steep, gravel-bed rivers. It would be useful to know the distribution and rates of movement on lowland streams flowing in wide flood plains.

Figure 17.3 at the *regional* scale shows a map of the location of channel changes revealed by a comparison of Ordnance Survey 1:10560 maps of two dates for the major rivers of Devon. The changes mapped range from about 10 m to 100 m. It is seen that large proportions of the streams of east Devon have altered their course over this period of 50+ years, but that elsewhere changes have occurred along short isolated reaches. As would be expected, the streams on the resistant granite of Dartmoor are relatively stable except where there has been erosion of superficial material. There is a lack of recent map information for the Exmoor area in the north and headwaters are too small to detect change on this scale. The streams of east Devon are mostly rather steep with gravel beds, the valleys are formed mainly in Keuper Marl with headwaters on Greensand, and the lower parts of the Creedy, Exe, and Culm are in Permian sandstone. The catchments are almost entirely rural.

To understand the scale and proportions of change and to elucidate the conditions under which different types of change take place, it is necessary to know the range and variability of the characteristics of the patterns. The stream traces were divided into sections, the number being dictated by the volume of data points required and by morphological considerations. In Figure 17.4 wavelength, calculated as twice the mean distance between points of inflexion, is plotted against catchment area, used as a surrogate of discharge. Alternatively the value of wavelength at which the power spectrum values showed the highest peak could be plotted. The correlation coefficient for Figure 17.4 is 0.837. The points for the River Otter mostly have high positive residuals from the least squares regression line, $Y = 15X^{0.45}$ and the variance is also high. Wavelength increases through the Axe basin at a lower rate than expected from the regression line. The

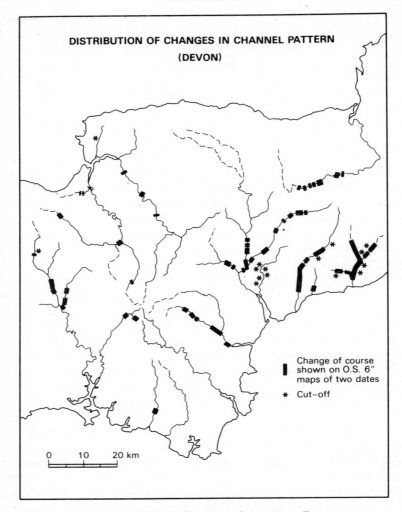

FIGURE 17.3 MAP OF CHANNEL CHANGES IN DEVON

Creedy shows a systematic increase in the upper part of the basin where there are many well-defined meanders but has relatively low wavelengths downstream. Few points are plotted for the River Culm because so much of that channel is divided with mill leats. The values for the River Yarty are more affected by spacing of the data points in the original digitising than the others, a problem discussed by Ferguson (1975).

The variation and trends in characteristics and mobility may be examined down through a *basin*. The Rivers Axe and Culm are used here to contrast the types of distribution of channel movement. On Figures 17.5a and 17.5b two types of movement have been distinguished: one, sections where the

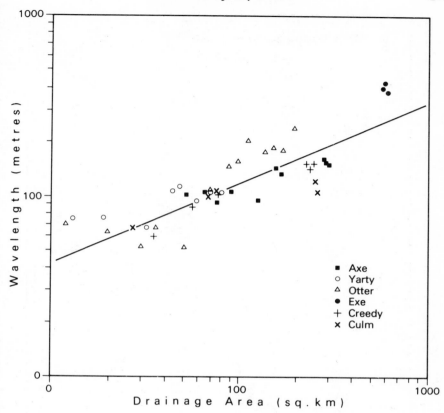

FIGURE 17.4 GRAPH OF WAVELENGTH VERSUS CATCHMENT AREA

channel has simply translated downstream maintaining its planform or where the proportion of movement is very low, and two, where movement accompanied by change in the form and characteristics of the pattern has occurred. Two special cases of the second class are distinguished: firstly, where a channel has been straightened and cut-offs formed, often artificially; secondly and conversely, where meanders have developed from a formerly straightened or canalized course. Other sections indicated are stable. The classification is arbitrarily based on the channel course divided by the double grid square lines of the 1:2500 maps and the channel is considered to have moved significantly if it has departed more than one width from its former course within that square, thus allowing for the different sizes of streams and scales of movement. It is apparent that a great proportion of the River Axe has altered its course, calculated as 40% of the channel length while 17% has also changed in form. This stream is unusual amongst those studied in that there are several prominent cut-offs, at least some of which are natural.

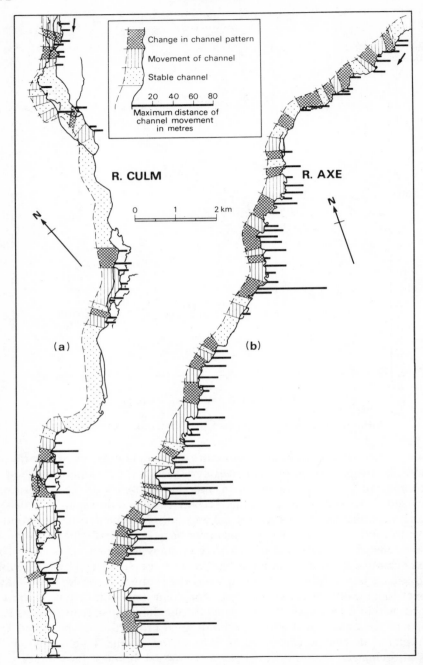

FIGURE 17.5 (a) DISTRIBUTION AMOUNT OF MOVEMENT ON RIVER CULM, (b) DIS-
TRIBUTION AMOUNT OF MOVEMENT ON RIVER AXE

On the Culm there are much longer stable reaches and much of the change occurs where there has been interference with the channel, especially in the upper part of the basin. Approximately 18% of the length has moved, of which about 40% has changed in characteristics. The maximum distance of movement in each square, indicated by the proportional bar, is larger where there has been direct cut-off rather than gradual movement, but the general level of movement is much higher on the Axe than on the Culm. Similar maps have been produced for other streams and it is calculated that 16% of the Creedy, 20% of the Otter, and 28% of the Yarty, show significant movement in the 50+ years between 1903 and the National Grid Series map publication. There is some increase in amount of movement with distance downstream but less than might be expected.

Section scale may include any reach from several hundred metres to a few thousand metres in length. The sections in the analysis are mostly $1\frac{1}{2}$–$2\frac{1}{2}$ km in length or 200+ data points at a spacing of about $1\frac{1}{2}$ channel widths. At this scale the characteristics of the pattern in a section can be identified from the spectra. Two sets of examples showing spectra for each of the map dates are illustrated in Figure 17.6. Well-defined meanders are characterized by a spectrum with a single dominant peak and low power at high frequencies.

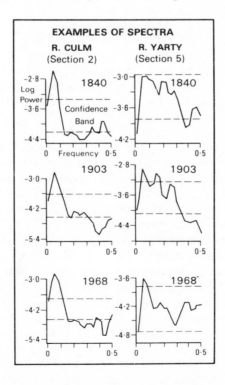

FIGURE 17.6 EXAMPLES OF SPECTRA

Straightened or irregular sections may have multipeaked spectra which approach random noise, no peaks reaching beyond the confidence bands which are plotted around the mean value to indicate significant peaks and gaps in the spectrum. To distinguish which sections have changed in characteristics of planform, the meander parameters obtained from the digitised data and measured between points of inflexion, were used to produce mean values for each section at each map date. T tests were made on these mean values of different dates for each meander parameter in a section. It was found that many sections on the Yarty and Axe have changed in characteristics. A greater number of sections produced a more significant change in size parameters than in shape parameters. The frequency of types of change was revealed by counting the increases and decreases in meander values of the parameters. Between the date of the tithe map and 1903 (period A), 75% of the sections decreased in mean wavelength and in period B, 1903–1958, 57% of the wavelength means decreased. There was an increase in sinuosity in 55% of the sections in period A and 68.8% of the sections in period B. Similarly maximum curvature increased in more sections, radius of curvature decreased, and the bends became sharper in shape. Measurements of change in meander breadth made directly from the maps of well-defined meanders revealed a mean increase in 77% of the sections in Period A and all sections in Period B. Map measurements of percentage changes in sinuosity varied from −0.33% to +31.4% and in meander breadth from −1.06% representing a mean distance of 4.5 m to +134.0% representing an average increase of 16 m. In terms of actual movement, taking the value measured for each grid square as shown in Figure 17.5b, the mean maximum movement is 36.4 m for the River Axe, including all reaches and 46.8 m excluding completely stable reaches, both measured over a period of about 60 years.

Figures 17.7, 17.8, and 17.9 demonstrate some of the types of change found in sections of these channels. Natural channels tend often to be rather irregular in movement but a few cases of regular movement were identified. Both examples 17.7a and 17.7b demonstrate the downstream translatory movement of the bends, with maximum erosion just downstream of the apex. Even in Figure 17.7c, where the form of the bend is much less regular, the same type of movement has taken place and is continuing. Erosion has been measured on this section over a 2-year period, yielding an average rate of 0.37 m/yr, an average maximum rate for five sections of 0.96 m/yr, and a maximum rate of 1.32 m/yr. The maximum rate of erosion measured from the maps at the same section is 0.54 m/yr. In the second set of examples change of a much more chaotic nature is illustrated. The River Sid has steep headwaters generating high peak flows, occasionally causing catastrophic changes as in 1968. The section of the River Coly is one of the few markedly unstable sections on that stream. The last examples show the effect of human interference and the natural tendency for streams to

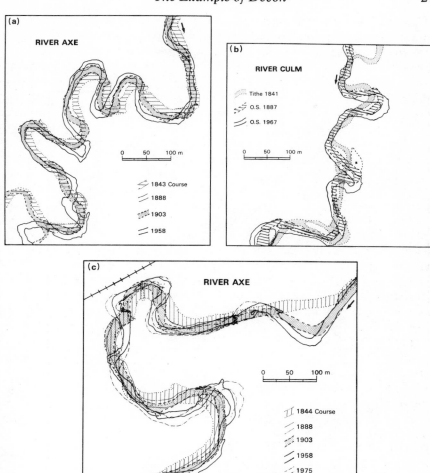

FIGURE 17.7 EXAMPLES OF CHANGES: (a), AXE—REGULAR BENDS; (b), CULM—
REGULAR: (c), AXE—EROSION MEASUREMENT SITE

meander. Both illustrate sections that were straight in 1903 but 50 or so
years later, when these reaches were no longer maintained and weirs had
collapsed, the development of bends had begun. The example on the Culm
exhibits a damped oscillation form.

At the scale of the individual *bend*, the distribution of erosion can be
analysed and the relation between the mode of movement and the form of
the bend examined. Graphs of percentage change against sinuosity (or
similar) reveal little because several types of change are incorporated. In
many cases the most sinuous bends will continue to move most rapidly but

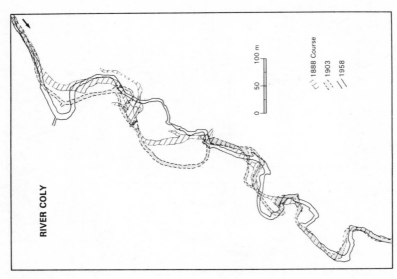

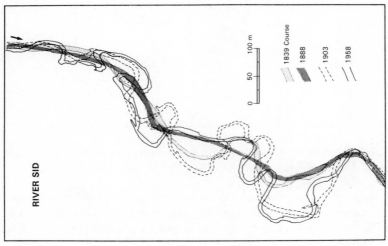

FIGURE 17.8 EXAMPLES OF CHANGES: (a), SID; (b), COLY

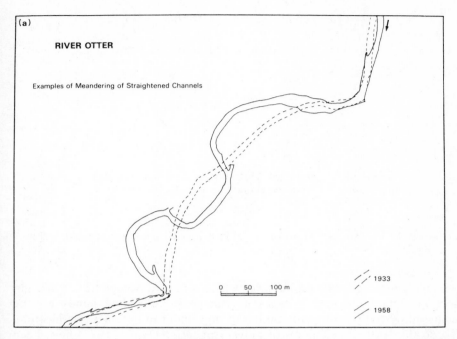

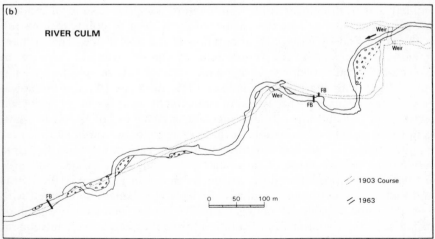

FIGURE 17.9 EXAMPLES OF CHANGES: (a), OTTER; (b), CULM

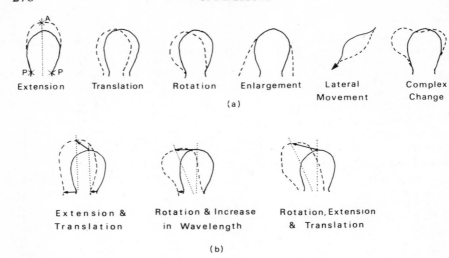

FIGURE 17.10 TYPES OF MOVEMENT: (a) PRIMARY ELEMENTS (b) COMBINATIONS

percentage change may also be large where a bend develops from a straight section. The frequency of types of change in individual bends can be analysed by comparison with models of movement to obtain a classification. To do this, primary elements of movement (after Daniel, 1971) have been identified and in double or triple combination compose a suite of models. Each is uniquely defined by the vectors of movement of points of inflexion and the apex, and by change in orientation of the apical line. The primary elements of movement are translation, extension, rotation, change in wavelength, lateral movement, and complex change. These are illustrated in Figure 17.10a with the points of inflexion (P), the apex (A) and the apical line (dotted). Each of the types of movement can be in one of two directions, i.e. up and downstream, increase or decrease, or to left or right. Examples of double combinations include extensions outwards and translation downstream, or rotation downstream and increase in wavelength. Triple combinations involve three of these elements such as extensions outwards and rotation and translation downstream, as shown in Figure 17.10b. It was found that of the 444 bends examined, 76% of the changes conformed with only eight of the possible 70 models. Fifty-five percent involved extension or translation or their combination. Rotation proved to be relatively infrequent but increase in complexity by development of a secondary lobe was quite common. In general, change around bends tends to be gradual and consistent in time, taking place by progressive erosion and deposition rather than by catastrophic change causing cut-offs or complete alteration of the direction of movement.

Conclusions

Thus it is seen from the sample of rivers studied, in the particular environment of east Devon, considerable change has taken place during the last 100–150 years and is continuing to occur. Distribution of lateral movement has been examined at five scales, each of which affords different information. The interregional and regional scales demonstrate how widespread or restricted channel mobility is in different areas and how it may be related to environmental variables such as rock type. At the larger scales, analysis provides detail of the relation between form and movement in a sequence of bends or in individual arcs, and reveals the types and directions of change and the distribution of amounts of change throughout a basin. Such a scheme of study could usefully be applied to streams in other regions, and the automated production of coordinate data from the map evidence facilitates rapid analysis of large amounts of information. As much as 40% of the length of a channel in a single basin was found to be laterally mobile and movement of up to 100 m in a 50-year period has been found from the map evidence. The rivers in this area are generally becoming more sinuous, moving by extension of bends, and decrease in wavelength and radius of curvature. The extent of human interference, and how much the movement and change is an adjustment to indirect changes in agricultural practice, field drainage, and urban activity, is difficult to assess. Some examples of direct interference have been demonstrated but they also substantiate the argument that streams such as the Otter, Yarty, and Axe are naturally mobile. The continuing activeness of these rivers is also demonstrated by the efforts made by the South West Water Authority to combat erosion. The normal means of prevention now used is the placement of block stones around the apex of the meander and the total cost of river maintenance in east Devon is over £10,000 per year although much more could be spent (Bridges, personal communication). Each year farmers lose acres of valuable farmland and although in the long-term, as observation and the map evidence reveals, deposition tends to balance erosion, it takes an estimated minimum of 10 years for deposited material to become usable agricultural land. Hence this is an economic problem which, though not on the scale of the Mississippi, is by no means negligible in Britain. Perhaps as more evidence is obtained on planform change we shall begin to understand the controls and therefore take action which, as Winkley (1972) has advocated, uses the river rather than works against it.

References

Bluck, B. J. 1971. Sedimentation in the meandering River Endrick, *Scott. Journ. of Geol.*, **7,** Pt. 2, pp. 93–138.

Brice, J. C. 1974. Evolution of meander loops. *Geol. Soc. Amer. Bull.*, **85,** pp. 581–586.

Carey, W. C. 1969. Formation of flood plain lands, *J. Hydraul. Div. Amer. Soc. Civ. Engrs.*, **95**, pp. 981–994.

Carlston, C. W. 1965. The relation of free meander geometry to stream discharge and its geomorphic implications. *Amer. J. Sci.*, **263**, pp. 864–885.

Chang, T. P. and Toebes, G. H. 1970. A statistical comparison of meander planforms in the Wabash basin, *Water Resour. Res.*, **6**, pp. 557–577.

Chitale, S. V. 1973. Theories and relationships of river channel patterns, *J. Hydrol.*, **19**, pp. 285–308.

Daniel, J. F. 1971. Channel movement of meandering Indiana streams. *U.S. Geol. Survey Prof. Paper 732-A.*

Duncanson, H. B., 1909. Observations on the shifting of the channel of the Missouri River since 1883. *Science*, **XXIX**, No. 752, pp. 869–871.

Dury, G. 1954. Bedwidth and wavelength in meandering valleys. *Nature*, **176**, p. 31.

Dury, G. 1958. Tests of a general theory of misfit streams, *Trans. Inst. Brit. Geog.*, **25**, pp. 105–118.

Ferguson, R. I. 1975. Meander irregularity and wavelength estimation. *J. Hydrol.*, **26**, pp. 315–333.

Fisk, H. 1952. Mississippi River valley geology relation to river regime, *Trans. Amer. Soc. Civ. Engnrs*, **117**, pp. 667–689.

Friedkin, J. F. 1945. *A laboratory study of the meandering of alluvial rivers*, U.S. Waterways Expt. Statn., Vicksburg, Miss.

Harley, J. B. 1975. *Ordnance Survey maps; A descriptive manual*, Ordnance Survey, Southampton.

Hickin, E. J. 1974. The development of meanders in natural river-channels. *Amer. J. Sci.*, **274**, pp. 414–442.

Hill, A. R. 1973. Erosion of river banks composed of glacial till near Belfast, Northern Ireland, *Z. Geom.*, **17**, pp. 428–442.

Hooke, J. M. and Perry, R. A. 1976. The planimetric accuracy of tithe maps. *Cartographic Journal*, December, pp. 177–183.

Inglis, C. C. 1947. Meanders and their bearing on river training. *Inst. Civ. Engrs. Papers*, Vol. V, Maritime and Waterways Paper No. 7, pp. 3–54.

Jefferson, M. 1902. Limiting width of meander belts. *Nat. Geog. Mag.*, **13**, pp. 373–384.

Leopold, L. B. and Wolman, M. G. 1957. River channel patterns. *U.S. Geol. Survey Prof. Paper 282-B.*

Lewin, J. 1972. Late-stage meander growth, *Nature, Physical Science*, **240**, p. 116.

Mosley, M. P. 1975. Channel changes on the River Bollin, Chesire. 1872–1973. *E. Mid. Geogr.*, **42**, pp. 185–199.

Speight, J. G. 1965. Meander spectra of the Angabunga River, *J. Hydrol.*, **3**, pp. 1–15.

Twidale, C. R. 1964. Erosion of an alluvial bank at Birdwood, South Australia, *Z. Geom.*, **8**, pp. 189–211.

Winkley, B. R. 1972. River regulation with the aid of nature, *Int. Comm. on Irrigation and Drainage*, **29**, pp. 433–457.

Wolman, G. 1959. Factors influencing erosion of a cohesive river bank. *Amer. J. Sci.*, **257**, pp. 204–216.

G. H. DURY

Professor of Geography and Geology
University of Wisconsin, Madison

18

Underfit Streams: Retrospect, Perspect, and Prospect

This review will be concerned retrospectively with the last 25 years, perspectively with the present state of the art, and prospectively with what might usefully be done in the fairly near future. In reviewing the retrospect, I shall attempt to keep documentation to the minimum. Considerable bulks of literature have been discussed elsewhere, both by myself and by others (for references up to 1960, see Dury, 1964a, 1964b, 1965). There will be occasion to correct some past errors. In dealing with the present time, I shall not undertake a survey of the literature on meanders in general, preferring to concentrate instead upon channel characteristics which can be identified and measured in respect of paleochannels, and upon chronological problems. As will appear in due course, I foresee future work as elaborating the known sequence of cut-and-fill, and as potentially providing a well-integrated scheme of paleoclimatology.

The Twenty-five Year Retrospect

It was in 1951 that a deliberate field test was first applied to the hypothesis that, if valley bends are authentic former meanders, former large channels should have been associated with them. As a matter of record, this first test was successful. It demonstrated the presence of a large meandering channel, up to ten times as wide as the existing channel, cut in bedrock and complete with pool-and-riffle sequence (Dury, 1952). Subsequent tests at other localities proved equally successful. By elimination, hypotheses of stream shrinkage by means of underflow through alluvium, deep percolation, or cessation of meltwater discharge or of lacustrine overspill, were readily disposed of. The capture hypothesis was also rejected as generally inapplicable. The first major test of this hypothesis related to the Cotswold Coln, for which Davis is now known to have invented field evidence.

In actuality, the capture hypothesis could have been ruled out from the beginning, had proper account been taken of its hydrologic implications. Empirical power-functional equations which relate discharge to drainage area have been available since the late nineteenth century (Chorley, Dunn, and Beckinsale, 1964, pp. 419–420). The exponent of area ranges from 0.5 in one equation to 0.75 in the other, this latter value being quite close to the 0.7 recently obtained by Alexander (1972). Even with no allowance made for the power-functional relationship of channel dimensions to discharge, and even if the exponent of area is taken as unity, the territorial implications of the capture hypothesis are drastic. Furthermore, the sensible constancy of the wavelength ratio between valley meanders and stream meanders, throughout entire regions, ought from the outset to have cast doubt on the general capture hypothesis. Where capture has in fact occurred, its effects may be readily separable from those of general stream shrinkage (Dury, 1963; Dury, 1964a, Fig. 45 and text; Kohchi, 1976).

Underfitness other than Manifest

A manifestly underfit stream combines large former with small present meanders. Most commonly, the large former meanders are recorded in the contours of the valley walls, although it is possible, as on the upper Evenlode, for the trace of the large former meanders to be preserved within a large former meander-trough. However, although manifestly underfit streams are sufficiently widespread, both in France and on the English Plain, to falsify the general capture hypothesis, they are not universally distributed there. This statement holds, when all due allowance has been made for over-generalized mapping and for artificial regularization. It may have been this set of circumstances which buttressed denials of the regional development of underfits. Moreover, such regional development demands a hypothesis of climatic change, which during the 1950s was as little fashionable in geomorphology as was continental drift in geology.

Laboratory and field studies combine to show that streams may become underfit, without being manifestly so. At least five sets of conditions may be imagined for ancestral streams, from which shrinkage leads by at least eleven paths to at least eight contrasted conditions of underfitness (Dury, 1964a, A45–48). Only two of the deduced conditions involve manifest underfitness. A third condition, which has certainly caused much past confusion, is that of Osage-type underfitness. This involves the combination of valley meanders with a shrunken stream which does not meander, but which possesses pools and riffles spaced appropriately to its bedwidth. That is to say, the pool-to-pool or riffle-to-riffle spacing is about five bedwidths, as in meandering streams in nature, and as in many flume studies of channels with mobile beds. Osage-type underfitness was first identified by name for the Colo River in New South Wales, Australia (Dury, 1966), but was originally recognized for the Osage River itself. Subsequently, a number of additional examples have been described (Shoobert, 1968; Dury, Sinker, and Pannett, 1972; Young, 1974; Kohchi, 1975).

The degree of abundance of Osage-type underfits is not known. Whereas manifest underfits can be identified immediately, from air photographs or from reliable maps, Osage-type underfits can only be suspected from their plan dimensions. With their existing channels perforce inflected round valley bends, they display an apparently very high wavelength/width ratio, a characteristic which, in combination with unreliable mapping, has sadly confused part of the total discussion. In order to test for Osage-type underfitness, one must determine the long-profile of the channel bed. On the other hand, until the channel bed has been profiled, and until the pool-and-riffle sequence has been demonstrated not to exist, or, if present, to be spaced inappropriately along the existing channel, the possibility remains open that an ingrown stream with an apparently very high wavelength/width ratio is an Osage-type underfit. The identification of a reach of the Severn as an Osage-type underfit touched off a lively debate upon meanders in general, as sundry issues of *Area* (1972–3) testify. The original description would perhaps have been more satisfactory, had the authors then been aware that, in the reach first examined, the Severn has reduced its channel to about three-quarters of expectable capacity. The spacing of pool-to-pool is, in fact, far closer to the expected five bedwidths than was originally claimed.

Meanders in Bedrock

The idea that meanders in bedrock are in some essential fashion different from meanders in alluvium goes back for many years. It still appears, either in explicit or in implicit form, in some recent current work. The only real difference seems to be that meanders in bedrock are commonly ingrown, whereas meanders on a floodplain, by definition, are not. Entrenched meanders, which have cut vertically down without enlarging themselves in the lateral and axial directions, are distinctly uncommon. They represent one end of a series, which extends through the intermediate range of normally ingrown meanders to meanders on a floodplain. The long-touted hypothesis that ingrown meanders are necessarily inherited from a flood-plain has seldom been tested, and, when tested, has been rejected (Dury, 1969; Young, 1974; see also Keller, 1973; Schumm and Khan, 1972; Shepherd and Schumm, 1974). It is of course true that the concept of bankfull flow is not applicable to ingrown meanders where the valley cross-section constitutes a V-shaped notch. Storm runoff of high magnitude and low frequency can there prove highly erosive (Dury, 1964a, A63–A64).

G. H. Dury

Young (1970, 1974) and Page (1972) have greatly extended the stock of information on the planform of ingrown meanders, successfully fitting to numerous examples the sine-generated curve of Langbein and Leopold (1966). In addition, Young has provided data on the wavelength and bedwidth of 71 meanders in bedrock, and on the radius of curvature for 66 of those meanders. When his data are grouped by meander trains, ranging from two to seven loops per train, and when one dimension is regressed upon another, the relationships which emerge are sensibly identical with those derived for alluvial meanders (Figures 18.1, 18.2 and 18.3). The wavelength/width ratio for Young's bedrock streams, at 10.844, fails to differ significantly from the approximate 11.0 obtained for alluvial streams (Dury, 1976). Equally, the mean of 10.36 for the ancestral channels of existing underfits fails to differ significantly either from the mean for Young's data, or from the mean for alluvial meanders. The probability P, in the relevant difference-of-means tests, ranges from about 0.4 to about 0.9. The wavelength/radius ratio, $4.7 R^{0.98}$ (or $4.15 R$) of Leopold, Wolman, and Miller (1964, Figure 7-41B) for alluvial meanders, is insignificantly different from the $4.27 R$ for Young's data. That is to say, meanders in bedrock, however great the distorting influence of structure, tend to be geometrically similar to alluvial meanders, in every respect except that of being ingrown.

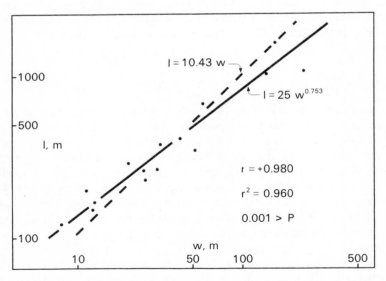

FIGURE 18.1 REGRESSION OF WAVELENGTH ON BEDWIDTH FOR R. W. YOUNG'S BEDROCK MEANDERS AVERAGED BY TRAINS

Based on data from Young (1974), Ph. D. Thesis, University of Sydney, by permission of the author

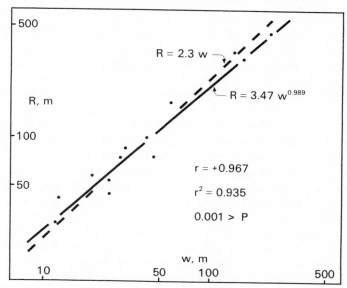

FIGURE 18.2 REGRESSION OF RADIUS OF CURVATURE ON BEDWIDTH FOR R. W. YOUNG'S BEDROCK MEANDERS, AVERAGED BY TRAINS

Based on data from Young (1974), Ph.D. Thesis, University of Sydney, by permission of the author

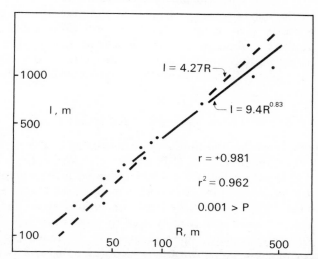

FIGURE 18.3 REGRESSION OF WAVELENGTH ON RADIUS OF CURVATURE FOR R. W. YOUNG'S BEDROCK MEANDERS, AVERAGED BY TRAINS

Based on data from Young (1974), Ph.D. Thesis, University of Sydney, by permission of the author

Some False Leads

My initial calculations of the discharges required to fill the former channels, and of the rainfalls capable of sustaining those discharges, were made to depend, inappropriately, on the Rational Formula for storm runoff. The establishment of power-functional relationships between channel characteristics on the one hand and discharge on the other (Leopold and Maddock, 1953; Leopold and Langbein, 1962) made it possible to attempt greater precision, especially when the significance of discharges of modest magnitude and high frequency was also demonstrated (Wolman and Leopold, 1957). However, information on bankfull discharge is by no means abundant, and the discussion about the relationship of such discharge to the magnitude–frequency scale has been long drawn out. The obvious difficulties resulting from tendencies to cut or fill need no more than a passing mention. Only recently has it been possible to show clearly that known bankfull flows concentrate themselves round the 1.58-year mark on the annual series (Dury, 1974). Later still, an equation relating meander wavelength to the square root of discharge (Dury, 1965, Figure 5 and text) has been rejected in favour of an equation wherein the exponent is 1.81 (Dury, 1976; also see below).

There is no need here to reexamine the extraordinary attempt by Geyl (1968) to explain valley meanders by former tidal action, nor the effort by Palmquist (1975) to account for the pools of paleochannels by scour on the part of existing streams; nor, again, the redundant rediscovery by Tinkler (1972) that narrow canyons can act as flumes. Refutations have been made elsewhere. We may pass these items over with only a brief word of comment. The three authors in question appear to have made trouble for themselves, by setting out not to falsify but to verify an hypothesis. In so doing, they have placed themselves in the situation where, in the words of Alexander Pope,

'The wild Mœander washed the artist's face'.

The full context, from the *Dunciad,* may seem worthy of the checking.

The work of Schumm (1968, 1972) deserves more serious notice, as attempting to incorporate the influence of sediment calibre upon meander geometry (for a thorough-going review of this topic, see Shahjahan, 1970). One may wonder, however, if Schumm's conclusions about the Murrumbidgee may have not been unduly influenced by those prominent Australian pedologists who rely implicitly upon pedogenesis in humid episodes, and upon upstream erosion and downstream deposition in arid periods. The postulate of aridity during glacial maxima owes something, and possibly much, to the writings of Galloway (e.g., 1970), who however has not gone without challenge (Dury, 1973). As noted elsewhere in this collection (Chapter 4), Schumm's equations for the characteristics of paleochannels fail to accord with the theoretical equations for stream channels in general; they are particularly poor in connection with downstream changes of channel slope and of flow velocity.

The State of the Art

Improved equations for the retrodiction of former bankfull discharge from former channel dimensions are now available (Dury, 1976). In m^3/s, and in m or m^2 as applicable, these equations are:

$$Q = (W/2.99)^{1.\overline{81}} \qquad (1)$$

or

$$= (L/32.857)^{1.\overline{81}} \qquad (2)$$

or

$$= 0.83 A_c^{1.09} \Omega \qquad (3)$$

where Q is former bankfull discharge, W is former bedwidth, L is former meander wavelength, A_c is former cross-sectional area, and Ω is the sinuosity of the existing stream. It is easy to write compound equations for the situation where values are known for more than one of the independent variables. Where values are known for all of these variables, the predicting equation becomes

$$Q = \{(W/2.99)^{1.\overline{81}} + (L/32.857)^{1.\overline{81}} + 0.83 A_c^{1.09} \ \Omega\}/3 \qquad (4)$$

The implied exponent of drainage area, when former bankfull discharge is taken as a power function of area, can be obtained by multiplying 1.81 by the exponents which relate former meander wavelength to area. I have previously determined wavelength/area relationships for 16 regions. The implied exponents in the discharge/area equation range from 0.45 to 1.03, averaging 0.76, which is not greatly different from the 0.7 of Alexander (1972). Obviously, however, it would be preferable to use a multiple predicting equation, and, as I have proposed elsewhere (Dury, 1976) to apply it to whole valleys rather than to individual reaches.

Some recent progress may be claimed in respect of the understanding of the change in form ratio which accompanies reduction of a stream to underfitness (Dury, 1974, 1976). Channel reaches on the Nene and the Great Ouse, reduced in discharge by the offtaking of water mills, have adjusted by reducing their form ratios. In a precisely corresponding manner, it seems usual that the ancestral channels of existing underfit streams had distinctly higher form ratios than those of present channels.

The Chronological Problem

Questions of the time of the last major stream shrinkage are bedevilled by the imprecisely dated climatic shifts of early deglacial times, by the insistence of some writers that the last glacial maximum was a time of aridity, by the scarcity of information upon the times when former large streams began and ceased to flow, by doubts about strict synchroneity of deglacial events in the northern and southern hemispheres, and by uncertainty about the linkage between events in middle latitudes on the one hand and in low latitudes on the other.

Morrison (1965) regarded lacustrine intervals in the Great Basin as synchronous with intervals of glaciation in the higher mountains. Saltzman and Verneckar (1975), modelling the northern-hemisphere climatic zonation for maximum-glacial conditions at 18,000 B.P., inferred that the full-glacial atmosphere was colder and drier than the atmosphere of today. At the same time, their Figure 15 shows a greater-than-present excess of precipitation over evapotranspiration than at present obtains. This excess can be read from the diagram as extending roughly between the tropic and 60° N latitude, and as meaning a 10% increase over the present-day excess.

Full-glacial conditions, however, do not necessarily come into question. For the Great Basin, Broecker and Kaufman (1965) concluded that pluviation lagged glaciation, especially in respect of the last very high stand of Lake Lahontan, which dates to 9500 radiocarbon years ago. The valley meanders of Black Earth Creek, Wisconsin, did not form until the ice had retreated from its last outermost stand, far enough for loess to cover the end-moraine. The shrinkage of the Creek to underfitness may have been deferred as late as 10,000 B.P. (Dury, 1964a), for which time Knox and Johnson (1974) report considerable channel disturbance at a Southwest Wisconsin site. The suggestion of a pluvial episode at about 10,000 B.P. is strongly supported by the results of Webb and Bryson (1972), although Davis (1974) has identified bogs, some of them in the cutoffs of large meanders, as already growing by 10.500 B.P. As the present author has suggested elsewhere, the last main shrinkage to underfitness may have been a double event. If so, its two parts may prove difficult to separate at many sites.

Studies in Siberia (Volkov, 1962, 1969, 1971, 1972) result in very good general agreement with the conclusions about humid/dry alternations which have been reached for Europe and North America. They are especially instructive in their treatment of the rise and fall of water tables. Efforts to calculate the precipitation required to sustain former stream discharge and former lakes have produced, in a number of cases, a factor of 1.5 to 2.0, whereby present-day precipitation should be multiplied (Butzer, Fock, Stuckenrath, and Zilch, 1973; Dury, 1964, 1968, 1973; Stoertz and Ericksen, 1974). However, precise dates on the times when the former streams flowed and the former lakes existed are urgently needed, in far greater quantity than at present. Especially is this so, if deglaciation in the southern hemisphere set in earlier than in the northern. Thus, Bowler (1971) dated high lake stands in part of southeastern Australia at 32,750, 23,500, and 15,500 B.P. This last date is late enough to equate with the last major ice-stand of northern-hemisphere continental glaciers, but does not necessarily correspond to the last stand in the southern hemisphere.

Low-latitude information comes mainly from Africa, for which Butzer, Isaac, Richardson, and Washbourn-Kamau (1972) have inferred a notable lack of correlation between high-latitude glacial advances and maximal stands of tropical lakes. They have identified transgressions which, beginning about 12,000 years ago and including a minor recession close to 10,000 B.P., culminated in the 10,000-year high stand. Williams and Adamson (1973) were in reasonable general agreement with these findings, and concurred in identifying an interval of high lake stands between about 6000 and 4000 years ago. By an analysis of Saharan radiocarbon dates, Geyh and Jäkel (1974) have narrowed one humid peak down to 10,500–10,000 B.P., and have also concluded that the Saharan climate became slightly more humid than immediately before, about 6000 B.P.

It is tempting to infer that the humid period in low latitudes, somewhere about 10,000 B.P., corresponded to the last episode, or one of the two last episodes, of pluviation and high stream discharge in middle latitudes. On the other hand, Williams (1970) dated calcretes and spring tufas, indicative of high water tables in the northern Sahara, at $11,640 \pm 350$ B.P., while van Geel and van der Hammen (1973) found for the near-equatorial Andes that lake level was high between about 13,000 and 10,800 B.P., in response to increased rainfall, fell between 10,800 and 9500 B.P., and once more stood high between 9500 and 7500 B.P. It might appear, either that the late-Glacial–early Holocene sequence of climatic shift differed from region to region, or that (as is entirely possible) it was diachronous, or that it was distinctly more complicated than has yet been generally demonstrated.

Perhaps because we know more about the mid-Holocene than about early cataglacial times, the Holocene picture seems somewhat the clearer. The African evidence of pluviality between about 6000 and 4000 B.P. seems closely matched by the partial re-clearance of alluvial fill by English rivers in about the same interval (Dury, 1964b: esp. Table 4), and by the onset of increased humidity and increased runoff in the U.S. Midwest (Knox, 1972). But still further information may serve in part only to increase uncertainty. The tabular summaries presented by Davis (1975, Figure 49) and by Knox (in press, Figure 5) leave scope for some degree of mismatch between region and region, and between one and another type of evidence of environmental shift. Johnson (1976) reports clusterings of radiocarbon dates for valley fills in Southwest Wisconsin, in about 6000–5000 B.P., 4400–4200 B.P., and 3000–1900 B.P., as indicating some form of environmental instability. In all cases, however, the relevant disturbances failed to scour clean the largest of the paleochannels.

Immediate Future Prospects

As has been clear for some time, the investigation of underfit streams forms part of the general enquiry into paleoclimatology and paleohydrology. Foreseeable future work includes the refinement of computations of former discharge, by the already-suggested investigation of complete valleys and by the use of improved equations; the refinement of chronology, by means of continued radiocarbon dating and palynological analysis, and the refinement of the analysis of dimensional ratios between former and present channels, both on a regional and on a temporal basis.

In this last respect, the Great Lakes region probably has much to offer. Discussion of former discharges has hitherto been concentrated on the discharges required to fill paleochannels at maximum former capacity. However, the reduction in the wavelength/area ratio, and thus implicitly in the discharge/area ratio, has been discussed for the situation where the Granta, Nene, and Great Ouse extended themselves across the consolidating peat fill of the Fenlands (Dury, 1964b, Figures 40, 42, and text). A similar exercise, or series of exercises, appears possible for the drainage which established itself on the floors of ancestral Great Lakes. Whereas at the mark of 250 km^2 the wavelength ratio between valley meanders and stream meanders is about 9.0 in unglaciated southwestern Wisconsin (Dury, 1965, Figure 9), for streams draining into the Green Bay arm of Lake Michigan the ratio is 5.6 (Dury, 1964b, Figure 32). For the Green Bay drainage below the Lake Nipissing shoreline the ratio is about 3.75.

Here seems to be a possible means of making comparisons among bankfull discharge at its cataglacial maximum, bankfull discharge at the present day, and bankfull discharge immediately subsequent to a datable stand of a former lake. Similar investigations are possible, but could well prove less easy to conduct, in respect of some ingrown meandering valleys. Young (1974, and references therein) has reported for his field area in eastern Australia that meanders cut into the K_3 terrace, which has been dated to about 29,000 B.P., are three to five times as large as are younger bends occurring at comparable drainage areas. The large bends generally contain, in their point bars, sediments of the K_2 terrace, which has been dated to about 3750 B.P. On this evidence, Young has inferred a pluvial episode in the general range of 4000 to 3000 B.P. The full significance of his findings cannot be assessed, until the possible latitudinal variation in degree of maximal stream shrinkage has been extensively studied.

A topic intimately related to shrinkage to underfitness is that of drainage-net retraction. This topic includes, but is by no means limited to, the study of dry valleys in permeable rocks. Gregory (1971) has reviewed the dry-valley question in some detail, and for basins in southwest England has established marked differences between sum valley length (greater) and sum channel length (lesser). Coleman (1973) has concluded that about one-third of the effective drying of valleys in the Chalk is ascribable to percolation, but that the large residual effect must be laid to climatic change. As in so many other contexts, many more firm dates are needed for changes in stream behaviour. One can only suggest here that the subsurface exploration of former stream channels, now obliterated in consequence of network retraction, might prove highly rewarding.

Additional information on Osage-type underfits can be expected to lead, in due course, to an understanding of the failure of streams of this kind to meander in their present channels. The equations for retrodicting discharge from channel dimensions might perhaps be profitably applied to former meltwater sluiceways, in order to establish the surface hydrology of the last major episode of ablation. In this connection, however, additional information on the hydraulic geometry of braided channels is sorely needed.

Possibly more urgent, but certainly more difficult, than any of this is the modelling of the general circulation, both at the time of the breakdown of full-glacial conditions, and at the time of main glacial onset. About cataglaciation we know at least something. About anaglaciation, including anaglacial shifts of precipitation, we know very little indeed.

References

Alexander, G. N. 1972. Effect of catchment area on flood magnitude. *J. Hydrol.*, **16,** pp. 225–240.

Bowler, J. M. 1971. Pleistocene salinities and climatic change. In D. J. Mulvaney and J. Golson (Eds.), *Aboriginal Man and Climate in Australia.* Australian Nat. U. Press, Canberra, pp. 47–65.

Broecker, W. S. and Kaufman, A. 1965. Radiocarbon chronology of Lake Lahontan and Lake Bonneville II, Great Basin. *Geol. Soc. Amer. Bull.,* **76,** pp. 537–566.

Butzer, K. W., Fock, G. F., Stuckenrath, R., and Zilch, A. 1973. Palaeohydrology of Late Pleistocene lake, Alexandersfontein, Kimberley, South Africa. *Nature,* **243,** pp. 328–330.

Butzer, K. W., Isaac, G. L., Richardson, J. L., and Washbourn-Kamau, C. 1972. Radiocarbon dating of East African lake levels. *Science,* **175,** pp. 1069–1076.

Chorley, R. J., Dunn, A. J., and Beckinsale, R. P. 1964. *The History of the Study of landforms, I; Geomorphology Before Davis.* Methuen, London.

Coleman, A. M., 1973. Personal communication.

Davis, A. M., 1974. Pollen analysis and climatic reconstruction: a Driftless Area example. In J. C. Knox and D. M. Mickelson (Eds.), *Late Quaternary Environments of Wisconsin.* AMQUA Field Guide, 1974 Meeting, pp. 79–85.

Davis, A. M. 1975. *Reconstructions of Local and Regional Holocene Environments from the Pollen and Peat Stratigraphies of some Driftless Area Peat Deposits.* Ph.D. dissertation, The University of Wisconsin-Madison.

Dury, G. H. 1952. The alluvial fill of the valley of the Warwickshire Itchen near Bishop's Itchington. *Proc. Coventry District Nat. Hist. and Sci. Soc.,* **2,** pp. 180–185.

Dury, G. H. 1963. Underfit streams in relation to capture: a reassessment of the ideas of W. M. Davis. *Trans, Inst. Brit. Geog.,* **32,** pp. 83–94.

Dury, G. H. 1964a. Principles of underfit streams. *U.S. Geol. Surv. Prof. Paper 452-A.*

Dury, G. H. 1964b. Subsurface exploration and chronology of underfit streams. *U.S. Geol. Surv. Prof. Paper 452-B.*

Dury, G. H. 1965. Theoretical implications of underfit streams. *U.S. Geol. Surv. Prof. Paper 452-C.*

Dury, G. H. 1966. Incised valley meanders on the lower Colo River, New South Wales. *Austr. Geog.*, **10**, pp. 17–25.

Dury, G. H. 1968. Introduction to the geomorphology. In G. H. Dury and M. I. Logan (Eds.), *Studies in Australian Geography.* Heinemann Educational Australia, Melbourne, pp. 1–36.

Dury, G. H. 1969. Tidal stream action and valley meanders. *Austr. Geog. Studies*, **7**, pp. 49–56.

Dury, G. H. 1973. Paleohydrologic implications of some pluvial lakes in northwestern New South Wales, Australia. *Geol. Soc. Amer. Bull.* **84**, pp. 3663–3676.

Dury, G. H. 1974. Magnitude-frequency analysis and channel morphometry. In Morisawa, M. (Ed.), *Fluvial Geomorphology*, State University of New York, Binghampton, pp. 91–121.

Dury, G. H. 1976. Discharge prediction, present and former, from channel dimensions *J. Hydrol.*, **30**, pp. 219–245.

Dury, G. H., Sinker, C. A., and Pannett, D. J. 1972. Climatic change and arrested meander development on the River Severn. *Area*, **4**, pp. 81–85.

Galloway, R. W. 1970. The full-glacial climate in the southwestern United States. *Ann. Amer. Geog. Assoc.*, **60**, pp. 245–256.

Geyh, M. A. and Jäkel, D. 1974. Late glacial and Holocene climatic history of the Sahara Desert derived from a statistical assay of ^{14}C dates. *Palaeogeogr., Palaeoclimatol., Palaeoecol.*, **15**, 205–208.

Geyl, W. F. 1968. Tidal stream action and sea level change as one cause of valley meanders and underfit streams. *Austr. Geog. Studies*, **6**, pp. 24–42.

Gregory, K. J. 1971. Drainage density changes in South-West England. In Gregory K. J. and Ravenhill W. (Eds.), *Exeter Essays in Geography.* Exeter, pp. 33–53.

Johnson, W. C. 1976. Personal communication.

Keller, E. A. 1973. New insight into the role of bedrock control of stream channel morphology. *Geol. Soc. Amer. Abstracts*, 1973 Annual Meeting, p. 689.

Knox, J. C. 1972. Valley alluviation in Southwestern Wisconsin. *Ann. Assoc. Amer. Geog.*, **62**, pp. 401–410.

Knox, J. C. In press. Concept of the graded stream (expected to appear in the S.U.N.Y.-Binghamton *Geomorphology* series).

Knox, J. C. and Johnson, W. C. 1974. Late Quaternary valley alluviation in the Driftless Area. In Knox J. C. and Mickelson D. M. (Eds.), *Late Quaternary Environments of Wisconsin*, AMQUA Field Guide, pp. 134–149.

Kohchi, N. 1975. Meandering valley and underfit stream of the Ota River. *Ann. Tohoku Geog. Assoc.*, **27**, p. 2 (in Japanese).

Kohchi, N. 1976. Incised meanders in the Chugoku Mountains, Southwest Japan. *Geog. Review of Japan*, **49**, pp. 43–53 (in Japanese).

Langbein, W. B. and Leopold, L. B. 1966. River meanders—theory of minimum variance. *U.S. Geol. Surv. Prof. Paper 422-H.*

Leopold, L. B. and Langbein, W. B. 1962. The concept of entropy in landscape evolution. *U.S. Geol. Surv. Prof. Paper 500-A.*

Leopold, L. B. and Maddock, T. 1953. The hydraulic geometry of stream channels and some physiographic implications. *U.S. Geol. Surv. Prof. Paper 252.*

Leopold, L. B., Wolman, M. A., and Miller, J. P. 1964. *Fluvial Processes in Geomorphology.* W. H. Freeman, San Francisco.

Morrison, R. B. 1965. Quaternary geology of the Great Basin. In Wright H. E. and Frey D. G. (Eds.), *The Quaternary of the United States.* Princeton U.P., pp. 265–285.

Page, K. J. 1972. *A Field Study of the Bankfull Discharge Concept in the Wollombi Brook Drainage Basin, N.S.W.* University of Sydney, M.A. Thesis.

Palmquist, R. C. 1975. Preferred position model and subsurface of valleys. *Geol. Soc. Amer. Bull.* **86,** pp. 1392–1398.

Saltzman, B. and Vernekar, A. D. 1975. A solution for the Northern Hemisphere climatic zonation during a glacial maximum. *Quaternary Research,* **5,** pp. 307–320.

Schumm, S. A. 1968. River adjustment to altered hydrologic regimen— Murrumbidgee River and paleochannels, Australia. *U.S. Geol. Surv. Prof. Paper 598.*

Schumm, S. A. 1972. Fluvial paleochannels. In J. K. Rigby and W. K. Hamblin (Eds.), *Recognition of Ancient Sedimentary Environments.* Soc. Econ. Paleontol. and Mineral, Spec. Pub. No. 16, pp. 98–107.

Schumm, S. A. and Khan, H. R. 1972. Experimental study of channel pattern. *Geol. Soc. Amer. Bull.,* **83,** pp. 1755–1770.

Shahjahan, M. 1970. Factors controlling the geometry of fluvial meanders. *Bull. Internat. Assoc. Sci. Hydrol.,* xve année (3), pp. 13–24.

Shepherd, R. G. and Schumm, S. A. 1974. Experimental study of river incision. *Geol. Soc. Amer. Bull.,* **85,** pp. 257–268.

Shoobert, J. A. 1968. Underfit stream of the Osage type: head of the Port Hacking River. *Austr. Geog.,* **10,** pp. 523–524.

Stoertz, G. E. and Ericksen, G. E. 1974. Geology of Salars in Northern Chile. *U.S. Geol. Surv. Prof. Paper 811.*

Tinkler, K. J. 1972. The superimposition hypothesis for incised meanders—a general refutation and specific test. *Area,* **4,** pp. 86–91.

van Geel, B. and van der Hammen, T. 1973. Upper Quaternary vegetational and climatic sequence of the Fuquene area (Eastern Cordillera, Colombia). *Palaeogeogr., Palaeoclimatol, Palaeoecol.,* **14,** pp. 9–92.

Volkov, I. A. 1962. The history of river valleys in the south part of western Siberian lowlands. *Works of the Novosibirsk Institute of Geology and Geophysics,* No. 27, pp. 34–46 (in Russian).

Volkov, I. A. 1969. Periods of flooding and aridization of the extraglacial zone. *Problems of the Quaternary Geology of Siberia,* U.S.S.R. Academy of Sciences, pp. 17–32 (in Russian).

Volkov, I. A. 1971. Fluctuations of climate of Late-glacial and early Holocene times in the south of the Western Siberian Valley. *U.S.S.R. Academy of Sciences, Geology and Geophysics,* No. 8, pp. 72–81 (in Russian).

Volkov, I. A. 1972. Late Quaternary time and principal continental sedimentary formations of temperate zone plains. *Problems of Quaternary Research,* U.S.S.R. Academy of Sciences, pp. 7–12 (in Russian).

Webb, T. and Bryson, R. A. 1972. Late and Postglacial climatic change in the Northern Midwest, U.S.A.—quantitative estimates derived from fossil pollen spectra by multivariate statistical analysis. *Quaternary Research,* **2,** pp. 70–115.

Williams, G. E. 1970. Piedmont sedimentation and Late Quaternary chronology in the Biskra Region of the northern Sahara. *Geomorph.,* Suppbd. **10,** pp. 40–63.

Williams, M. A. J. and Adamson, D. A. 1973. The Physiography of the Central Sudan. *Geog. J.,* **139,** pp. 498–508.

Wolman, M. G. and Leopold, L. B. 1957. River flood plains: Some observations on their formation. *U.S. Geol. Surv. Prof. Paper 282-C.*

Young, R. W. 1970. The patterns of some meandering valleys in New South Wales. *Australian Geog.,* **11,** pp. 269–277.

Young, R. W. 1974. *The Meandering Valleys of the Shoalhaven River System: A Study of Stream Adjustment to Structure and Changed Hydrologic Regimen.* University of Sydney, Ph.D. Thesis.

Section IV

NETWORK CHANGE AND THEORY

K. J. GREGORY

Professor of Geography
University of Southampton

19

Network Adjustments and Progress towards Theory

Whereas earlier review chapters (2, 6, 10) have been able to review progress already realized, studies of change placed in the context of the drainage network will be elaborated in future research so that succeeding chapters point the ways in which progress may be achieved. These chapters are sufficient by themselves and this introduction merely endeavours to collect several strands together in the belief that an extensive review is premature. It has been argued that closer liaison is required between network, pattern, geometry, and sedimentary sequences, and that such an improved liaison will cement the gulfs which have existed between separate branches of study (Gregory, 1976a). Approaches to palaeohydrology (Schumm, 1965) have been concentrated at the drainage basin level, whereas river metamorphosis (Schumm, 1969) has focused upon the reach or geometry of the river channel. Between these two approaches is the drainage network and it is necessary to attempt analysis of empirical data and model building which reconciles the two approaches; an attempt is made in Chapter 25. To this end it is desirable for empirical data to be collected from the network and related to the channel pattern and geometry as exemplified by Chapters 20 and 26; to gain an improved understanding of geometry variations and changes throughout the network (21, 22); to obtain further knowledge of the downstream consequences of upstream changes (23, 24), and to improve theoretical arguments (27).

Eventually we need a theoretical basis for modelling network processes which can be the basis for analysis of change. Simons and Gessler (1971) contended that 'Theory on hydraulic processes is years ahead of theory on geomorphic processes and there is a pool of knowledge in the field of boundary layer theory which could be tapped for answers in relation to geomorphic problems.' Progress has been realized with the increased use of theory in geomorphology since 1965, but such theoretical developments have concentrated upon river channel cross-sections or reaches because networks are necessarily more complex. In fact networks have been studied in at least three principal ways by geomorphologists. First, have been studies of network topology which have only recently been directed towards the problem of the relationship between network topology and stream flow hydrograph formation as attempted by Surkan (1974). Second, have been studies of drainage network densities in relation to climatic characteristics and streamflow (Gregory, 1976b), but only since 1968 has the nature of the relationships become apparent so that we are in a position to model change in the manner previously attempted for river channel metamorphosis. Thirdly, there has been great progress in the study of network extension by gullying (Cooke and Reeves, 1976) but only recently has it become usual to include ancillary consideration of associated geometry changes downstream. Unfortunately, studies of network contraction, by the production of dry valleys, have not usually been investigated for recent short periods of time but have been placed in a longer time scale context.

Progress towards theory must depend upon a greater use of flood routing models. Many geomorphological studies, encouraged by the use of data from small drainage basins, have relied upon statistical multivariate models which have related climatic, and water and sediment discharge parameters to measures of drainage basin character. The need for more physically based models has become evident and this has been attempted for example by Calver, Kirkby, and Weyman (1972). An early initiative was available from the progress achieved by Wooding (1966) towards a hydraulic model; more recently Diskin (1973) has elaborated a quasi-linear system and emphasized the desirability of a regional model, and Onstad (1973) has provided a review of flood routing using distributed parameters. It is increasingly possible to evaluate empirical results against deductive models and Smith (1974) has constructed an analytical model of channel growth by assuming that sediment mass is conserved, that the channel form is sufficient to transport the water discharge, and that the channel form is also sufficient to transport the sediment supplied. Smith suggests that further models could be built for different channel types and it would seem that these could then be placed in a network context. An attempt to model the fluvial system in this way was attempted by Rzhanitsyn (1963) but has not yet been refined to reduce the dependence upon ordering techniques. In future work it may be expedient to express network form in a more complete and more significant way than has been previously attempted. Hence Coulson and Gross (1967) followed Sribnyi (1961) in describing the channel system distribution by a drainage factor which was obtained from measurements of total channel length in relation to basin area, and which can index the 'hydrologic shape' of a basin. An alternative approach was devised by Rogers (1971) who used the frequency distribution of first-order stream lengths and related such frequency distributions to hydrograph shapes. The opportunity for further progress is shown by Simons and Gessler (1971), who commented in a review of research needs in fluvial processes that 'geomorphic processes have been primarily descriptive and for many years the approach to actual problem solving has been highly empirical. This discrepancy makes it quite evident that research effort should concentrate on geomorphic aspects.' The potential for the application of results obtained by further progress was sketched by Simons (1967) who contended that ' . . . by education and the proper utilisation of existing knowledge of hydraulics, sedimentation, channel geometry, river mechanics and fluvial morphology one can predict the responsiveness of river systems through varying degrees of development.' This statement appeared in a paper entitled *River Hydraulics* but the sentiment could equally well apply to network hydraulics.

References

Calver, A., Kirkby, M. J., and Weyman, D. R. 1972. Modelling hillslope and channel flows. In Chorley, R. J. (Ed.), *Spatial Analysis in Geomorphology.* Methuen, London, pp. 197–218.

Cooke, R. V. and Reeves, R. W. 1976. *Arroyos and Environmental Change in the American South West.* Clarendon Press, Oxford.

Coulson, A. and Gross, P. N. 1967. Measurement of the physical characteristics of drainage basins. *Inland Waters Branch Department of Energy, Mines and Resources, Tech. Bull. No. 5,* Ottawa.

Diskin, M. H. 1973. The role of lag in a quasi-linear analysis of the surface runoff system. In Schulz, E. F., Koelzer, V. A., and Mahmood, K. (Eds.), *Floods and Droughts.* Water Resources Publications, Fort Collins, pp. 133–144.

Gregory, K. J. 1976a. Changing drainage basins. *Geogr. J.,* **142,** pp. 237–247.

Gregory, K. J. 1976b. Drainage networks and climate. In Derbyshire, E. (Ed.), *Climate and Landforms.* Wiley, London, pp. 289–315.

Onstad, C. A. 1973. Watershed flood routing using distributed parameters. In Schulz, E. F., Koelzer, V. A., and Mahmood, K. (Eds.) *Floods and Droughts,* Water Resources Publications, Fort Collins, pp. 418–428.

Rogers, W. F. 1971. Hydrograph analysis and some related geomorphic variables. In Morisawa, M. E. (Ed.) *Quantitative Geomorphology: Some aspects and applications.* State University of New York, Binghamton. pp. 245–257.

Rzhanitsyn, N. A. 1963. *Morphological Regularities of the structure of the River Net.* (Trans.: Krimgold, D. B.) Soil and Water Conservation Research Division, U.S. Dept. Agric. and Water Resources Div. Geol. Surv. U.S. Dept. Interior.

Schumm, S. A. 1965. Quaternary Palaeohydrology. In Wright, H. E. and Frey, D. G. (Eds.), *The Quaternary of the United States,* Princeton University Press, Princeton, pp. 783–794.

Schumm, S. A. 1969. River metamorphosis. *J. Hydraul. Div. Amer. Soc. Civ. Engrs.,* **95,** pp. 255–273.

Simons, D. B. 1967. River hydraulics. *Proc. Twelfth Cong. Internat. Assoc. Hydraulic Research,* Vol. 5, Colorado State University, Fort Collins, pp. 376–398.

Simons, D. B. and Gessler, J. 1971. Research needs in fluvial processes. In Shen, H. W. (Ed.), *River Mechanics,* Vol. II, Colorado State University, Fort Collins, pp. 32-1 to 32-13.

Smith, T. R. 1974. A derivation of the hydraulic geometry of steady-state channels from conservation principles and sediment transport laws. *J. Geol.*, **82**, pp. 98–104.

Sribyni, M. F. 1961. Geomorphological characteristics of catchment drainage basins (drainage areas). In *Problems of river runoff control*, Academy of Sciences, USSR Trans. Israel Program for scientific translations.

Surkan, A. J. 1974. Simulation of storm velocity effects on flow from distributed channel networks. *Water Resour. Res.*, **10**, pp. 1149–1160.

Wooding, R. A. 1966. A hydraulic model for the catchment-stream problem. *J. Hydrol.*, **3**, pp. 254–267; 268–282; 4; 21–37.

A. M. HARVEY

Lecturer in Geography
University of Liverpool

20

Event Frequency in Sediment Production and Channel Change

It is generally accepted that within the overall framework of the major climatic regime, and major relief and slope characteristics of the drainage basin, stream channel morphology is controlled primarily by sediment and water supply to the stream system. Channel size, shape, and pattern are dependent on flood hydrology, sediment calibre and amount. These relationships, particularly in the context of channel stability, may be further influenced by the frequency of major discharges and of sediment-producing events.

This paper deals with the frequency of these events on a small upland stream in an area of active slope erosion and substantial sediment supply. Previous studies of the magnitude and frequency aspects of fluvial systems have tended to deal with lowland streams or larger rivers, either in the context of sediment transport (Wolman and Miller, 1960) or in the context of channel adjustment to discharge as expressed by the bankfull discharge (Dury, 1961; Harvey, 1969; Woodyer, 1968). Both types of study suggest that alluvial channel morphology tends to adjust to relatively frequent events, those with a recurrence interval of between one and several years.

The purpose of this paper is to examine the frequency characteristics of major sediment production events, and major channel changes in what appears to be a more dynamic and unstable environment than those previously studied. The area studied is the catchment of Grains Gill, a tributary of Carlingill in the Howgill Fells, Cumbria (Figure 20.1). Grains Gill drains a catchment of approximately 0.9 km^2 with a range of relief from 185 m to 475 m, which experiences a cool damp climate with rain all year averaging approximately 1350 mm/year. In winter some of this falls as snow, usually in several major falls each winter, but snow cover rarely persists for more than a few days on any one occasion. The underlying bedrock is Silurian shale

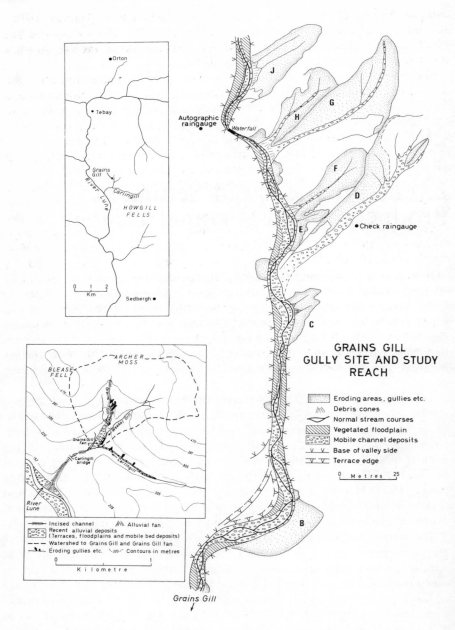

and grit overlain by Pleistocene glacial till and soliflucted till (Harvey, 1974), and by post-glacial terrace gravels and fan deposits. The catchment lies wholly on unenclosed land, predominantly grass moorland but with heather covered peat on Archer Moss (Figure 20.1).

Midway along Grains Gill are a series of deep active gullies dissecting the till slopes on the valley sides. These gullies act as a major sediment source for the stream. Indeed, their importance is reflected in the contrast in channel morphology of Grains Gill above and below the gully site. Above the gully site the channel is narrow and bordered in places by grass-covered, stabilized alluvial gravels. Below the gully site, the channel is generally much wider and much less stable, occupying most of the valley floor.

These tendencies are illustrated in Figure 20.2 where channel bed width is plotted against distance along the valley floor. Grains Gill rises on Archer Moss at a point just over 900 m upstream of its confluence with Carlingill, where pipes and runnels issue from the peat to form an identifiable stream channel. For the first 100 m the channel has grass-covered, alluvial margins, but then for 100 m the channel is steep, in places cut into bedrock and is interrupted by small waterfalls. The bedrock-controlled sections are indicated on Figure 20.2. Over the next 140 m downstream to the gully site the channel again has alluvial margins. Throughout this upper course width shows a slow but steady increase downstream, with relatively little variation and only at one point does it exceed 2.5 m.

Downstream from the gully site, not only does width suddenly increase, but it also shows tremendous variability from site to site, such that further

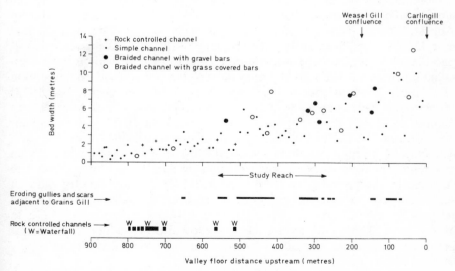

FIGURE 20.2 BEDWIDTH VARIATIONS ALONG GRAINS GILL (BASED ON FIELD SURVEY)

downstream the effects of the increments of the small right bank tributary and the larger left bank tributary of Weasel Gill are totally obscured by the variability in the width of Grains Gill itself. Furthermore from the gully site downstream much of the channel has a braided character.

The coarse gravel and cobbles which form the mobile channel bed deposits are derived mainly from the boulder clay being eroded by the gullies. During heavy rain the gullies supply both bed sediment and suspended sediment to the stream. The fines are washed straight into the stream, and during runoff periods suspended sediment concentrations are visibly much higher below the gullies than above. During only moderate rainfall measured suspended sediment concentrations below the gullies are as much as two-and-a-half times those above. The coarse material eroded from the gullies tends to accumulate in debris cones or fans below the gullies, only to be removed and incorporated with the stream bed deposits by a major flood. Such floods remove the accumulated debris, rework the mobile stream bed material and bring about major changes in channel configuration throughout the reach from the gully site down to the confluence with Carlingill.

This paper examines the frequency of sediment production, sediment removal, and channel change, particularly along a 300 m study reach through and below the gully site (Figure 20.1).

Observations and Methodology

Observations relating to gully erosion, sediment accumulation and removal and channel change have been maintained over a $6\frac{1}{2}$-year period from August 1969 to March 1976. Changes in gully, debris cone, and channel morphology have been recorded photographically, from a standard set of photograph sites. Over the whole period 6-monthly observations have been made (Spring and Autumn) and over most of the period (with the exception of October 1969 to March 1970 and August 1972 to September 1973) observations have been made at 1- to 2-month intervals. From May 1971 to August 1972 and from September 1975 to March 1976 more frequent observations (10-day to 2-week intervals) have been made of specific sites. The data on morphological change provided by the photographs have been augmented in the context of gully erosion by data from erosion nails and stakes over the period May 1970 to March 1976 and by detailed data on sediment yield to sediment traps at the base of the gully systems for the periods April 1971 to July 1972 (Harvey, 1974) and September 1975 to March 1976. These data have enabled a record to be maintained of major geomorphological events both in terms of gully erosion and sediment supply to the stream and of channel change.

The major controls of sediment production and sediment transport are surface erosion by mass movement and by running water (Harvey 1974), and stream discharge, both primarily controlled by precipitation. No data on stream discharge are available but daily precipitation data for the whole period of field observations are available from the raingauge at Sedbergh, on the margin of the Howgill Fells, 9 km to the south of Grains Gill. These data are augmented by daily data from Orton raingauge, 8 km to the north, for the period 1971–1976, and by a continuous record from an autographic raingauge installed at Grains Gill for the periods May 1971 to July 1972 and September 1975 to March 1976. The raingauge at Sedbergh provides daily data from 1930.

From the record of geomorphological events, classifications of the magnitude of channel change and sediment supply events have been devised and examined in the context of the available record of daily precipitation events. Because precipitation data for Grains Gill are available only for the sediment supply study periods, precipitation at Grains Gill for the rest of the channel change study periods had to be estimated from the data for Sedbergh and Orton rain gauges. Correlation and regression relationships were established between storm rainfall at Sedbergh and Orton and storm rainfall at Grains Gill. In order to minimize the correlation problems associated with the exact timing of precipitation, especially around 0900 hours, discrete rain periods of durations up to 3 days were identified from the autographic record from Grains Gill and correlated with the equivalent 1-, 2-, or 3-day totals from Sedbergh and Orton. To minimize the distorting influence of localized light, and geomorphologically ineffective, rains only those days or periods with rainfall totals greater than 5 mm at the independent station were used in the analysis. No separate account was taken of snowfall as on almost all occasions snow melted soon after falling. The results may be summarized as follows:

$$P_G = 1.03 \, P_S^{1.00} \qquad (n = 87, r = 0.90, \text{s.e.}_G = 0.15) \tag{1}$$

$$P_G = 1.53 \, P_O^{0.85} \qquad (n = 66, r = 0.88, \text{s.e.}_G = 0.14) \tag{2}$$

where P is precipitation (G at Grains Gill, S at Sedbergh, O at Orton), n is sample size, r is the correlation coefficient, and s.e. is the standard error of the estimate (log units).

The near linearity of the Sedbergh relationship for storm rainfall would suggest that a reasonably accurate estimate of daily rainfall over the long term may be made from the daily totals at Sedbergh. The relationship of rainfall at Sedbergh to that at Grains Gill was also calculated, so that return periods of daily rains at Grains Gill could be estimated from the Sedbergh data (see below). This relationship is as follows:

$$P_S = 1.21 \, P_G^{0.90} \qquad (n = 86, r = 0.90, \text{s.e.}_S = 0.13) \tag{3}$$

The long term frequency characteristics of the daily rainfall totals at Sedbergh for the 40 year period 1930–1969 were examined by calculating the recurrence intervals of daily rainfalls (Chorley and Kennedy, 1971) using methods similar to the partial duration series method of flood frequency analysis (Langbein, 1949). The results are illustrated in Figure 20.3.

Event Frequency and Sediment Production

A detailed analysis of sediment yield was carried out over the period 1971–1972 (Harvey, 1974) based on sediment yields over 5–14 day periods to traps installed at the base of rill systems on Gully D (Figure 20.1). In that study five classes of sediment yield event were identified, based on sediment amount and particle size characteristics. The rainfall intensity and duration characteristics determining sediment yield type were identified, using data from the autographic raingauge installed at Grains Gill. In this present study there is a need to generalize the rainfall data to daily totals so that the long term frequency of sediment-yielding events may be considered. In addition the sediment yield data for 1971–1972 have been augmented by data, classified by the same system (Table 20.1), collected at intervals of up to 14 days during 1975–76.

A summary of the period maximum daily rains associated with each sediment yield class is given in Table 20.2. It is apparent that a reasonable distinction may be made between Class A (major) yields and the minor yields, classes C, D, and E, with Class B yields occupying transitional positions. Of the 33 Class A yields, 29 resulted from daily precipitation in excess of 15 mm. None of the 36 minor yields resulted from precipitation in excess of 12 mm. Class B yields were transitional with two of the nine resulting from rains in excess of 15 mm. Of the four Class A yields resulting from daily precipitation less than 15 mm, two (both in December 1971) resulted from snowmelt from a thin snow cover, itself resulting from precipitation equivalent to 8–9 mm. The other two Class A exceptions (July 1971 and July 1972) were unusual in that they resulted from very short duration intense showers: both were only marginally Class A yields.

TABLE 20.1 SUMMARY OF CLASSIFICATION OF PERIOD SEDIMENT YIELDS
(BASED ON HARVEY, 1974, 52–3)

Class A	Major sediment yield, indicative of widespread surface erosion by mass movement or running water
Class B	Transitional yield, indicative of very limited surface erosion
Class C	Stony yields, indicative of frost heave or intense desiccation, but no other surface erosion evident
Class D	Small yields, indicative of rill wash only
Class E	Negligible erosion

TABLE 20.2 SUMMARY OF MAXIMUM PERIOD DAILY RAINFALL TOTALS FOR
PERIOD SEDIMENT YIELD CLASSES

Sediment yield class	Total	Maximum period daily rainfall (mm)				
		Maximum	UQ	Median	LQ	Minimum
A	33	98	33	21	18	8
B	9	20	15	10	7	4
C+D	23	12	6	4	2	0
E	13	6	3	1	0	0

(UQ, LQ are upper and lower quartile respectively)

The normal minimum rainfall required for a Class A sediment yield, in other words the daily rainfall threshold for major slope erosion, appears to be between 12 and 15 mm. However, apart from the influence of snowmelt no obvious seasonality in this threshold appears. By applying the least squares method to the deviations provided by the exceptions (i.e. Class A yields from lower rainfalls and Classes B, C, D, E yields from higher rainfalls) the 'best fit' threshold appears to be 13 mm when the snow yields are included and 14 mm when they are excluded, with the best overall fit at *c.* 13.8 mm. Then, by applying equation (3) to this figure, the most probable equivalent rainfall at Sedbergh may be estimated at *c.* 12.8 mm, which according to Figure 20.3 has a long term recurrence interval of about 0.035–0.04 years. It would appear, then, that major sediment producing events occur approximately 25–30 times a year. Much of the sediment produced, particularly the coarse sediment, then accumulates in debris cones or fans at the base of the gully systems awaiting removal by stream processes.

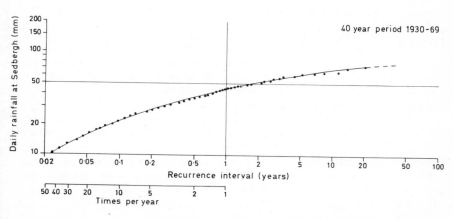

FIGURE 20.3 LONG-TERM DAILY RAINFALL FREQUENCY AT SEDBERGH 1930–1969

Event Frequency and Channel Change

The approach to the analysis of channel change was similar to that adopted for sediment supply but with the lower frequency of channel change; the observations were maintained over a longer period, and the nature of the field observations was different. The field evidence consisted of direct field observations and photographic recording of channel morphology along the study reach between Gullies B and J (Figure 20.1). Observations were maintained over the full field period, August 1969 to March 1976.

A hierarchical classification of channel change was devised and applied to each of the periods between observations (Table 20.3), and the results considered against the maximum daily rainfall during the period. Rainfall data from Grains Gill were used, where available, but otherwise data from Sedbergh were used. For each period the assumptions were made that the maximum class of channel change observed resulted from the maximum flood occurring during the period and that the period maximum daily rainfall is a reasonable surrogate for flood discharge. These assumptions appear to be acceptable in that the classification of channel change is hierarchical rather than cumulative, and that stream response to rainfall is immediate

TABLE 20.3 CLASSIFICATION OF PERIOD CHANNEL CHANGES

Class A	Complete channel change	Total removal of one or more debris cones. Marked erosion of valley sides and vegetated floodplain. Total destruction of pattern of channel bars. New pattern established.
Class B	Considerable channel change	Partial removal of one or more debris cones. Observable erosion of valley sides and vegetated floodplain. Major modification to pattern of channel bars, etc. Major shifts in channel position rather than creation of a totally new channel.
Class C	Moderate channel change	Some trimming of base of debris cones. Some erosion and trimming of valley sides and vegetated floodplain. Some modification to pattern of channel bars, etc. Minor shifts in channel position.
Class D	Minor channel changes	Minor change in the form and pattern of channel bars. Movement of loose particles from the base of debris cones. No real change in channel position.
Class E	Some sediment movement	No observable change in channel, debris cone or bar morphology but some movement of individual particles identified.
Class F	No change	No observable channel change. No observable sediment movement.

TABLE 20.4 SUMMARY OF MAXIMUM PERIOD DAILY RAINFALL TOTALS FOR
PERIOD CHANNEL CHANGE CLASSES

Class of channel change	Total	Maximum period daily rainfall (mm)				
		Maximum	UQ	Median	LQ	Minimum
A, B, C	8	98	58	50	41	31
D	24	51	31	25	20	12
E	15	35	25	20	13	4.0
F	17	20	12	8.5	4.1	2.9

(UQ, LQ are upper and lower quartile respectively)

and direct, especially in the case of the relatively rare events that produce sediment movement and channel change.

In the analysis channel change Classes A, B, and C were amalgamated since there were so few occurrences of these major channel changes. Certainly these classes are distinct from the minor changes (Classes D, E, F) in that they all show evidence of erosional change, to consolidated channel margins, whereas Class D shows only minor changes to the form and pattern of depositional features.

A summary of the maximum period daily rainfall totals associated with each class of channel change in given in Table 20.4 and as with the sediment yield classification there is relatively little overlap between classes, particularly at the upper end. Apparent thresholds between classes have again been derived by identifying the daily rainfall total producing the fewest exceptions and testing by the least squares deviation method described above. The results are shown in Table 20.5. A daily rainfall total of the order of 41 mm appears to represent the 'best fit' threshold distinguishing major changes, involving marginal erosion (Classes A, B, C) from minor changes (Classes D, E, F), involving no more than minor modifications to existing depositional forms. Two major changes resulted from daily rains of less than this figure (31 mm and 37 mm) but on both occasions the rainfall figure is derived from

TABLE 20.5 SUMMARY OF DAILY RAINFALL, BEST FIT THRESHOLDS

	Threshold between		Daily rainfall (mm) best fit threshold	Long term frequency (times/year)	N_1	N_2	Exceptions	
	Classes (1)	Classes (2)					(1)	(2)
Channel changes	ABC	DEF	41	1.05	8	56	2	2
	ABCD	EF	21	11	32	32	7	5
	ABCDE	F	13	30	47	17	4	4
Sediment production	A	BCDE	13	28	33	45	4	2

the Sedbergh records not from Grains Gill. Given the scatter about the relationship summarized by equation (1), it is quite possible that rainfall at Grains Gill on these occasions was well above 41 mm. Such deviations would be less than one standard error above the best fit regression. One of these periods occurred immediately after a Class B change (September 1970) when the channel may have been particularly unstable. Again both Class D exceptions, above 41 mm, are for periods when the rainfall data are from Sedbergh and the Grains Gill figure could have been less. In each case (51 mm and 43 mm) deviations from the best fit regression of less than one standard error would bring these figures below 41 mm. The other thresholds are a little less clear but the figures 21 mm for D/E and 13 mm for E/F are the 'best fit' estimates. The best available rainfall data for the whole period involve Grains Gill as well as Sedbergh data. Correction has not been made for the different stations and estimates of frequency are based directly on Figure 20.3.

In summary it would appear that conditions suitable for major channel changes, involving erosion of the channel margins occur approximately 1.05 times per year, for minor channel changes approximately 11 times per year, and for movement of individual cobbles roughly 30 times per year, the same frequency as major sediment supply events.

Discussion

The varying event frequency in sediment production and channel behaviour influences the dynamic nature of the relationship between stream and slope processes at Grains Gill. Coarse sediment, eroded from the gullies builds up in debris cones or fans at the base of the gullied slopes until the occurrence of a flood of sufficient magnitude to remove all or part of the debris accumulation and incorporate this material into the mobile bedload of the channel. In doing so it brings about major changes in channel configuration below and downstream from the gully site. Sediment production events occur on average about 30 times per year and sediment removal and major channel changes about once per year. There is consequently a cyclic build up of debris and a periodic removal. This trend can be illustrated by the sequence of cone development and channel change recorded at the base of Gully C over the period 1969–1975 (Figure 20.4).

In 1969 the lower parts of Gully C were mantled by debris which was removed by a flood the following winter. Small cones built up during the summer of 1970 only to be cleared again in September. There was a similar sequence through winter 1970–71 and summer 1971 before a major flood in August 1971. The growth of the main cone or fan over this period is illustrated on Figure 20.5 and characteristically the major period of growth can be seen to be during the winter, reflecting the seasonality of sediment yield totals, as identified in an earlier study (Harvey, 1974). After 1971 three further cycles may be identified with partial or total clearances: these

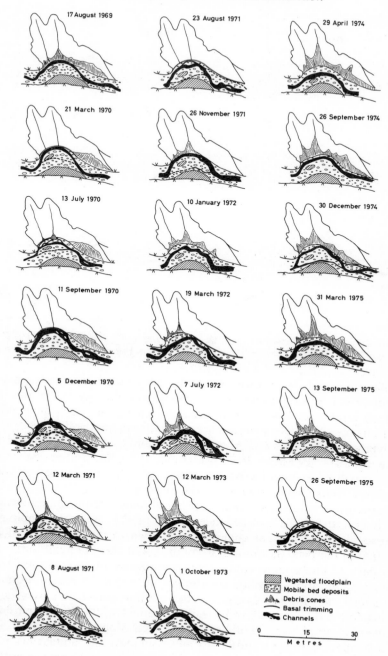

FIGURE 20.4 DEBRIS CONE AND FAN DEVELOPMENT AND CHANNEL CHANGE AT THE BASE OF GULLY C, GRAINS GILL (BASED ON FIELD SURVEY AND PHOTOGRAPHIC EVIDENCE)

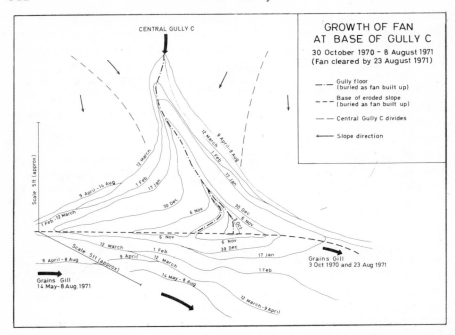

FIGURE 20.5 GROWTH OF FAN AT BASE OF GULLY C. 30 OCTOBER 1970 TO 8 AUGUST, 1971 (PERSPECTIVE DRAWING BASED ON PHOTOGRAPHIC EVIDENCE)

were before March 1972, October 1973, and September 1975. Figure 20.4 also illustrates the major changes in channel morphology at this site. Some of the apparent changes are simply due to changes in stage or at most to minor changes in bar configuration and minor channel shifts (Class D) but the changes illustrated by 21 March, 11 September, and 5 December 1970, 23 August 1971, 19 March 1972, 1 October 1973, and 26 September 1975, are all classified as major changes (Classes A, B, and C), on the basis of changes at Gully C or elsewhere in the study reach.

The occurrence of major floods, capable of removing the debris accumulation, is crucial for the maintenance of the present gully morphology and processes, and the present channel form downstream from the gullies. A reduction in the incidence of major floods would allow gradual burial of the lower parts of the eroding slopes, their colonization by vegetation, and their eventual stabilization. It would also reduce the supply of coarse sediment to the channel allowing stabilization of the existing gravel bars by vegetation colonization and an overall reduction in channel width. It is apparent that the overall morphology, both of the gullied slopes and of the stream channel, appears to be adjusted to approximately the annual event. This situation is similar to that in larger rivers and lowland streams reported on many occasions in the literature (e.g. Leopold, Wolman, and Miller, 1964, p.

319; Gregory and Walling, 1973, pp. 21–24). The detailed morphology, however, is closely related to more frequent events, both on the slopes and in the stream channel, particularly if there has been some time interval since the last major flood.

In this context it is interesting to compare the period of field observation (1969–1976) with the long term period of available rainfall data (1930–1969). The nineteen-seventies have so far been exceptionally dry compared to the previous 40 years. This is evident when both the mean annual total rainfalls and the number of days with totals equalling or exceeding the thresholds identified on Table 20.5 are compared (Table 20.6). In only one year (1974) did the total rainfall reach the long term mean and similarly the incidence of each of the threshold rains recorded at Sedbergh was far below the long term mean.

This comparison is illustrated in Table 20.6 where mean annual number of days with precipitation greater than the thresholds identified earlier are listed for the long term period 1930–1969 and for the study period, 1970–1975. In addition the data for days with precipitation greater than 37 mm have been added to take into account variation in daily rainfall totals between Sedbergh and Grains Gill rain gauges. This figure was arbitrarily selected as a possible Class ABC channel change threshold below which there was only one exception for major channel changes during the period of field observation.

There are some discrepancies between the Sedbergh 1970–1975 data and those for the same period using both Sedbergh and Grains Gill gauges (Table 20.6) but there are marked contrasts between either data sets for 1970–1975 and that for the long term period. This is particularly true for the incidence of low magnitude events, i.e. sediment production and sediment movement events. Of more significance perhaps is the spacing of the major channel-changing events. Between August 1971 and September 1975

TABLE 20.6 COMPARISON OF DAILY RAINFALL DATA FOR SEDBERGH 1930–1969 WITH SEDBERGH 1970–1975 AND WITH SEDBERGH AND GRAINS GILL 1970–1975

		Sedbergh 1930–1969	Sedbergh 1970–1975	Sedbergh with Grains Gill 1970–1975
Mean annual (mm) rainfall		1342 (224)	1198 (150)	—
Mean annual no. of days with rainfall	> 13 mm	32 (7.9)	25.5 (4.7)	26.7 (6.0)
	> 21 mm	11.5 (3.3)	9.3 (1.4)	9.2 (1.3)
	> 37 mm	2.0 (1.4)	0.8 (1.0)	1.2 (0.8)
	> 41 mm	1.3 (1.2)	0.8 (1.0)	1.2 (0.8)

(Standard deviations are given in brackets)

there were no Class A channel-change events and only two major channel-change events (Class B in March 1972 and Class C in August 1973) neither of which caused total debris removal from the base of gullies J, G/H, E/F and C (Figure 20.1). This has meant that many of the slopes have been protected at the base by debris over the whole of the period 1971–1975. Even on gullies G, H, F, and D the incomplete removal of debris by Grains Gill has constrained incision of the gully floor channels and allowed debris to build up at the base of the slopes. Photographic evidence suggests that there has been a marked increase in vegetation cover on the gully slopes over this period, particularly on gullies C, E, and G. Should the present trend continue, especially if major floods become rarer, one might expect the gully slopes to stabilize, sediment yield to be reduced, stream bedload to be reduced, and the channel downstream from the gullies to become more stable. An examination of the long-term records would suggest that such a sequence of events has not previously occurred since 1930, with the possible exception of the mid and late 1950's. There were no intervals of much more than $2\frac{1}{2}$ years between daily rains of greater than 41 mm, suggesting that throughout the period 1930–1969 the morphology and processes of the gullies and stream channels at Grains Gill were maintained in a state of dynamic equilibrium (Hack, 1960), this equilibrium being maintained by events with a return period of approximately 1 year. Whether the apparent departure from that equilibrium over the period 1971–1975 represents a long term trend will depend on the incidence of major storms over the next few years.

Acknowledgements

I am grateful to the University of Liverpool for a research grant enabling the field work to be carried out; to the Meteorological Office, Bracknell for permission to consult their records, and to the staff of the Drawing Office and Photographic sections of the Department of Geography of the University of Liverpool for producing the diagrams and for general photographic work connected with the research.

References

Chorley, R. J. and Kennedy, B. A. 1971. *Physical Geography. A Systems Approach*, Prentice-Hall International Inc., London.
Dury, G. H. 1961. Bankfull discharge: an example of its statistical relationships. *Int. Assoc. Sci. Hydrol. Bull.*, **5**, pp. 48–55.
Gregory, K. J. and Walling, D. E. 1973. *Drainage Basin Form and Process*, Edward Arnold, London.
Hack, J. T. 1960. Interpretation of erosional topography in humid temperate regions. *Amer. J. Sci.*, **258**, pp. 80–97.

Harvey, A. M. 1969. Channel capacity and the adjustment of streams to hydrologic regime. *J. Hydrol.*, **8**, pp. 82–98.

Harvey, A. M. 1974. Gully erosion and sediment yield in the Howgill Fells, Westmorland. Gregory, K. J. and Walling, D. E. (Eds), *Fluvial Processes in Instrumented Watersheds*. I.B.G. Sp. Publ. **6**, pp. 45–58.

Langbein, W. B. 1949. Annual floods and the partial duration series. *Trans. Amer. Geophys. Union*, **30**, pp. 879–881.

Leopold, L. B., Wolman, M. G., and Miller, J. P. 1964. *Fluvial Processes in Geomorphology*. Freeman, San Francisco.

Wolman, M. G. and Miller, J. P. 1960. Magnitude and frequency of forces in geomorphic processes. *J. Geol.*, **68**, pp. 54–74.

Woodyer, K. D. 1968. Bankfull frequency in rivers. *J. Hydrol.*, **6**, pp. 114–142.

J. B. THORNES

Lecturer in Geography
London School of Economics

21

Channel Changes in Ephemeral Streams: Observations, Problems, and Models

With almost one third of the world having arid or semi-arid climates and another third having seasonally concentrated flow, a large proportion of the natural channels of the world flow only periodically or infrequently. These characteristics are exaggerated in most areas by the abstraction of surface and subsurface water for irrigation and domestic consumption, and the fact that ephemeral channels tend to be relatively small. Often these difficulties are linked to high sediment yields resulting from a rather special set of channel hydraulic conditions and a phase of upland erosion.

A notable lead in the investigation of semi-arid channels was undertaken in the United States arising from conflicting opinions about the causes of historical changes in the channels of the South West. In this work three papers, by Leopold and Miller (1956), Schumm and Hadley (1957), and Schumm (1961), stand as landmarks. They emphasized the spatial and temporal variability of ephemeral channel morphology and this work extended to the study of channel changes (Schumm and Lichty, 1963; Schumm and Beathard, 1976).

The development and investigation of the hydrology and hydraulics of ephemeral channels is a rather later development, at least in formal terms. The violently non-uniform, unsteady behaviour of these channels, coupled with the transmission loss problem, has meant that progress has been slow. The uncertainty of flow forecasting arising from the rainfall characteristics and severe practical problems of sedimentation have meant that the instrumentation of channel systems has been remarkably slow to develop compared with that in perennial catchments. These difficulties have not deterred the observation of historic changes nor, indeed, attempts to explain them (Cooke and Reeves, 1976). In recent years two important developments have meant that real progress can now be made, with techniques for solving the equations of unsteady open channel flow (Kibler and Woolhiser, 1970) and the instrumentation of ephemeral streams, notably by the United States Department of Agriculture (Renard, 1970, 1972a).

This chapter begins by illustrating the problems and difficulties of the geomorphology of semi-arid catchments by reference to some studies currently taking place in southeast Spain. This is followed by a more general review of transmission losses, channel erosion and sedimentation and the morphological responses. The available strategies for solving these problems are discussed and finally the application of one of these solutions is outlined.

A Case Observation: Southeast Spain

South of the Sierra Nevada in Spain and to the west of Murcia is an extensive area of low rainfall, high temperatures, and ephemeral channels. Several small catchments in this area were selected for detailed study of morphology, sedimentology, hydrology, and geomorphic processes. The background to these studies is described in Thornes (1976a). The channels occur in a mountainous environment and have slopes between 1 and 10%, runoff occurs very infrequently from the winter rainfall, which is highly variable, but averages about 400 mm. The channels comprise large gully systems in which, characteristically, they show two sections: an upper zone similar to conventional perennial channels with steeper slope and slow increase in width, and a lower zone with rapidly increasing width in a downstream direction, and sharp oscillations in width associated with markedly unstable gravel and sand beds. Inset into the wide sandy beds are low flow channels which deepen, and also oscillate in width and change in character spatially. They disappear and reappear in a downstream direction, reflecting phases of scour and alluviation in a somewhat periodic fashion (Thornes, 1976b). Similar characteristics have been observed by Packard (1974) on alluvial fans in Arizona.

These characteristics raise the question as to whether the features of the valley gravel spreads and those of the inset channels represent essentially different mechanisms having different time bases, and if so how these mechanisms are expected to vary spatially in the channel. A further question would then be how do the temporal relations between forms vary spatially and, in particular, if events of different magnitudes have different relaxation times, what is the effect of these likely to be on channel morphology? This highlights the lead–lag problem in geomorphological systems recently discussed by Chorley and Kennedy (1971) and by Allen (1974). It was suggested that three phases of flow could be recognized: fully integrated flow, in which the whole channel is occupied by flowing water after the fashion of perennial channels; axial flow, in which flow occurs in the main valley axis to progressive distances downstream, and asynchronous tributary flow, in which the tributaries would contribute to a dry main channel (Thornes, 1974). Each of these represent threshold type of behaviour in geomorphological terms as a result of different types of flow and sediment conditions.

Schumm and Lichty (1963), and especially Burkham (1970), observed the important role played in this process by large events, though lately Schumm (1973) has inclined to a cyclical type of explanation in which sediment build up and erosion are mainly responsible for such changes. All three flow types have now been observed in the Spanish channels. In 1973, after the first survey of the smallest channel, a major flood occurred in Spain which has offered an opportunity to observe some of the relationships speculated on above.

Since 1258 there have been over 63 major recorded floods in the area, with those of 1834 and 1879 being particularly severe. The causes of these disasters have been essentially climatic and almost all of them relate to storms at the end of the summer anticyclonic conditions. A map of the intense storm of 18–19 October 1973 is shown in Figure 21.1. The heaviest rains fell over the Almanzora catchment in the east and over the Alpujarras, near Ugijar, in the west. It is in the latter area that the channels described in

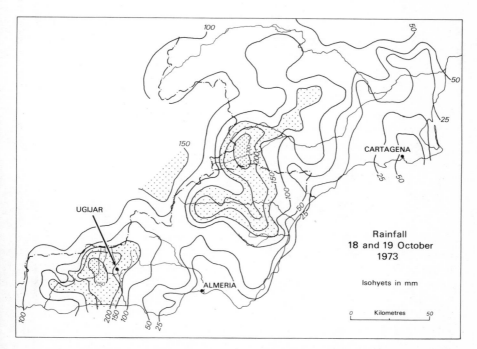

FIGURE 21.1 DISTRIBUTION OF RAINFALL IN THE STORM OF 18–19 OCTOBER, 1973

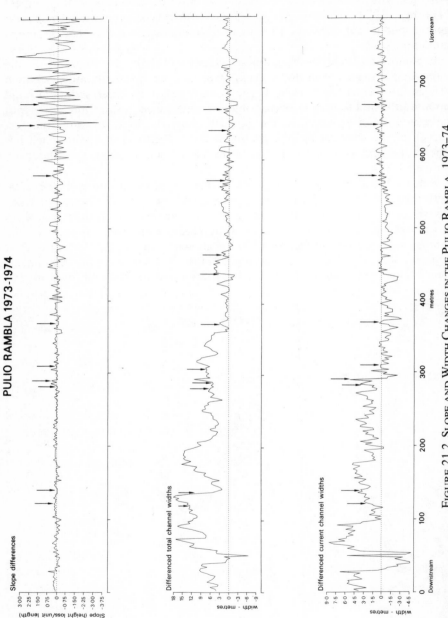

FIGURE 21.2 SLOPE AND WIDTH CHANGES IN THE PULIO RAMBLA, 1973–74

this chapter are to be found. The estimated flow at the local guaging station was 349 m^3/s, twice the expected 100-year flood (Cirugeda, 1973). As with all storm runoff in summer in this area, the hydrograph showed an almost vertical rise and a very sharp recession. Estimates of ground loss from the depth of incision of fresh gullies are from 56–420 mm on the micaschists, with a general estimate of 150 mm (Ruiz de la Torre, 1973).

Figure 21.2 shows changes which occurred in the Pulio channel in terms of slope and width and Figure 21.3 shows the changes in sediment size. In both diagrams upstream is to the right, and the arrows show entry points of

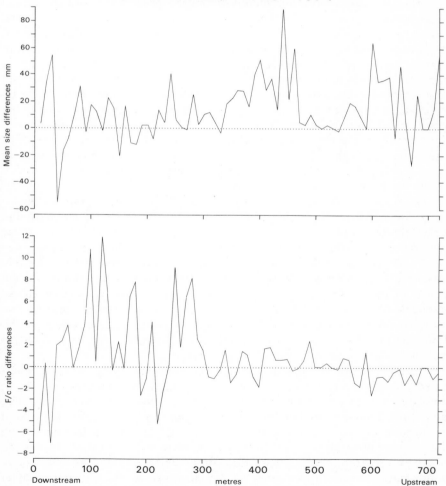

FIGURE 21.3 CHANGES IN MEAN COARSE PARTICLE SIZE AND FINE–COARSE RATIO, PULIO RAMBLA, 1973–74

major tributaries. Changes in slope were remarkably small along the length
of the channel downstream from the point of flood plain onset. Upstream
from this, especially in the headwaters, are violent oscillations representing
valley collapse, aggradation behind boulders, and similar types of feature. In
total width (valley-train) changes, there is a very clearly defined pattern. In
the upper 300 m the changes are relatively small; this is the upper zone,
above the point of alluviation. Below this there is a distinct tendency for the
channel to have widened by increasing amounts downstream. In other
words, valley-train widening has occurred essentially at the widest points
and the degree of widening is proportional to width. This most unusual
situation is a case of positive feedback and the process explains the funnel
shaped character of many of the channels. The current channel widths in the
lowest graph (Figure 21.2) represent the behaviour of incision *after* the
flood event and represent the channel cut by winter storms. An important
feature here is that the changes have been negative in many places. These
locations are essentially where the pre-storm channel has been obliterated
and replaced in the post-storm period, either by a narrower channel, or by
no channel at all. Three zones can be recognized, an upper zone, above the
point of alluviation, in which there is very little change; an intermediate
zone in which the changes are mainly negative and a lower zone in which
oscillations of change can be observed. This lower zone occurs downstream
from three important tributaries, and the large zone of constriction is a
newly built alluvial fan.

Data on sediment size changes is less frequent. The upper graph of Figure
20.3 shows changes in the coarse fraction (>2 mm), which appear sharp and
unsystematic. In the lower diagram the change in the ratio of fine to coarse
material at the surface reveals that in the upper half of the channel there has
been little change. This is hardly surprising, since the material is relatively
coarse in both periods. The rapid shifting between adjacent stations (10 m
apart) in the lower half of the channel suggests the movement of coarse
unevenly spread bars of the pre-flood period and their replacement by fine
material during and after the flood. The two large negative values at the
lower end of the system can be specifically related to large bars of coarse
material which were relocated during the storm.

These observations suggest that the whole system operates in an integrated manner during flood, with the main channel morphology related to,
coextensive with, and controlled by, flow there and in the tributaries. The
end of the flood was marked by extensive deposition of a near plane bed and
restoration of the channel to its original slope. The width pattern of the
valley train must largely be attributed to scour and deposition during the
flood event. From the point of alluviation there has been an increase in
width more-or-less proportional to the prestorm width. Current channel
widths are formed by asynchronous flows in the tributaries and the main
channels. In the middle reaches, scour and deposition by main channel flows

leave an oscillating current channel width which is smaller than that existing prior to the storm. In the lowest portion there has been a widening of the pre-storm width, though again some oscillatory behaviour is observed. This is also thought to represent scour and fill deposition sequences which are spatially distributed.

It is implicit in what has been said that (i) channel morphology overall is a function of the major flood; (ii) detailed morphology reflects the history of scour and fill since the flood, and (iii) this history is spatially variable so that morphology in one part of the channel is the response to a different history of flow events from that in another part of the channel. This problem seems rather specific to ephemeral channels. A similar sequence of events has been observed in the area again by my colleagues, Mr G. C. Butcher and Miss H. Scoging, but the changes are not yet fully evaluated. In most channels, and especially much larger ones, the simple sequence observed here is complicated by many further large events which, though yielding integrated flow do not obscure the whole channel. Also, if there is an upstream migration of the wide-sections, which is required by the positive feedback mechanism, the pattern of stream power availability will itself change.

Controls of Runoff, Sediment Movement, and Morphology

The stream power available for sediment transport fluctuates with the discharge through its relationship with velocity. Understanding the volume and characteristics of runoff events is a prerequisite to a full appreciation of the morphology of ephemeral channels. This problem involves generation of the flows and their translation through the channel system.

Runoff

In terms of flow generation, two features appear to be particularly important. First, precipitation is sporadic in time and is often highly concentrated in space. It is usually relatively intense compared with that in perennial areas. For example, in the Ugijar area the distribution of duration is steeply negative exponential with very few cases of rain being recorded on more than two successive days. Second, the strong seasonal variations in infiltration rates add to the complexity of rainfall effects because they affect surface runoff (Schumm and Lusby 1963) and the development of seasonal crusts is particularly important in this respect. As a result of these controls, the specific yield (runoff/unit area) is highly variable, both from year to year and between storms within the year; the time to rise of hydrographs is characteristically very short, and runoff may be generated over quite small areas, so that tributary inflow may occur when the rest of the system is quite dry.

In semi-arid channels, particularly with small watershed areas specific yields relate extremely well to precipitation. Osborn and Lane (1969), and Osborn, Lane, and Kagen (1971) have demonstrated these relationships and found the following relationship between peak discharge, Q in acre feet, and the 15-minute peak rainfall (P).

$$Q = 2.8P_{15} - 0.53$$

For larger watersheds (above 15.54 km^2) the relationship was changed only by allowance for a fixed storage. Both expressions yield low standard errors and high correlation coefficients. The inferences from regression in this context seem reasonable in view of the independence of the events. The implications are that antecedent soil moisture plays a relatively unimportant role, and this has been the source of recent debate. Chery (1972) shows conclusively that for the tested watershed, runoff volumes, ratios of runoff volumes to rainfall depths, and ratio of peak flow rate to 5- and 15-minutes rainfall rates were greater when the initial state was moist.

Transmission Losses

Water flow in the channels is characterized by the passage of well defined peaks, often of only a few hours duration in the summer season, separated by long periods without flow. In winter, storm flows become more attenuated and there may be base-flow for short periods of time. Because waves of water pass through the channel quickly and recharge is relatively low in both the channels and on the hillslopes, the piezometric surface below the channel is often at some depth, especially where the thickness of sediments is very deep. The channel may be likened to a reservoir (Figure 21.4) overlain by a perforated lid which allows infiltration. Infiltration into this reservoir is called transmission loss and is usually flux controlled. If the discharge inflow (both upstream and laterally) is less than the total transmission loss, then there will be no flow out of the end of the reach, so it is geomorphologically quite important to determine the magnitude of and controls on, these losses. Moreover, since many routing models rely on continuity, transmission losses are highly significant for these models.

Direct observations of transmission losses reveal that the volumes may be quite large. Keppel and Renard (1962) for example showed from measurements in successive flumes that over a 6.4 km reach 50% of the maximum discharge was lost and the total flow volume was reduced by 35%. Turner and coworkers (1943) stated that infiltration losses of 75% in 15–25 miles were common, and even in perennial channels a study of 57 flood events on 18 rivers in the Great Plains revealed an average loss of 40% on an average channel length of 53 miles, with as much as 75% of the flood volume being lost. Several workers have attempted to relate these losses to the velocity of the flowing stream, arguing that low velocities would allow the precipitation

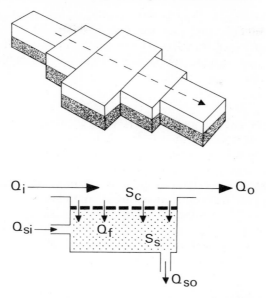

FIGURE 21.4 SCHEMATIC REPRESENTATION OF CHANNEL STRUCTURE WITH DEPTH
IN TERMS OF WATER DISTRIBUTION

The shaded area represents the bed material. S_c and S_s are subsurface and channel
storage. Q_i, Q_{si} are inflow to this storage and Q_o and Q_{so} the outflows. Q_f is
infiltration. The arrow denotes downstream

of fines at the surface or in the upper few centimetres (Matlock, 1965).
Others have examined the role of bed materials. Behnke and Schiff (1963)
investigated the hydraulic conductivity of sands with size ranges found in
alluvial fans and reported that the velocity and hydraulic conductivity
increased linearly with grain size for the range 0.1–1.0 mm in uniform sands.
The position and thickness of layers of low hydraulic conductivity significantly influence the velocity, hydraulic gradient and overall hydraulic conductivity. Matlock (1965) observed that because of infiltration losses there is a
point of no flow which fluctuated over a mile or more of channel according
to the daily flow cycle. He found that a silt layer 1.5–3 mm thick reduced
infiltration by 80%, while a 75–101 mm thick silt layer had no measurable
infiltration over a 4-hour period. In our own experiments in the Pulio
Rambla, sustained very high rates were a simple direct function of available
head and showed no decrease with time.

The relative importance of transmission losses is obviously related to the
inflow volume, at least up to some fixed discharge because of the stage
discharge characteristics of the cross-sectional form. These channels have
very high width–depth ratios, so that, for a plane bed channel, small
increases in depth yield rapid increases in width and consequent increases in

transmission loss. Where the current channel is inset, there may be a sharp break in the increased losses as the flow spreads out onto the floodplain. Murphey, Diskin, and Lane (1972) were able to demonstrate a high correlation ($r^2 = 0.94$) of transmission losses with the volume of the inflow hydrograph of the form:

$$L = 0.36V + 2.29$$

with V the inflow volume and L the yield in cfs/acre. The standard error of the estimate of this equation was found to be 0.29 cfs/acre. If loss is a function of inflow volume, without further lateral inflow down a reach, total volume will fall exponentially. This idea is expressed in Figure 21.5 which attempts to express the relationships in a diagrammatic manner. The oscillatory behaviour of transmission losses is intended to express the overall increase of losses in relation to width. For the smaller storm this curve should move down to reflect the smaller cross-sectional area. The concept of survival of flows between tributaries as a function of specific yield, transmission losses and yield is also made clear from the diagram.

Scour and Deposition

At the foot of Figure 21.5 are sketched expected locations and relative amounts of erosion and deposition for the lower of the two flows. This diagram does not try to illustrate the effects of systematic changes in bed-material size or channel roughness. Because the channels examined have only a very mild degree of concavity at most and normally have uniform slopes, it is thought that width reponses are most important with the passage of flood wave. This is partly a reflection of the fact that it is virtually impossible to observe bedform changes under flood conditions, and the residual form most easily measured is width. It also reflects, however, the generally low stability of the banks of these ephemeral channels. The pattern of widths is essentially a response to the passage of a flow event. As the channel widens the depth of flow increases at roughly $\frac{2}{3}$ the power of the width ratios, infiltration rates increase, and particle shear increases and then decreases leading to deposition as the flow diminishes. Because the observed channels have rapid fluctuations in width at low flows as well as high flows, the role of transmission losses in this process must be important. Sometimes, usually in winter but also in summer, the recharge actually resurges down channel in the form of a spring and although this cannot account for periodic pulsing, it may explain some of the scour and sedimentation pattern. Schumm (1961) observed also that the pattern of aggradation was non-uniform and stated that 'aggradation in these channels is associated with segments of the stream that receive only small contributions from tributaries.' He later says that 'deposition is occurring in reaches where the increase in drainage area is much less per unit length of channel than on

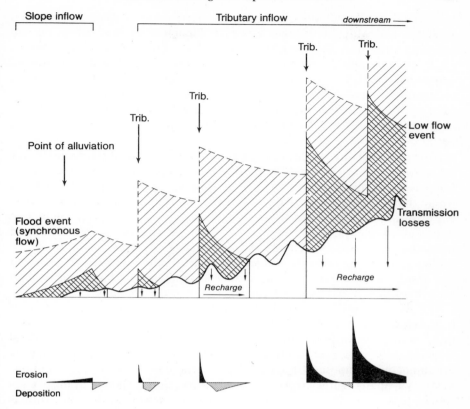

FIGURE 21.5 SCHEMATIC REPRESENTATION OF TRANSMISSION LOSSES AND ITS RELA-
TION TO CHANNEL FLOW AND EROSIONAL PROCESSES FOR STORMS OF TWO DIFFER-
ENT MAGNITUDES. SHADED AREA REPRESENTS DISCHARGE

stable reaches.' It might, of course, be argued that in the small channels referred to here the whole system is dynamically unstable by virtue of size (Bennett, 1976), but it seems more likely that ephemeral channels are inherently unstable.

Not only does the pattern of erosion and deposition present problems. The changes in sediment concentration with discharge are conditioned by the high availability of fine materials on the bed with each new flow, a circumstance uncommon to perennial flows. Consequently the hysteresis effects commence with high concentrations which are sustained until discharge begins to fall resulting in a '7' shape (Renard, 1972b). The result is that sediment movement is in a series of short jerks and the conventional pool and riffle sequence may be replaced by a horizontal sequencing of coarse and fine sediments. The same conclusion was reached by Leopold, Wolman, and Miller (1964). It is a fact that quite shallow flows can still move material of a diameter twice that of the flow (Leopold and Miller, 1956) and this must in part be attributable to the very high concentrations of suspended solids. Leopold and Miller (1956) found a systematic increase in concentration in a downstream direction which they considered to be related to transmission losses and which they thought would be offset by the observed increase in velocity downstream compared with perennial channels.

A further difficulty is that the coarse bedforms of ephemeral channels may be somewhat more complex than those occurring in perennial channels. Movement of specific 'bars' has been observed in the Spanish channels where 'bars' means low coarse bedforms in the channel. Some of these bars are artifically created when tributary inflows deposit very large volumes of sediment in the main channel as a result of changes of gradient and transmission losses. Sometimes the sequencing of flows can actually be evaluated from the deposits. Under low flows, these bars become dissected and yield low flow anastomosing channels, separated by lag deposits; with high flows they move as kinematic waves down the channel.

Morphology itself exerts an important role. In perennial channels there is a constant, time-continuous, adjustment between form and flow. Even with flood waves the response to discharge can be fairly rapid. In ephemeral channels a flow of completely different magnitude and frequency may be instantaneously imposed on an existing morphology. Instead of a continuously evolving, slightly lagged, highly autocorrelated system, transient response to a range of discrete pulses of varying intensity is the rule. How is this view of the system consistent with Leopold and Miller's (1956) view that ephemeral streams conform to the classical relationships established by Leopold and Maddock (1953)? The answer may lie in the scales, both spatial and temporal, at which the systems are observed. In the long-term, morphology is conditioned by a series of integrated flows; at the large scale or in the topological sense of the classical hydraulic geometry, the high frequency transient response is avoided. Hydraulic geometry is a low-pass filter, which explains its great unifying character, but also helps to account for the difficulties in coming to grips with a process explanation of the phenomena. What is needed is an explanatory model of ephemeral channel morphology which is consistant at both levels. In the last section of this chapter some of the work which moves in this direction is outlined.

Dynamic Models for Ephemeral Channel Changes

In the last decade the extensive development of numerical solutions to the unsteady flow equations of open channel flow has led to the adoption of much more appropriate, but also technically much more sophisticated, models for channel routing (Mamhood and Yevjevich, 1976). The application of such techniques to the ephemeral channel-flow problem, although still in its infancy, has already made some progress. This work, developed originally from that of Wooding (1965), by Smith (1972) seeks to describe the hydrological response of a semi-arid watershed by a kinematic wave which is flowing over an initially dry infiltrating plane. The interaction of surface flow and infiltration loss is considered to be a function of time since wetting (opportunity time). This function is combined with the continuity equation for unsteady open channel flow and the rating equation, and the whole solved numerically. The wavefront travels as a kinematic shock, and point infiltration is a function of time since the shock passed the point and so is dependent on time and the shock velocity. The infiltration function is essentially the Kostiakov formula. The main problems are in specifying the complex geometry of a real catchment in a manner which can be satisfied in computational terms at the level of resolution desired. For the oscillatory behaviour of the Spanish channels this of the order of 200-m reaches. Some success has already been achieved by Butcher using these techniques. Smith (1976) has followed Bennett (1974) in combining the (non-infiltrating) model with a continuity equation and sediment rating equation to obtain a deterministic distributed watershed model of erosion dynamics. In this case the model is far ahead of the data which could conceivably parameterize it, with the consequence that the results are at the will of the author.

Because of the uncertainty in runoff relationships several attempts have been made to simulate runoff, sediment yield or both. At the simplest level the runoff series is generated from an empirical function and related to an empirical sediment rating function. In the area of probablistic sediment modelling, Woolhiser and Todorovic (1974) have developed a stochastic model using a counting process in which sediment yield is treated as a random number of random events. Smith, Fogel, and Duckstein (1974) took the basic Universal Soil Loss Equation, and using a joint distribution of rainfall and storm duration, obtained the cumulative distribution of sediment yield for a single event. This work parallels that of Duckstein, Fogel, and Kesiel (1972) in that the seasonal sediment yield is computed as the sum of a random number of random events mutually independent and identically distributed. From the computed values of mean sediment yield and its variance, along with the mean and variance of the number of events per season, the mean and variance of the seasonal distribution may be computed directly. Unfortunately, most of these models are 'lumped' in the sense of taking no account of spatial variability. Lane (1972) has a distributed model for runoff in which transmission losses are considered to be a function of channel storage, but this has to be estimated from inflow–outflow data and assumes a considerable historic record availability.

It was argued earlier that the morphology of erosion is related to the unit stream power of surviving flows particularly as they change with no lateral inflow. Figure 21.5 suggests, for low flows, that scour and deposition is related to some flow criterion which in turn is related to transmission losses. If one supposes an average spacing of input tributaries (which may increase downstream), for any part of the channel a surviving flow is one which flows for longer than the average tributary spacing. Area increase in a downstream direction along the main channel conforms to Hack's (1957) area increment function, which is expressed by

$$l = pA^{0.6} \tag{1}$$

where L is the distance down the main channel and A the area upstream from that point. Corresponding exponents for the Ugijar and Cherin channels are 0.8 and 0.6 respectively. Assuming that discharge is a linear function of area, i.e.

$$Q = gA^{\alpha} \tag{2}$$

where α is unity and Q discharge and so g is the gain or specific yield in volume per unit area per second. Combining these two equations yields discharge as a function of distance down the main channel, though in fact the increase in area is a strongly stepped function (Thornes, 1976a; Horton, 1976). Moreover, this relationship is essentially over integrated channel flow,

$$Q = g(l/p)^{1.66} \tag{3}$$

Now Burkham (1970) expresses channel infiltration/unit width (I) together with width B in the continuity equation

$$\frac{\delta Q}{\delta L} + IB = B\frac{\delta D}{\delta t} \tag{4}$$

where D is depth of flow and L the length down a reach. To obtain a manageable solution, he has to assume steady flow

$$\frac{\delta D}{\delta t} = 0 \tag{5}$$

which leads to

$$\frac{\delta Q}{\delta L} = -IB \tag{6}$$

Substituting into this the hydraulic geometry relations of Leopold and Maddock (1953) and by integration and manipulation he obtains the expression

$$Q_{out} = Q_{in}^m - macC_lL \tag{7}$$

where m, a, and c are the coefficients in the expressions $B = aQ^b$, $D = cQ^f$ and $v = KQ^m$, the hydraulic geometry equations; C_l is the relationship of infiltration to depth in his key transmission loss equation

$$I = C_l D \qquad (8)$$

By further manipulation, we may obtain the following expression, which combines Burkham's equation (7) with equation (3) and assumes $Q_{out} = 0$;

$$L = \frac{g^m}{macC_l} \cdot \left(\frac{l}{p}\right)^{1.66m} \qquad (9)$$

This equation operates in an interesting fashion to yield for a given distance downstream (and hence basin area), l, and a specific yield (g) a survival length L for the flow. If at any distance L is greater than the average tributary spacing, the axial flow will survive. Integrated flow occurs when L is sufficiently large at each point to sustain flow to the next tributary. The equation also indicates the survival length for tributary inflow.

From Figure 21.6, survival length is seen to increase very rapidly for a small increase in specific yield; as g varies survival length increases and the rate of increase is greater downstream. From a morphological point of view, a more distinct morphology will be produced the sharper the distribution of

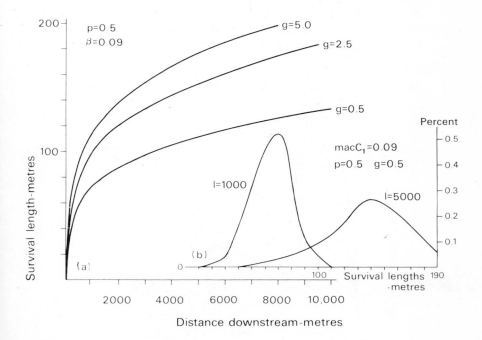

FIGURE 21.6 (a) SURVIVAL LENGTH AS A FUNCTION OF SPECIFIC YIELD (g) AND DISTANCE DOWN CHANNEL (b) INSET SHOWS THE INFLUENCE OF CATCHMENT SIZE ON THE DISTRIBUTION OF SURVIVAL LENGTHS ($\alpha = 0.8$)

L. If we assume that specific yield has a negative exponential distribution
with parameter λ, i.e.

$$f(g) = \lambda e^{-\lambda g} \tag{10}$$

then by expressing *g* as a function of *L* and taking the Jacobian, we obtain
for the distribution of *L*

$$f(L) = \left[\frac{1}{m}\left(\frac{L}{\beta}\right)^{1-m/m}\right] \cdot \left[\lambda \, \exp(-\lambda L^{1/m}/\beta)\right] \tag{11}$$

in which

$$\beta = \frac{1}{macC_l}\left(\frac{l}{p}\right)^{1.66m} \tag{12}$$

The distribution for two basin sizes is shown in Figure 21.6b. In the
headwaters the survival lengths are more strongly peaked, and generally in
this distribution have a steeper 'recession' limb, i.e. for headwater flows the
boundary of all flows is likely to be quite sharp. Further downstream flows
are more dependent on specific yield variations, as Figure 21.6a indicates.

Equation 9 is very sensitive to *m* and C_l. In increasing width more slowly
downstream than the classical values for semi-arid channels, but keeping
depth constant, *m* increases. The result is to produce a steep rise in the
length of survival. Likewise by reducing the term C_l, the infiltration coefficient, the survival length increases rapidly. These two effects are illustrated

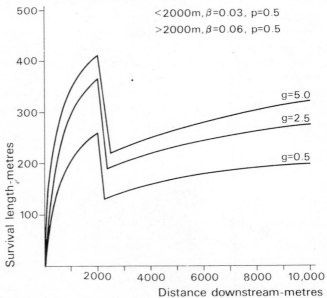

FIGURE 21.7 THE EFFECT OF ONSET OF ALLUVIATION ON THE DISTRIBUTION OF
SURVIVAL LENGTHS ($\alpha = 0.8$)

in Figure 21.7. A hypothetical increase in width and infiltration rate at the point of alluviation dramatically shortens the survival lengths.

This development is intended to indicate some possible directions for investigation of the impact of transmission losses on channel behaviour. The perennial channel analogue of transmission losses is attenuation of the hydrograph, though this also occurs in semi-arid channels. For different specific yields in perennial channels, the channel-forming discharges vary spatially through the area increment function (Thornes, 1974) though these may also be expressed through the properties of channel network (Kirkby, 1976). A further step is to obtain spatially varying discharge volumes in semi-arid channels as a function of network geometry, and some attempts along these lines have been made by Renard and Lane (1975). Unfortunately, their analysis is based on the Strahler system of stream ordering: the area-increment function or geometric path length would perhaps provide better bases.

Conclusion

In outlining some of the principle controls on the complex morphology of dynamically unstable ephemeral channels an attempt has been made to move away from the large time and space scale implied by hydraulic geometry back to the more localized process-oriented scale studied by Schumm (1961). The problems will ultimately be solved by the type of model described by Smith (1976) but at the moment this is a distant prospect. Because the spatial effects of time-generated flows are important in understanding the geomorphology on a historical basis, a model which reproduces these space–time effects is required. Burkham's (1970) model solved for outflow volumes and survival lengths in the manner described above seems to offer some interesting possibilities. If linked to either a formalized, or empirical area-increment function or to the type of network generator described by Kirkby (1976), it could be highly suited to the problem in hand.

References

Allen, J. R. L. 1974. Reaction, relaxation and lag in natural sedimentary systems: general principles, examples and lessons. *Earth Sci. Rev.*, **10**, pp. 263–342.

Behnke, J. J. and Schiff, L. 1963. Hydraulic conductivity of uniform, stratified and mixed sands. *J. Geophys. Res.*, **68**, pp. 4769–4775.

Bennett, J. P. 1974. Concepts of mathematical modelling of sediment yield. *Water Resour. Res.*, **10**, pp. 485–492.

Bennett, R. J. 1976. Adaptive channel geometry. *Earth Surface Processes*, **1**, pp. 131–150.

Burkham, D. E. 1970. A method for relating infiltration rates to stream-flow rates in perched streams. *U.S. Geol. Surv. Prof. Paper 700-D*, pp. D266–D271.

Chery, D. L. 1972. Significance of antecedent soil moisture to a semi-arid watershed rainfall–runoff relation. *Hydrology and Water Resources in Arizona.* Ariz. Acad. Sciences, Vol. 2, pp. 385–406.

Chorley, R. J. and Kennedy, B. A. 1971. *Physical Geography, A Systems Approach,* Prentice-Hall, London.

Cirugeda, J. 1973. *Informe relativé las crecidas de Octubre de 1973 en el Sureste: Estudio de caudales.* Centro de Estudios Hidrograficos, Madrid, Spain.

Cooke, R. U. and Reeves, R. W. 1976. *Arroyos and Environmental Change in the American Southwest.* Oxford University Press.

Duckstein, L., Fogel, M., and Kesiel, C. C. 1972. A stochastic model of runoff-producing rainfall for summer type storms. *Water Resour. Res.* **8,** pp. 410–421.

Hack, J. T. 1957. Studies of longitudinal stream profiles in Virginia and Maryland. *U.S. Geol. Surv. Prof. Paper 294-B.*

Horton, M. 1976. *A study of Area Distributions in Small Drainage Basins.* Unpublished B.Sc. dissertation, Geography Department, King's College, London.

Keppel, R. V. and Renard, K. G. 1962. Transmission losses in ephemeral stream beds. *J. Hydraul. Div. Amer. Soc. Civ. Engrs.,* **88,** pp. 59–68.

Kibler, D. F. and Woolhiser, D. A. 1970. The kinematic cascade as a hydrologic model. Colorado State University, *Hydrology Papers No. 39.*

Kirkby, M. J. 1976. Tests of the random network model and its application to basin hydrology. *Earth Surface Processes,* **1,** pp. 197–213.

Lane, L. J. 1972. A proposed model for flood routing in abstracting ephemeral channels. Arizona Academy of Science, Hydrology Section, *Proceedings May 5–6 Meeting,* Prescott, Arizona, II, pp. 439–453.

Leopold, L. B. and Maddock, T. 1953. The hydraulic geometry of stream channels and some physiographic implications. *U.S. Geol. Surv. Prof. Paper 252.*

Leopold, L. B. and Miller, J. P. 1956. Ephemeral streams—Hydraulic factors and their relation to the drainage net. *U.S. Geol. Surv. Prof. Paper 282-A.*

Leopold, L. B., Wolman, M. G., and Miller, H. P. 1964. *Fluvial Processes in Geomorphology.* Freeman & Co., San Francisco.

Mamhood, K. and Yevjevich, V. (Eds). 1976. *Unsteady Open Channel Flow.* 3 vols. Water Resources Publications. Ft. Collins.

Matlock, W. G. 1965. *The Effect of Silt-Laden Water on Infiltration in Alluvial Channels.* Ph.D. Dissertation, Univ. Arizona, Tucson, Arizona.

Murphy, J. B., Diskin, M. H., and Lane, L. J. 1972. Bed material characteristics and transmission losses in an ephemeral stream. Arizona Academy of Sciences, Hydrology Section, *Proceedings May 5–6 meeting,* Prescott, Arizona, II, pp. 455–472.

Osborn, H. B. and Lane, L. 1969. Precipitation-runoff relations for very small semi-arid rangelands watersheds. *Water Resour. Res.,* **5,** pp. 419–425.

Osborn, H. B., Lane, L. J., and Kagan, R. S. 1971. Determining significance and precision of estimated parameters for runoff from semi-arid watersheds. *Water Resources Bull.,* **7,** pp. 484–494.

Packard, F. A. 1974. *The Hydraulic Geometry of a Discontinuous Ephemeral Stream on a Bajada near Tucson, Arizona.* Ph.D. dissert., Tucson, Arizona Univ.

Renard, K. G. 1970. The hydrology of semi-arid rangeland watersheds. United States *Dept. of Agric. Public. ARS 41–162,*

Renard, K. G. 1972a. Sediment problems in the arid and semi-arid South-west. *Proceedings of the 27th Annual Meeting of the Soil Conservation Society of America,* August 6–9, Portland, Oregon.

Renard, K. G. 1972b. *The Dynamic Structure of Ephemeral Streams.* Ph.D. Dissertation, Univ. Arizona, Tucson, Arizona.

Renard, K. G. and Laursen, E. M. 1975. Dynamic behaviour model of ephemeral streams. *J. Hydraul. Div. Amer. Soc. Civ. Engrs.*, **101**, pp. 511–528.

Renard, K. G. and Lane, L. J. 1975. Sediment yield as related to a stochastic model of ephemeral runoff. Proceedings of Sediment Yield Workshop, *United States Department of Agriculture Public. ARS–S–40*, pp. 253–263.

Ruiz de la Torre, J. 1973. *Informe sobre los effectos de la lluvias de los dias 18 y 19 de Octubre 1973 en el sureste de Espana, desde putito de vista sedimentologico.* Centre de Estudios Hidrograficos, Madrid, Spain.

Schumm, S. A. 1961. Effect of sedimentation on erosion and deposition in ephemeral stream channels. United States Geol. Survey, *Prof. Paper 352-C.*

Schumm, S. A. 1973. Geomorphic thresholds and complex response of drainage systems. *Fluvial Geomorphology*, Ed. Marie Morisawa, State University of New York, Binghamton., pp. 299–310.

Schumm, S. A. and Beathard, R. M. 1976. Geomorphic thresholds: an approach to river management. *Rivers 76*. Symposium on Inland Waterways for Navigation, Flood Control and Water Diversions. *Amer. Soc. Civ. Eng. V.I.*, pp. 707–724.

Schumm, S. A. and Hadley, R. F. 1957. Arroyos and the semi-arid cycle of erosion. *Amer. J. Sci.*, **255**, pp. 161–174.

Schumm, S. A. and Lichty, R. W. 1963. Channel widening and flood-plain construction along Cimarron River in Southwestern Kansas. *U.S. Geol. Surv. Prof. Paper 352-D.*

Schumm, S. A. and Lusby, G. C. 1963. Seasonal variations of infiltration capacity and runoff on hillslopes in Western Colorado. *J. Geophys. Res.*, **68** (12), pp. 3655–3666.

Smith, J. H., Fogel, M., and Duckstein, L. 1974. Uncertainty in sediment yield from a semi-arid watershed. *Proceedings 18th Annual Meeting, Arizona Academy of Sciences*, Flagstoff, Arizona, April.

Smith, R. E. 1972. Border irrigation advance and ephemeral flood waves. *Proc. Amer. Soc. Civ. Eng., Jo. Irrig. Div.*, **98**.

Smith, R. E. 1976. Simulating erosion dynamics with a deterministic distributed watershed model. *Third Interagency Conference*, Denver, Colorado, March. Vol. 1, pp. 163–173.

Thornes, J. B. 1974. *Speculations on the behaviour of stream channel width.* London School of Economics, Department of Geography, *Discussion Paper No. 49.*

Thornes, J. B. 1976a. Semi-Arid Erosional Systems: Case studies from Spain. London School of Economics, *Geography Department Papers No. 7.*

Thornes, J. B. 1976b. Autogeometry of semi-arid channel systems. *Rivers 76.* Symposium on Inland Waterways for Navigation, Flood Control and Water Diversions. *Amer. Soc. Civ. Eng.* Vol. II, pp. 1715–1726.

Turner, S. F. *et al.* 1943. Groundwater resources of the Santa Cruz Basin, Arizona. United States Geol. Survey *Report*, Tucson, Arizona, pp. 35–53.

Wooding, R. A. 1965. A hydraulic model for the catchment-stream problems, II. Numerical solutions. *J. Hydrol.*, **3**, pp. 268–282.

Woolhiser, D. A. and Todorovic, P. 1974. A stochastic model of sediment yield for ephemeral streams. Proceedings U.S.D.A.-IASPS Symposium on Statistical Hydrology, Tucson, Arizona, U.S. Department of Agriculture, *Misc. Publications No. 1275.*

S. J. RILEY

Lecturer in School of Earth Sciences
Macquarie University, New South Wales

22

Some Downstream Trends in the Hydraulic, Geometric, and Sedimentary Characteristics of an Inland Distributary System

Inland distributary channel networks have largely been ignored in the literature on channel systems. There are few detailed studies of these systems, yet they constitute important systems in terms of their occurrence and their frequent use in semi-arid and arid agricultural projects.

The location of many of the larger distributary systems in areas with low population and poor access has resulted in a dearth of reliable hydraulic and hydrologic data for these systems. While river heights have been monitored for long periods at a few sites on distributaries in eastern Australia, many of these sites have ill-defined rating curves, and information on sediment load, water quality, channel morphometry, and sedimentology is meagre and in most cases, non-existent. These deficiencies are being corrected at present, but it may be several years before information is available. In the meantime, the increasing utilization of distributary systems for agriculture demands some investigation of their character and behaviour using the information that is available or can be readily obtained.

This paper describes the downstream hydraulic, morphologic, and sedimentary changes that occur in several streams of the Namoi-Gwydir distributary system of eastern Australia. These changes are examined in order to assess the nature of some of the processes and behavioural characteristics of the Namoi–Gwydir system.

It has been recognized for a long time that distributary systems are a particular class of channel pattern (Lane, 1957). The author has discussed elsewhere the implications of the various terms used to describe distributary systems (Riley, 1975). The basic feature that defines a distributary network is that it has bifurcating channels that persist relatively independent of the parent stream, the stream from which they offtake, for some length far in excess of their width. However, they may rejoin the parent channel or each other. The Namoi–Gwydir distributary system conforms to the author's definition.

The Namoi–Gwydir Distributary Network

The lower section of the Namoi and Gwydir Rivers (Figure 22.1) is one of several distributary systems in eastern Australia that drain westwards from the Great Dividing Range. The Namoi–Gwydir system was selected for this study for several reasons including the presence of distinct north and south boundaries that preclude any possible confusion that may arise from over-

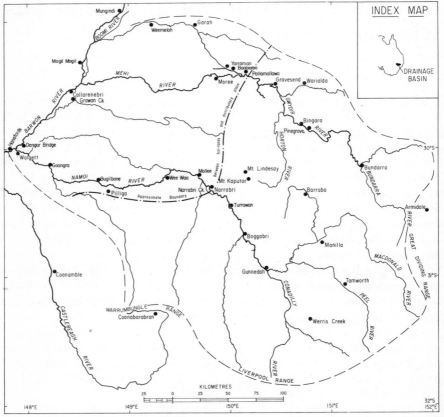

FIGURE 22.1 THE NAMOI–GWYDIR DRAINAGE BASIN AND LOCATION OF DISTRIBUT-
ARY SYSTEM

lapping networks; the several gauging stations with long periods of records and reliable rating curves, and the fact that there is a minimal man-made interference with the channel network. Furthermore the streams are easily accessible, and the fact that the system is one of the smallest distributary systems in eastern Australia facilitated its study.

The Namoi–Gwydir distributary system commences at a well defined north–south topographic boundary slightly east of Moree and Narrabri (Figure 22.1) the plains to the west are of low relief and have a westward gradient, as shallow as 1×10^{-4}. Annual average rainfall varies between 585 mm in the east to 480 mm in the west.

The majority of inputs into the distributaries are received from the eastern highlands. Local storms appear to replenish depression and soil storage but do not usually contribute significantly to runoff. The Namoi and Gwydir streams have daily flows usually of the order of 1–50 m³/s, but droughts are not unknown. The small streams are characterized by long periods of zero flow (Figure 22.2). Most flow in the small distributary streams is related to high stages of flow in the parent streams.

Four channels representing a selection of the major and minor streams in the study area have been selected in order to examine downstream changes

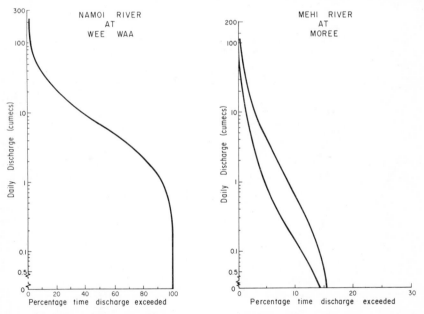

FIGURE 22.2 DURATION OF DAILY FLOWS AT WEE WAA, BASED ON PERIOD 1.12.1956 TO 31.3.1971, AND MOREE

Curve A based on period 1.1.1943 to 31.5.1956 and curve B on period 1.2.1964 to 30.3.1970

S. J. Riley

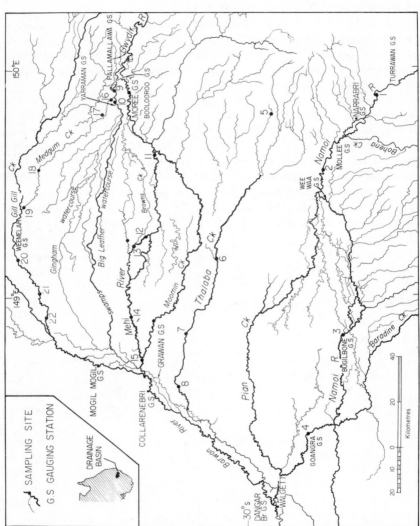

FIGURE 22.3 THE NAMOI-GWYDIR DISTRIBUTARY SYSTEM AND LOCATION OF GAUGING STATIONS AND STREAM SITES

in hydraulic, morphometric, and sedimentary parameters. The *Namoi River* is a major parent stream in the system and exists as a continuous channel from the eastern highlands to the Barwon River (Figure 22.3). Many distributaries offtake along its course, particularly in the area between Narrabri and Wee Waa. *Thalaba Creek* is a small stream in the system. While it is interrupted by swamps, it can be continuously traced from the eastern highlands to the Barwon. Few streams offtake from Thalaba Creek while several distributaries join it. The *Mehi River* is a major distributary of the Gwydir River and represents the main continuous channel between the Gwydir and Barwon Rivers. *Gil Gil Creek* is a distributary of the Gwydir River, but differs from the Mehi in that several of its tributaries arise in the eastern hill lands and are not distributaries. It is smaller than the Mehi in terms of cross-sectional areas and bankfull discharge. Together the four streams encompass a wide variety of sizes and distributing characteristics and give a comprehensive view of the types of downstream changes that are found in the Namoi–Gwydir distributary system.

Several sampling sites were randomly selected along each stream, although accessibility did have an influence on the exact location of each site. Sites were selected near gauging stations (Figure 22.3). At each site several cross-sections were surveyed with the A-Frame (Riley, 1969) and the bed profile was surveyed by dumpy level at all but four sites. Grab samples of bed sediment were taken at 18 sites. The parameters selected in order to study downstream changes (Table 22.1) are considered by the author to be those whose trends will suggest trends in the dominant processes.

Hydraulic Changes

Several gauging stations are situated on the Namoi River between the eastern hill lands and the Barwon River (Figure 22.3). The Gwydir, on the other hand, probably because of the extent to which it distributes, is poorly gauged on the plain. Nevertheless, several common trends may be noticed along the two streams.

The 1.58 and 2.33 year discharges decline downstream, probably as a result of removal of water from the parent channels by distributaries and by seepage into aquifers (Figure 22.4). The ratio of $Q_{2.33}$ to $Q_{1.58}$ also decreases across the plain, the decrease being greatest for the Gwydir where distributaries remove most of the flow above a certain stage. A notable exception to the downstream decrease is the high value for the Mehi River at Moree. The 1.58-year flood in the Gwydir contributes little of its flow to the Mehi while the larger 2.33-year Gwydir flood contributes a larger proportion of its flow to the Mehi distributary.

Velocity is constant along the streams except at the point where the Namoi and Gwydir Rivers enter the plain. At these points there is an increase in velocity (Figure 22.4), probably as a result of readjustments of streams to altered slope, and hence sediment load and channel pattern. The high velocity in the Mehi at Moree, which is near the offtake of the Mehi from the Gwydir, and the low velocity for Gil Gil Creek at Weemelah, which is some distance from the distributary offtake and eastern hill lands, supports the above explanation for the increase in velocity at the junction of hill lands and plain.

TABLE 22.1 HYDRAULIC, MORPHOMETRIC, AND SEDIMENTARY PARAMETERS AND
THEIR METHOD OF DERIVATION

Parameter	Method
Hydraulic:	
Discharge (m³/s), $Q_{1.58}$ $Q_{2.33}$	Annual floods with recurrence intervals of 1.55 and 2.33 yr respectively. Determined from Gumbel distribution of annual floods plotted on gumbel paper and line of best fit fitted by eye.
Velocity (m/sec) $V_{1.58}$	Velocity determined from velocity–discharge curve for discharge of 1.58 yr. Velocity discharge data obtained from Water Conservation and Irrigation Commision of N.S.W.
Morphometric: Width (m), W	Bankful width, averaged for several cross-sections at a site, where bankfull stage is defined by the bench index (Riley, 1972).
Hydraulic radius (m) R	Ratio A/P, where P is perimeter length and A cross sectional area at bankfull stage. Average for several cross sections at a site.
Cross sectional area (m²), A	Cross sectional area, averaged for several cross sections at a site, at the bankfull stage.
Bed slope, S	Slope of line of best fit fitted to a longitudinal bed profile surveyed by dumpy level over a distance of at least 300 m.
Sedimentary: Silt-clay percentage, Sc	Percentage of bed sediment that passes through a sieve of mesh diameter 64 μm.
Median diameter (mm), d_{50}	50 percentile of grain size frequency curve of bed sediment. Silts and clays analysed for size distribution by hydrometer method as recommended by British Standards Institution (1963), sands analysed by visual accumulation tube, as recommended by Guy et al. (1969), and gravels analysed by seiving.

TABLE 22.1 *(cont'd)*

Parameter	Method
95 percentile (mm), d_{95}	Percentile for which 95 percent of bed sediment is finer. Grain size frequency curve determined as outlined above. The sediment sample for the analyses outlined above was a grab sample collected from the lowest point in the cross section and from the top 10 cm of the bed sediment. The cross sections surveyed for derivation of the variables listed above were located at points of inflexion of stream meanders for reasons outlined by Riley (1972). Cross sections were surveyed with the A-Frame (Riley, 1969).

Downstream modification to floods along the Namoi suggests that the downstream decrease in flood magnitudes and ratio of 2.33 to 1.58 year floods is also a result of flood attenuation through channel storage (Figure 22.5). There is no simple relation between flood peak and distance travelled, and the hydrographs suggest that water is lost from the channels, particularly for the large floods (see flood of January, 1968). The pattern is complicated by the contribution of southern tributaries, e.g. the May 1968 flood is larger at Bugilbone than at Wee Waa. Nevertheless, part of the decline in flood peak is a result of attenuation. Since, for all the gauging

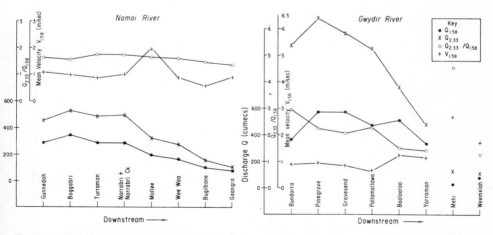

FIGURE 22.4 DOWNSTREAM CHANGES IN 1.58- AND 2.33-YEAR ANNUAL FLOODS, BANKFULL VELOCITY, AND RATIO OF 2.33- TO 1.58-YEAR FLOODS ALONG THE NAMOI AND GWYDIR RIVERS

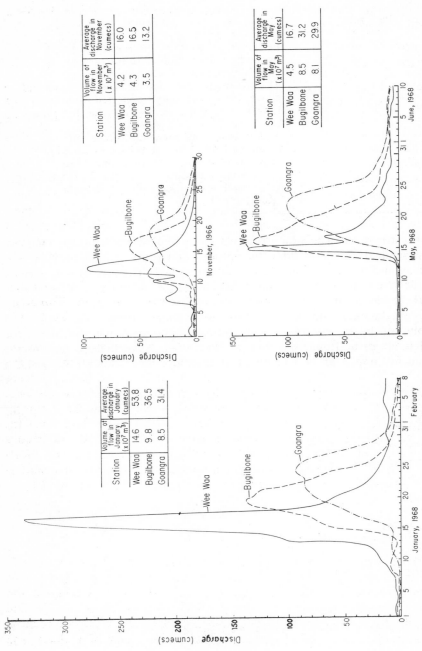

Station	Volume of flow in November (x 10⁷ m³)	Average discharge in November (cumecs)
Wee Waa	4.2	16.0
Bugilbone	4.3	16.5
Goangra	3.5	13.2

Station	Volume of flow in May (x 10⁷ m³)	Average discharge in May (cumecs)
Wee Waa	4.5	16.7
Bugilbone	8.5	31.2
Goangra	8.1	29.9

Station	Volume of flow in January (x 10⁷ m³)	Average discharge in January (cumecs)
Wee Waa	14.6	53.8
Bugilbone	9.8	36.5
Goangra	8.5	31.4

FIGURE 22.5 SELECTED FLOOD HYDROGRAPHS FOR WEE WAA, BUGILBONE AND GONGRA GAUGING STATIONS ON THE NAMOI RIVER

stations, there is a positive relation between discharge and velocity, the attenuation will also contribute to a downstream decrease in velocity for a particular flood peak (Figure 22.4).

Morphometric Changes

There is a rapid decrease in channel capacity, width, and hydraulic radius for the Gwydir and Namoi Rivers and Medgum Creek in the eastern area of the plain (Figure 22.6). The large number of distributaries and consequent loss of flow accounts for this decrease (Figure 23.3). The increase in width, hydraulic radius, and capacity along Thalaba Creek may be attributed to its tributaries.

Channel width appears to be the least variable downstream with respect to percentage variation about the mean. In general the three variables, width, hydraulic radius, and cross sectional area tend to parallel each other. However, there are notable exceptions, such as site 1 which is characterized by coarse sediment, and sites 8, 12, and 22. Non-parallelism suggests change either in channel boundary sediments or in hydraulic behaviour of the streams.

In the absence of distributary offtakes and major tributary inflows, there is a tendency for width, radius, and area to decrease downstream. This trend may be attributed to loss of water by evaporation and to flood wave attenuation.

Channel slope is low along the streams on the plain although it increases for Gil Gil Creek and Thalaba Creek near the western extremity of the plain. The influence of the Barwon River on the slope of the smaller streams near their confluence with that river is the probable reason for the increase. Large streams such as the Namoi and Mehi, which tend to have steeper slopes than the small streams, have probably adjusted their beds over the plain to the base level of the Barwon River. Certainly a stream with a hydraulic radius of 3 m does not have as great a fall into the Barwon with a radius of 2 m. The small slopes of streams in the distributary system suggest that a difference in elevation of 2 m can represent a considerable drop in elevation.

Sedimentary Changes

The percentage of silt-clay in the bed of the Namoi River varies little along its course (Figure 22.7). This lack of change reflects the competence of the stream to transport fine materials even at low stages. An increase in the size of the median and 95 percentile grain size is explained by the several tributaries from the south which deliver sand to the river. Westward of these tributaries the grain size declines, probably as a result of a decreasing discharge and competence.

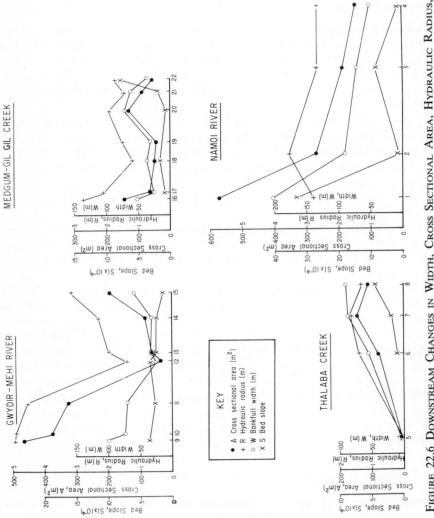

FIGURE 22.6 DOWNSTREAM CHANGES IN WIDTH, CROSS SECTIONAL AREA, HYDRAULIC RADIUS, AND BED SLOPES ALONG FOUR STREAMS IN THE NAMOI-GWYDIR DISTRIBUTARY SYSTEM

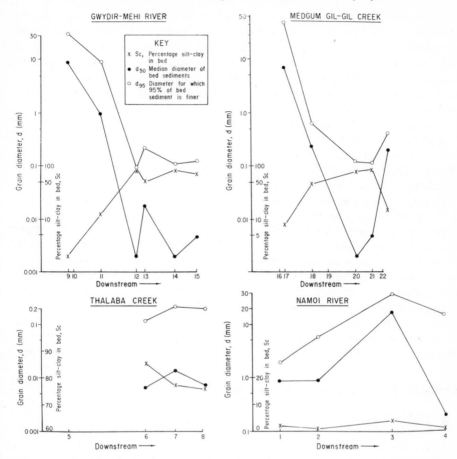

FIGURE 22.7 DOWNSTREAM CHANGES IN PERCENTAGE SILT-CLAY IN THE BED, MEDIAN, AND 95 PERCENTILE DIAMETERS OF BED SEDIMENT FOR FOUR STREAMS IN THE NAMOI–GWYDIR DISTRIBUTARY SYSTEM

Thalaba Creek has a high percentage of silt-clay in its bed, a result of its draining areas of clay rich soils and long periods of zero flow (Figure 22.7). There is some sand in the stream, as indicated by the 95 percentile. An absence of a downstream trend in the parameters may be a result of the small variation in channel capacity, slope, and hydraulic radius between sites 6 and 8 (Figure 22.6). The small variation suggests minimal downstream changes in sediment-transporting capacity and mode of transport.

There is a rapid decline in grain size and increase in percentage of silt-clay in the transition zone where the Gwydir–Mehi River flows from the highlands onto the plain (Figure 22.7). Beyond this zone the silt-clay and 95 percentile grain size are uniform downstream. Variation in median grain size may be attributed to local sediment characteristics such as presence or absence of dunes on the bed and variations in input from distributaries. The downstream uniformity of the 95 percentile, which is a measure of stream competence (Passega, 1957) suggests uniform flow conditions with respect to sediment-transporting capacity. The latter suggestion is supported by the minimal variation in slope, cross sectional area, width, and hydraulic radius between sites 12 and 14.

Sediment trends along Gil Gil Creek resemble those of the Mehi River except near the Barwon River, where there is a rapid increase in the median and 95 percentile size and a decrease in the percentage silt-clay in the bed (Figure 22.7). The influence of the Barwon and its sediment on Gil Gil Creek near the confluence cannot be ignored.

In general, for the parent and distributary streams, there is a trend for the percentage silt-clay in the bed to increase downstream or to remain constant, and a tendency for the 95 percentile diameter to decrease downstream across the distributary system. The median diameter is influenced by local conditions and peculiarities of flow and sediment supply, and hence does not appear to be as uniform downstream as the other two sediment parameters. The constancy of the silt-clay percentage and 95 percentile diameter on the plain is a result of a stream assuming a stable flow condition, with only small distributaries offtaking and rejoining. A downstream decrease in the 95 percentile diameter may be attributed to a downstream decrease in discharge and hydraulic radius and hence transporting capacity and competence.

Similarity Patterns among Stream Sites

In order to examine downstream changes the degree of similarity among stream sites must be assessed as this will reveal the extent of downstream change. Because the degree of correlation among variables influences most grouping procedures, the several variables (Table 22.1) are reduced to an uncorrelated set of factors before the stream sites are grouped. An agglomerative procedure is used herein to assess the degree of similarity among stream sites.

Factor analysis of the seven variables, transformed where necessary to normalize them, reveals that only four factors explain more than 5% of the variance (Table 22.2). Factor I is a channel sediment parameter, Factor II a bed slope parameter, Factor III a hydraulic radius–area parameter, and Factor IV a width–area parameter.

The four factors confirm the choice of the seven variables. The variables clearly describe different aspects of channel regime as they are not highly correlated (Table 22.3). The high variance explained by the sedimentary factor (Factor I) may be a function of the degree of correlation among the three sediment variables Sc, d_{50}, and d_{95} and of the sampling prodedure. A high variation in the bed sediment at a site increases the total variance of the

TABLE 22.2 FACTOR LOADINGS AND PERCENTAGE VARIANCE
EXPLAINED

Variable	Factors			
	I	II	III	IV
R	0.43	0.01	0.88	0.19
W	0.12	0.30	0.11	0.94
Sc	−0.92	−0.12	−0.21	−0.17
$\log d_{50}$	0.96	0.04	0.20	0.09
$\log d_{95}$	0.90	−0.11	0.30	0.14
$\sqrt{A}$	0.32	0.20	0.59	0.71
$\log S$	0.01	0.95	0.04	0.31
% Variance explained	59.1	23.9	9.3	5.4
Cumulative % variance explained	59.1	83.0	92.3	97.7

N.B. Factor analysis performed on the correlation matrix and varimax rotation used to maximize factor loadings.

sediment samples for the distributary area. Analysis of factor score trends in the downstream direction for each of the four streams reveals similar trends to those of the individual variables that are highly loaded on each factor.

Clustering the stream sites by means of a sum of squares distance measure on standardized variables and centroid agglomeration (Dixon 1975) reveals five groups of sites at the level of agglomeration of 2 (Figure 22.8).

The largest group of sites are those which occupy the central portion of the Mehi–Gil Gil–Thalaba stream system. Sites 21 and 18 on Gil Gil Creek, sites 15 and 12 on the Mehi River, and sites 8 and 6 on Thalaba Creek bracket the area (Figure 22.4). Within the area, at a lower level of agglomeration (i.e. greater similarity) two subgroups can be defined. Sites 14, 13, and 12 on the Mehi River are one of the subgroup. These sites occupy an

TABLE 22.3 CORRELATION MATRIX, MEANS, AND STANDARD DEVIATIONS
OF THE SEVEN REGIME VARIABLES

Variable	1	2	3	4	5	6	7
1. R	1.00						
2. W	0.33	1.00					
3. Sc	−0.62	−0.33	1.00				
4. $\log d_{50}$	0.60	0.23	−0.92	1.00			
5. $\log d_{95}$	0.67	0.24	−0.87	0.92	1.00		
6. $\sqrt{A}$	0.79	0.83	−0.57	0.49	0.55	1.00	
7. $\log S$	0.10	0.58	−0.17	0.08	−0.03	0.43	1.00
Mean	2.4	67.	46.	−1.2	−0.08	12.2	−3.6
s.d.	1.1	42.	39.	1.3	1.07	5.1	0.5

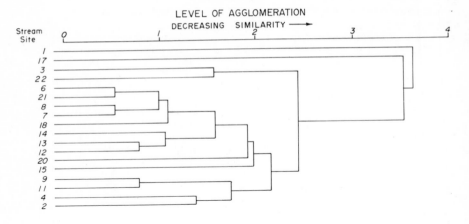

FIGURE 22.8 DENDROGRAM OF GROUPING AMONG STREAM SITES

area of considerable distributary inflow and outflow. Sites along Thalaba Creek and sites 21 and 18 on Gil Gil Creek form a second subgroup. The existence of the two subgroups may be accounted for by several basic differences between the Mehi River and the two Creeks. Thalaba and Gil Gil Creeks differ from the Mehi River in terms of their size, drainage area, and tributary contribution to flow and sediment. The three streams are ephemeral and have similar bed sediment characteristics. However, Thalaba Creek flow is entirely dependent on runoff from the eastern hill lands near the plain whereas the distributaries receive water and sediment from streams that drain larger areas further east (Figure 22.2) with consequent differences in sediment load and runoff. Gil Gil Creek, although supplied by distributaries, is also supplied by tributaries from the near eastern hill lands.

The second largest group of sites is composed of sites 9 and 11 on the Gwydir–Mehi River, and 2 and 4 on the Namoi River. The group appears to represent the parent streams on the plain. The Bugilbone site is not included in the group because it is the one site at which bedrock crops out into the channel bed. The bedrock bed of the stream will differ from sand and dry-bedded streams in response to flow and sediment load. The rock also represents a control on base level, which may explain why sites 3 and 22 form a group. The resistant bed behaves in a manner similar to the base level control of the Barwon.

Sites 1 and 17 cluster at a high level with the other 16 sites. Site 1 is probably distinct because it is on the junction of the hill-lands and plains and above the distributary offtake areas. Site 17 should be related to the group of sites 11 and 9 and no explanation can be offered at the present for its difference.

Discussion

Several trends for the Namoi–Gwydir distributary systems are evident from the preceding analysis.

(i) The parent streams distribute in the region where they enter the plain, have higher bankfull velocities than upstream or downstream of the marginal areas. The high velocity is probably a result of readjustment of the streams to conditions of low slope, reduced discharge, and sediment load—conditions different from those in the adjacent hill lands.

(ii) The parent streams rapidly lose water downstream from the region of entry onto the plain, but the rate of loss declines downstream. The loss is a result of distributary offtakes removing water at high stages of flow in the parent, a fact confirmed by the decrease in the ratio $Q_{2.33}/Q_{1.58}$, and of attenuation of the flood hydrograph.

(iii) Sediment size in terms of silt-clay percentage and 95 percentile diameter tends to remain constant downstream for the parent streams except where tributaries contribute a large proportion of sediment to the total load.

(iv) Streams smaller than the parent streams of the distributary system generally resemble each other in terms of most regime variables, no matter whether they are tributary or distributary streams. However, there are several differences among small tributary and distributary streams, particularly with respect to flow and sediment.

(v) Channel reaches with considerable numbers of distributing offtakes and tributary/distributary inflow differ in terms of channel regime from those reaches of the same stream with no or very few distributary outflows and inflows.

(vi) Sediment and morphological parameters tend to vary least in the downstream direction for both the large parent and small distributary and tributary streams.

(vii) Base level influences stream geometry and sediment, as evident in the interaction between the Barwon River and several streams of the Namoi–Gwydir system, and this influence is most pronounced in the case of the smaller streams. However, the direct consequences of the influence rapidly decreases upstream from the point of confluence.

The trends noted above have been defined for only four streams in the Namoi–Gwydir distributary system, and there is some possibility that the trends are not reproduced for other streams in the system. However, although the variance of the several variables is large, the trends are clearly defined and the possibility of their being due to random fluctuation is remote.

On the whole the trends suggest that the rate of change of hydraulic, sedimentary, and morphometric characteristics of streams crossing the plain decreases and may be adopting equilibrium values. That is, sediment, velocity, discharge, and geometry are tending either towards a constant value for each stream or towards some constant rate of change. In all probability the latter is the case as discharge declines through evaporation and infiltration.

Considerable disruption occurs to the character of streams in those regions where the parent streams enter the plain, where the distributary streams offtake and distributaries/tributaries rejoin, and where the distributary system joins the Barwon.

The patterns noted above are in the first instance only relevant to the Namoi–Gwydir distributary system. However, the general morphological pattern of the Namoi–Gwydir system resembles that of most distributary systems in eastern Australia. Hence, there are some grounds for accepting the fact that the trends defined for the Namoi–Gwydir system apply to the distributary systems of eastern Australia.

References

British Standards Institution. 1963. *Methods of Testing Soils for Civil Engineering Purposes.* British Standards House, London.

Dixon, W. J. (Ed.) 1975. *BMDP Biomedical Computer Programs.* Univ. California, Berkeley.

Guy, H. P. (A.S.C.E. Committee for preparation of manual on sedimentation). 1969. Sediment measurement: F. Laboratory procedures. *J. Hydraul. Div. Amer. Soc. civ. Engrs.,* **95,** pp. 1515–1543.

Lane, E. W. 1957. A study of the shape of channels formed by natural streams flowing in erodible material. *U.S. Army Corps* of Engineers, *Missouri River Div., Omaha, Nebr. Sediment Ser. 9.*

Passega, R. 1957. Texture as characteristic of cluster deposition. *Bull. Amer. Assoc. Petrol. Geol.,* **41,** pp. 1957–1984.

Riley, S. J. 1969. A simplified levelling instrument: the A-Frame. *Earth Sci. J.,* **3,** pp. 51–53.

Riley, S. J. 1972. A comparison of morphometric measures of bankfull. *J. Hydrol.,* **17,** pp. 23–31.

Riley, S. J. 1975. Some differences between distributing and braiding channels. *N.Z. J. Hydrol.,* **14,** pp. 1–8.

J. LEWIN
Lecturer in Geography

B. E. DAVIES
Lecturer in Geography

and

P. J. WOLFENDEN
Formerly Department of Geography
University College of Wales, Aberystwyth

23

Interactions Between Channel Change and Historic Mining Sediments

Mining activity and stream activity may interact in the fluvial environment in a number of ways. In the first place mining wastes in the form of solutes and sediments may enter streams where they are dispersed, and are then redeposited in what constitutes a special case of fluvial sedimentation; secondly, the input of materials and the modification of discharge characteristics may lead to repercussions in terms of channel morphology and dynamics. Modern mining practice and environmental legislation both tend to curtail the input of mining wastes into fluvial environments, but this was not the case with historic mining from which much waste has been and continues to be incorporated into fluvial systems active today.

For Britain, the Fifth report of the 1868 Rivers Pollution Commission (1874) pertinently identified the major causes for late nineteenth century concern. The Commission noted especially the problems caused by the discharge of large quantities of fine sediment (from coal washing, tin mining, and china clay working) and the dispersal of toxic metals (notably from lead, zinc, and arsenic mines). This fluvial dispersal of mining wastes is graphically summarized in Figure 23.1. Mobilized wastes may be transported as clastic sediments or as solutes to be deposited or precipitated in a range of

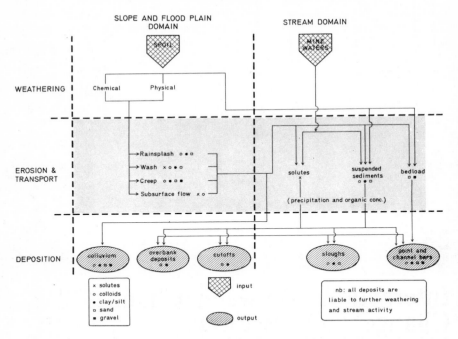

FIGURE 23.1 A MODEL FOR THE DISPERSAL OF MINING WASTES

sedimentary environments. For different types of ore, mining practice and indeed environment, the nature and quantity of dispersal are likely to vary; in this paper we are concerned to examine the dispersion patterns manifest in the *depositional* phase resulting from historic metal mining in mid-Wales, and to consider what the possible modifications to river channel patterns have been.

Metal mining in mid-Wales has a long history (Lewis, 1967) but is now inactive. Mining for lead and later zinc was intensively undertaken from the beginning of the nineteenth century when new techniques were adopted: large quantities of wastes were indiscriminately fed into local streams until preventative legislation was enacted (1876). Mining activity also declined as prices fell to a less than economic level in the last decade of the century. Ores were hand-sorted, crushed, and then jigged or 'buddled' to separate metals by water flotation. These processes were inefficient and both the heaps of tailings and the finely divided wastes fed directly into streams from flotation still contained a high proportion of ore particles and metal-rich spoil. In the relatively narrow and steep-sided valleys of mid-Wales, inputs of sediment in the form of country rock and metal particles, and solutes (from mine waters and the subsequent input of soluble salts from the weathering of tailings dumps) were considerable. The Commission reported the rivers near Aberystwyth as being the most polluted:

'All these streams are turbid, whitened by the waste of the lead mines in their course; and flood waters in the case of all of them bring down poisonous 'slimes' which, spreading over the adjoining flats, either befoul or destroy the grass, and thus injure cattle and horses grazing on the dirtied herbage, or, by killing the plants whose roots have held the land together, render the shores more liable to abrasion and destruction on the next occasion of high water. It is owing to the latter cause as well as to the immense quantity of broken rock which every mine sends forth that the small rivers *Rheidol* and *Ystwyth* present such surprising widths of bare and stony bed.' (Rivers Pollution Commission, 1874, p. 15)

As we shall see later, the quantity of 'broken rock' is here perhaps a little overdrawn in view of the normal state of local gravel rivers, but the waste heaps have certainly continued to provide sediment in the century that has followed. The spoil is commonly permeable, lacking in nutrients and rich in toxic metals to an extent which prevents a vegetation cover; both weathering and the physical entrainment of metals and crushed country rock makes the extensively gullied heaps a continuing source of solutes and sediments.

It has long been appreciated that this situation leads to serious agricultural and biological problems (Griffith, 1918; Newton, 1944; Alloway and Davies, 1971); most research has understandably been biological in focus, and only recently have attempts been made to identify patterns of fluvial dispersal and channel activity (Davies and Lewin, 1974; Grimshaw, Lewin and Fuge, 1976). This paper adopts such an approach, taking the River Ystwyth as a case study.

River Ystwyth Study Sites

The River Ystwyth drains a catchment area of 193 km^2; along the lowest 20 km of the river, sections of alluvial valley floor are separated by short steeper-gradient rock-cut channels (Brown, 1952). The river can be classified as a coarse bed load stream, deriving much of its present sediment load from channel sources (Grimshaw and Lewin, 1976). The river channel actively migrates across the alluvial sections except where confined by road, railway or valley side. Into this active fluvial system, mining wastes have been injected from a number of sources, notably the crushing mills, dressing floors, flotation plant, and spoil tips at Cwm Ystwyth, Pontrhydygroes, Frongoch, and Grogwynion (Fig. 23.2). Details of mine histories appear in Jones (1922), Lewis (1967) and Bick (1974).

We have analysed sediment samples for Pb, Zn, Cu, and Cd using atomic absorption spectrophotometry (air/acetylene flame) at three floodplain sites down-river from the former mine locations, together with some reconsideration of a site previously discussed in Alloway and Davies (1971). Analytical

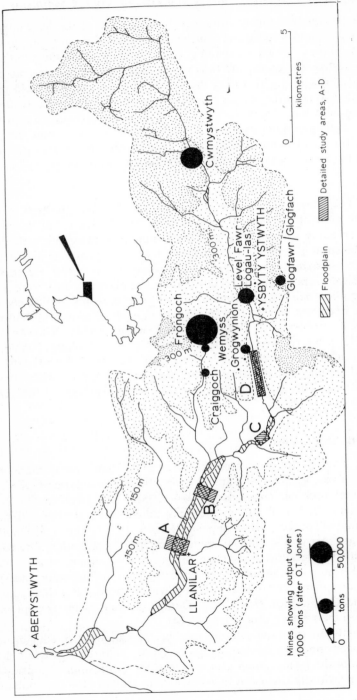

FIGURE 23.2 THE YSTWYTH CATCHMENT, MID-WALES

The main floodplain sites are shown in relation to the larger mines

methods were as described in Davies and Lewin (1974), except that in some cases particular size fractions of sediment have also been analysed (see Figure 23.6).

Study sites were selected in the light of information on the dynamics of channel change derived both from recent field observations (Lewin, 1976) and from analysis of available manuscript and printed maps and air photography (Lewin and Hughes, 1976). Each site will be discussed in turn, the metal data being listed in Table 23.1.

Llanilar (SN 6275)

The floodplain morphology and contemporary channel activity at this site have been examined previously in some detail (Lewin and Manton, 1975; Lewin, 1976): metal values taken on a floodplain cross-profile are shown diagrammatically in Figure 23.3 whilst further data for these and other samples are given in Table 23.1.

Three points are of immediate interest. Firstly, metal values are hardly anomalous (as defined in Alloway and Davies, 1971, Table II) unless on the floodplain within reach of river activity. Secondly, the *highest* values occur to the south of the railway embankment in an area now protected from channel activity, but on the site of the active channel prior to railway construction in the 1860s. Thirdly, contemporary river sediment has comparatively low metal values, though this situation is liable to vary locally where the river impinges on nineteenth-century channelway sediments.

In Table 23.1 (column 3), an attempt is made to classify sample sites according to the sedimentary environment prevailing and relevant to the deposition of mining waste. On this basis three categories of metal levels are clear: valley slopes, areas which have been or are flood liable, and areas of channel activity dating to the nineteenth century. The present active channel (sample number 367), an old cutoff formed prior to the nineteenth century (366), and the other floodplain soils (370, 371, 377), all group together as having anomalous metal contents, though with Pb concentrations an order of magnitude less than those with channels active in the mining heyday. Hence only parts of the floodplain have very high metal values: these parts cannot be distinguished by morphology or location, but only by the dating of active sedimentation at channel margins.

Dol Fawr (6574)

The same pattern emerges at Dol Fawr in data previously reported by Alloway and Davies (1971): the 1845 tithe map shows that samples 2001 and 2002 are from areas that were part of the active channel at that time; samples 2003 and 2004 have Pb concentrations an order of magnitude less (and not so high as in equivalent locations further downvalley); slope soils are only marginally anomalous.

Wenallt (6771)

Here a meander loop, with a radius of approximately 250 m and arc length 130°, has oscillated within a band of about 100 m width without major development for at least 150 years; to the north the valley side rises steeply, but a terrace exists to the south (Figure 23.4). High values occur in areas marginal to channel activity during mining (382, 383); lower values occur in contemporary sediments (381) but with the highest values of all in a disused mill race (384). This is mapped as having a functioning race and dam across the river in 1886, but the dam had gone by 1904. There can be little

TABLE 23.1 SAMPLE ANALYTICAL DATA

Sample	Grid Reference (SN)	Site type	Ignition loss (%)	Metal (p.p.m.)			
				Pb	Zn	Cu	Cd
Llanilar							
363	624744	S	9.9	51	110	15	0.6
364	627749	S	9.4	83	137	23	0.6
365	627751	T(solif)	8.7	69	111	18	0.6
366	625753	Ch	7.2	443	483	25	1.5
367	628753	Ca	0.8	274	480	27	1.3
368	628754	Cm	5.1	6017	730	129	1.7
369	628752	Cm	4.6	6290	592	65	1.9
370	628753	Cm	3.2	718	505	21	1.7
371	629756	F	8.2	667	197	22	1.0
372	630758	S	8.1	85	160	23	0.7
373	629760	S	8.4	39	105	19	0.5
374	629763	S	11.0	55	109	13	1.0
375	638758	S	8.2	58	129	19	0.8
376	637757	S	7.8	58	124	19	0.7
377	636753	F	9.2	384	225	20	0.9
Dol Fawr							
2001	650744	Cm	8.4	1406	305	20	1.0
2002	651743	Cm	7.5	1980	460	25	1.4
2003	651742	F	10.7	238	305	22	0.8
2004	651741	F	12.4	210	260	18	0.8
2005	651739	T	9.0	95	235	16	0.7
2006	652737	S	11.1	80	226	18	0.6
2007	650736	S	13.9	70	170	15	0.8
2008	649735	S	13.1	58	168	13	0.8
Wenallt							
378	676713	S	9.2	71	111	20	0.7
379	672715	T	6.4	670	226	22	0.9
380	676717	S	11.4	65	145	17	0.7
381	674717	F	0.6	941	826	36	2.0
382	674716	F	3.6	2321	707	80	1.9
383	675718	F	8.2	1193	321	26	0.9
384	676716	Cm	25.0	7610	2275	56	13.9

TABLE 23.1 (*cont'd*)

Sample	Grid Reference (SN)	Site type	Ignition loss (%)	Metal (p.p.m.)			
				Pb	Zn	Cu	Cd
Grogwynion 2							
151	697718	S	12.3	240	120	15	1.1
152		F	2.8	1507	461	36	1.9
153		Ca	4.6	1593	587	36	2.0
154		F	1.7	821	377	31	1.3
155		F	3.0	1422	753	30	2.2
156		S	7.9	104	140	17	1.3
157		F	0.6	1024	501	37	1.9
Grogwynion 1							
158	713721	S	14.4	435	86	19	1.0
159		Cm	6.9	2105	379	29	2.2
160		Ca	1.8	1047	607	30	1.9
161		Cm	6.5	3423	793	45	2.3
162		F	6.0	1550	566	28	2.1
163		S	10.0	573	1905	20	1.1

S, valley slope; T, terrace; F, area currently or historically flood liable; Ca, active channel; Cm, channel active since 1800; Ch, old channel active before 1800.

doubt that finer metal sediments were especially liable to sedimentation here.

Sample 379 is particularly interesting since it comes from a terrace above contemporary flood limits (cf. Alloway and Davies, 1971, Table IV): values are, however, comparable with those previously quoted for floodplain areas not within the bounds of active channel migration in the mining era. This raises the whole question of the role of floods, for previous writers have often implied that flood discharges deposit a flood drape of polluted sediments right across areas liable to flood. In practice, however, most flood sediments, together with those transported near bankfull stage, are deposited at the channel margins, and suspended sediments from mining processes may be unlike 'natural' sediments in that inputs and transport can be

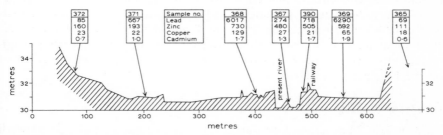

FIGURE 23.3 A PROFILE OF THE YSTWYTH FLOODPLAIN AT LLANILAR (SN 628754)

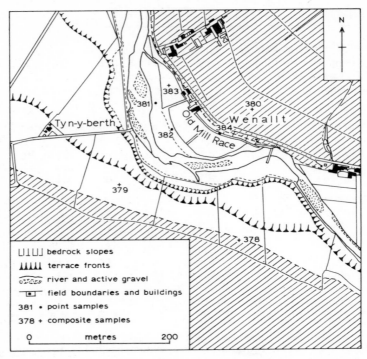

FIGURE 23.4 THE YSTWYTH FLOODPLAIN AT WENALLT (SN 6771)

high on quite moderate discharges. Such sediment may be inserted directly into streams, not entrained on slopes following precipitation; at less than overbank flows they can only be redeposited within the channel margins. Hence fairly high metal values elsewhere on the occasionally-inundated floodplains could also result from predocumentary *channel* deposition during many centuries of less-intensive mining activity prior to the nineteenth century. Alternatively (and this must apply to river terrace deposits) erosion of exposed lodes may have contributed metals to alluvial sediments for millenia. Thus whilst the role of floods during and after mining activity has certainly been to entrain and transport mining sediments, these sediments have been redeposited especially in and near active channels. Some metals may have been transported as far as the limits of flood inundation (and the waters may have been toxic with dissolved metal salts subsequently precipitated), but there are other explanations for the moderately high metal values on the floodplain and terraces.

Grogwynion (7071)

For nearly 2 km above Llanafan Bridge (D in Figure 23.2), the Ystwyth passes through a steep-sided valley, 200 m deep with a relatively narrow floor, which has been swept across by the river in the last 130 years (Figure 23.5). Much of this floor consists of bare gravel cut into by active and abandoned channels.

Several lodes parallel the valley on the north, and adit mining has been operated intermittently for centuries at Gwaithgoch (710723) and more importantly at Grogwynion (714723): mining here was particularly active from the 1860s to the early 1880s. In addition, the Gwaithgoch crushing mill was installed in the First World War to rework the Frongoch waste tips for lead and zinc; an aerial ropeway was completed after the Great War to the tips 3 km away and these were worked for their ore content, particularly in the late 1920s when prices improved, but operations ceased at the end of that decade (Bick, 1974).

It is tempting to regard the locally exceptional braided channel as a response to the input of mining sediment: the 1845 tithe map (Figure 23.5), and the series of sketched maps of the mine dated 1741, 1792, and *c.* 1800 held in the *Gogerddan Collection* in the National Library of Wales, all show a single channel. However, the quality of these pre-nineteenth century maps is not such that great reliance can be placed on them.

Furthermore, there is little real evidence that large quantities of *coarse* sediment (of which the valley floor is composed) were fed locally into the river during mining: the spoil dumps at Grogwynion do not appear to have been undercut, whilst the size and roundness characteristics of the floodplain sediments in the supposedly affected reach are not anomalous compared with trends in the river system as a whole.

The input of toxic metals has probably been more important: sampling undertaken on two traverses across the valley (Profiles 1 and 5; Table 23.1) show high metal values across the valley floor, and it seems likely that the lack of vegetation contributes to greater bank and floodplain erodibility.

This particular type of multiple channel is also not generally characteristic of a high sediment input (Smith, 1974; Fahnestock and Bradley, 1974). The relatively low gradient (0.015), the pattern of long diagonal channels connected by shorter crossovers, and the rate of channel change, closely resemble the Knick River in Alaska, in kind if not in size (Fahnestock and Bradley, 1974). Whilst channel change rates are certainly more rapid than on most local meandering streams, short-term bedform changes do not compare with other braided rivers where a multiplicity of lowstage channels reflects the intricacy of the numerous mobile unit bedforms active at higher stages. Instead, changes observed over a 6-year period (1969–75) have been in response to relatively few high flows, involving point-bar and confluence bar sedimentation and the reactivation of an abandoned channel by crevassing to reform a multiple system during a flood in August 1973. Otherwise much of the braided appearance of the valley results from the flooding of inactive channels, and a lack of vegetation. It may also be noted in passing that the channel here is not noticeably aggrading nor are the sediments

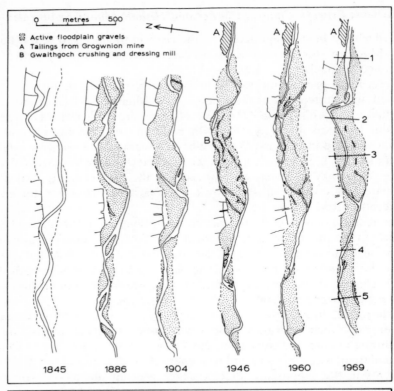

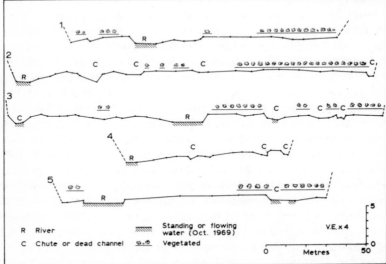

FIGURE 23.5 THE YSTWYTH FLOODPLAIN AT GROGWYNION (SN 7971)

Top: The Channel at various dates; Bottom: Floodplain profiles in 1969

fine: both criteria have been previously identified with the development of channel patterns of this type (Smith, 1974; Chitale, 1970).

Discussion

We can conclude on the basis of this and our other studies, that historic mining in mid-Wales led to the rapid dispersal of toxic waste, and that sedimentation at channel margins, which probably covered approximately 1.7 km^2 of the Ystwyth catchment during the mining heyday 1820–1880, was a preferred environment for the redeposition of metal sediments. The remainder of the floodplain (2.5 km^2) and some gravel terraces also have high metal concentrations as a result of flood dispersal or previous channel margin activity.

Since the mining phase, there has been a fall-off in metal levels in floodplain sediments (Davies and Lewin, 1974) until contemporary channel sediments are relatively but by no means absolutely 'clean.' Nineteenth-century floodplain sediments now themselves form sources of metal (possibly the most active source at present), in addition to the still-remaining waste tips, so that a continued input of metals is assured. With further data we feel it will be possible to establish temporal decay models giving fall-off rates for metal values applicable to varying types of floodplain environment.

This can be balanced by a model for downstream changes in the metal contents of materials of different particle size, as illustrated by active sediments on a tributary of the Ystwyth draining from Wemyss mine and spoil heaps (Figure 23.6). Both lead and zinc contents decrease downstream until about 4000–5000 m, when the lead curve shows a small peak, whereas zinc appears to be increasing more uniformly. Furthermore, the fall off in zinc concentration is less than lead, particularly in the 63 μm fraction. The increase in metal concentration at 4000–5000 m can be attributed to sediment lying upstream of the sampling point in an old mill race. A small pulse of material injected into the river channel at the latter would increase the proportion of 'polluted' to 'clean' sediment, thus giving rise to an overall higher metal level downstream. There may, however, be other possible processes and chemical changes involved.

If the downstream dispersal is a simple clastic process then several possibilities can be considered. The metal-rich material entering the stream is broadly of two kinds, namely crushed gangue minerals and wall rock. In addition there are small particles of galena (PbS) which were not separated during the dressing process, and sphalerite (ZnS) particles of varying size which either also escaped separation or were deliberately discarded for lack of economic value. Galena is much heavier than sphalerite (densities 7.5 and 4.1 g/cm^3, respectively) and both are denser than the rest of the detritus. Consequently, running water might be expected to cause differential downstream movement with the lightest material travelling furthest from the

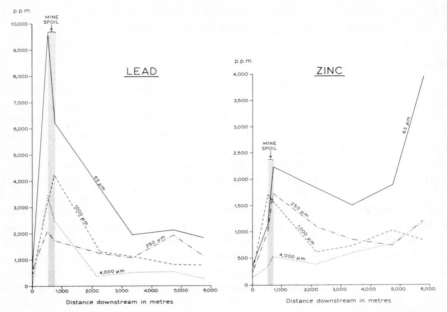

FIGURE 23.6 METAL VALUES (P.P.M.) BY SIZE ON THE NANT CWM-NEWYDION DOWN-
STREAM OF THE WEMYSS MINE (SN 716741)

source, and the rate of transport might have been such that an increase in
concentration since insertion had proceeded only a restricted distance downstream by the time of observation. Insertion is, however, a continuing slow
process not a massive singularity, and the rate of transport of finer sediments
is much too great for this model to be appropriate, in general, for the fall-off
in metal values. More generally, the input of 'clean' sediment from downstream slopes and channel banks must imply that metal sediments form a
decreasing proportion of the total sediment load in any size range, and this
will lead to a fall off in concentrations down-valley.

A further complication arises from the fact that the ore particles are not
inert: they dissolve and react in water and the resulting weathering products
migrate in different ways. Under acid conditions such as moorland waters of
low redox potential (E_h) or within spoil heaps containing pyrite (FeS) the
sulphide ores oxidize to sulphates. If carbonic acid is present as in some
heaps (calcite rather than quartz is the chief gangue mineral in some mines),
or in soils, metal carbonates are likely to form. These different metal salts
have very different solubilities (Table 23.2) and those of zinc are all more
soluble, i.e. more mobile, than the corresponding lead compounds. Garrels
and Christ (1965) give a detailed review of the controls of E_h and pH on
metal compound solubilities.

However, it is unlikely that metals migrate to any appreciable extent as

TABLE 23.2 SOLUBILITY OF SALTS
OF LEAD AND ZINC

	Solubility of salt (mol/l)	
	M = Pb	M = Zn
MS	1.78×10^{-14}	1.26×10^{-12}
MSO$_4$	1.26×10^{-4}	5.37
MCO$_3$	2.18×10^{-7}	3.97×10^{-6}
MCl$_2$	1.58×10^{-2}	18.01

simple ions. They are adsorbed by fine particles, especially those <2 μm, and it is noteworthy that, as illustrated in Figure 23.6, metal contents increase with decreasing particle size. Lead is more strongly adsorbed than zinc. The metals are also readily complexed (chelated) by organic substances which can considerably modify their geochemical dispersal (see Davies, 1976) and most chelating agents complex lead more strongly than zinc.

Thus it can be seen that downstream dispersal is a complex process and detailed investigations are needed. Spatial and temporal decay functions of metal concentrations are composite and result from several possible physical and chemical processes as well as from varying levels of mining activity. Quite apart from the relevance to the practical problem of the dispersal of toxic mine waste, clarification of the processes may allow metal sediments to be used as tracers in long-term sedimentological experiments (cf. Gross, 1972). The input of dated volumes of 'labelled' sediment could allow volumes and sources for other sediment to be derived. This possibility will be developed elsewhere.

Although downstream fall-off in lead values occurs on smaller streams, the same is not the case for active sediments on streams with floodplains, nor for floodplain sediments of the same date for comparable environments. Here again a number of explanations may be advanced. Previous work suggests that comparatively little fine sediment is provided by the lowland parts of west Wales catchments, so that 'clean' *fine* sediment is not so available to diminish the proportional significance of metal sediments downstream. Alternatively, erosion of 'old' metal-rich floodplain sediments can lead to local metal enrichment. Finally, and concerning floodplain sediments in particular, metals are liable to pedogenic translocation so that the value now observed may no longer be those that obtained when sediments were emplaced.

A final point for discussion concerns channel dynamics: here again the input of toxic fine sediments seems to have been of more significance than the input of voluminous but inert waste. This conclusion contrasts with that of some earlier writers such as Jones (1940) who suggested that at Llanilar (8 km downstream from the nearest mine) the channel was 'silted up with vast quantities of stone, rubble and gravel that have come down from the mine workings.' The nature of this gravel, and the alternative sources available for it, is not in fact that unusual for local rivers, and these are likely to have been affected locally rather than regionally by the input of inert coarse mining spoil.

Conclusions

It is clear from the above discussion that within the framework of a general model (Figure 23.1) which has both spatial and temporal dimensions, very specific and individual patterns of metal dispersal and channel modification may take place in particular locations. In this case, a syngenetic secondary dispersal pattern (Hawkes and Webb, 1962) dominates, involving dispersal of metal fines rather than large volumes of country rock. Partly this is a reflection of the nature of the mining operation and ores involved, but the nature of the local fluvial process system is also crucial. It is highly unlikely that even identical mining operations in contrasting fluvial environments would produce similar patterns of environmental interaction, and a specific input of solutes or sediments could prove quite acceptable in one environment but not in another. An understanding of channel processes and dynamics in a variety of environments therefore appears highly desirable from an environmental management point of view.

Acknowledgements

We thank the Natural Environment Research Council for support: as research grants to Dr Brian E. Davies and Dr John Lewin, and a Research Studentship to Mr Paul J. Wolfenden.

References

Alloway, B. J. and Davies, B. E. 1971. Trace element content of soils affected by base metal mining in Wales. *Geoderma*, **5**, pp. 197–208.

Bick, D. E. 1974. *The old metal mines of Mid-Wales, Part I Cardiganshire South of Devil's Bridge.* The Pound House, Newent, Glos.

Brown, E. H. 1952. The River Ystwyth, Cardiganshire: a geomorphological analysis. *Proc. Geol. Ass. Lond.*, **63**, pp. 244–69.

Chitale, S. V. 1970. River channel patterns. *J. Hydraul. Div. Amer. Soc. Civ. Engrs.*, **96**, pp. 201–221.

Davies, B. E. 1976. The role of organic matter in heavy metal problems in soil and water. *Welsh Soils Disc. Group Rept.*, **16**, (in press).

Davies, B. E. and Lewin, J. 1974. Chronosequences in alluvial soils with special reference to historic lead pollution in Cardiganshire, Wales. *Environ. Pollut.*, **6**, pp. 49–57.

Fahnestock, R. K. and Bradley, W. C. 1974. Knick and Matanuska Rivers, Alaska. In Morisawa, M. (Ed) *Fluvial Geomorphology*, State University of New York, Binghamton, pp. 226–250.

Garrels, R. M. and Christ, C. L. 1965. *Solutions, minerals and equilibria.* Har Row, New York.

Griffith, J. J. 1918. Influence of mines upon land and livestock in Cardiganshire. *J. Agric. Sci.,* **9,** pp. 241–71.

Grimshaw, D. L. and Lewin, J. 1976. In preparation.

Grimshaw, D. L., Lewin, J., and Fuge, R. 1976. Seasonal and short-term variations in the concentration and supply of dissolved zinc to polluted aquatic environments. *Environ. Pollut.,* **11,** pp. 1–7.

Gross, M. G. 1972. Waste discharges as sedimentological experiments. *Geol. Soc. America Memoir,* **132,** pp. 623–630.

Hawkes, H. E. and Webb, J. S. 1962. *Geochemistry in mineral exploration.* Harper and Row, New York.

Jones, J. R. E. 1940. A study of the zinc polluted river Ystwyth in north Cardiganshire, Wales. *Ann. Appl. Biol.,* **27,** pp. 368–378.

Jones, O. T. 1922. The mining district of north Cardiganshire and west Montgomeryshire. *Mem. Geol. Surv.,* 20.

Lewin, J. 1976. Initiation of bedforms and meanders in coarse-grained sediments. *Geol. Soc. Amer. Bull.,* **87,** pp. 281–285.

Lewin, J. and Hughes, D. 1976. Assessing channel change on Welsh rivers. *Cambria,* **3,** pp. 1–10.

Lewin, J. and Manton, M. 1975. Welsh floodplain studies: the nature of floodplain geometry. *J. Hydrol.,* **25,** pp. 37–50.

Lewis, W. J. 1967. *Lead mining in Wales,* Cardiff Press, Cardiff.

Newton, L. 1944. Pollution of the rivers of west Wales by lead and zinc mine effluent. *Ann. Appl. Biol.,* **31,** pp. 1–11.

Rivers Pollution Commission (1868). 1874. *Fifth report of the commissioners appointed in 1868 to inquire into the best means of presenting the pollution of rivers.*

Smith, D. G. 1974. Aggradation of the Alexander-North Saskatchewan Rivers, Banff Park, Alberta. In Morisawa, M. (Ed.) *Fluvial Geomorphology,* State University of New York, Binghamton. pp. 201–219.

K. S. RICHARDS

Lecturer in Geography
University of Hull

and

R. WOOD

Severn-Trent Water Authority

24

Urbanization, Water Redistribution, and their Effect on Channel Processes

The morphology of alluvial channels continually responds to varying hydrological conditions, despite the possible existence of equilibrium forms when controlling environmental variables are constant. Schumm and Lichty (1965) suggest that geomorphological phenomena reflect the influence of differing independent controls at various time scales, although in practice separation of the time-continuum into discrete phases is impossible. Three scales of change in river channels may be identified. Within a single year, short-term random fluctuations in channel morphology occur following bank erosion during floods, as the channel migrates over its flood-plain (Knighton, 1975). The location of a section varies, but its morphology is stable if the environment remains unchanged (Leopold, Wolman, and Miller, 1964, p. 201). Over periods of several thousand years, palaeohydrological conditions respond to climatic trends. Consequently, long-term changes in channel morphology have occurred, reflecting the existence of dynamic equilibrium rather than steady state. These effects are evident in studies of underfit streams (Dury, 1965) and palaeochannels (Schumm, 1968). At this scale, process is inferred from form, and quantitative expressions of the impact of changing climate on channel form and process (Rango, 1970) are rare. Human interference causes considerable medium-term hydrological change over decades, particularly following urbanization. Increases in sediment yield during intense building activity are well documented (Wolman, 1967; Walling and Gregory, 1970), and channel capacity changes have been investigated (Hammer, 1972; Leopold, 1973). Nevertheless, considerable potential remains for research into changes in hydrological and geomorphological processes at this scale.

Although these components can be distinguished qualitatively, quantitative analysis cannot readily separate effects at different scales. Long term changes are so gradual that their effects are obscured by short-term, random variations. The magnitude of changes following urbanization simplifies identification of such effects. Nevertheless, the hydrological and geomorphological changes resulting from urbanization are complex, and the paucity of data encourages oversimplification.

In densely populated catchments, local surface and groundwater reserves cannot meet domestic and industrial demand for water, and a substantial proportion of water supply must be imported from wetter, less-populated basins. This introduces a complication to urban hydrology which has received little attention. The scale of inter-basin water transfer is exemplified by the Tame catchment. Between 1950 and the early 1970s demand for water increased by approximately 3% per annum (Trent River Authority, 1968). In 1965, the amount of imported water averaged 4.9 m³/s, or 423 mld (Figure 24.1a) and the projected 1975 importation was 6.9 m³/s or 596 mld (Figure 24.1b). Original projections beyond 1975 have been revised downwards because of recent population trends and the slowing of industrial expansion, and the growth rate of water supply is now about 1.2% per annum. Nevertheless, water imports to the Tame basin will continue to increase to offset the permanent deficiency in Birmingham. This reliance of populated areas on remote water supplies is a common feature of water supply policies for urban centres (Simons, 1969; O'Riordan and More, 1969; Rees, 1969). The hydrological and geomorphological effects of massive water transfers are intermixed with other effects of urbanization, and have received little direct attention. The effects are widespread since areas of water supply undergo alterations in process regime as well as recipient areas. This paper illustrates some effects of water transfer, and highlights some problems which arise in the analysis of the effects of urban growth on hydrological processes and channel forms.

Hydrological Effects of Water Transfer

The hydrological effects of water storage and redistribution to other watersheds can be conveniently examined by considering water supply and water receiving areas separately.

Supply Areas

A common result of reservoir storage is a reduced amplitude of the annual streamflow oscillation downstream. Reservoir storage frequently involves complementary flood protection; flood peaks are substantially

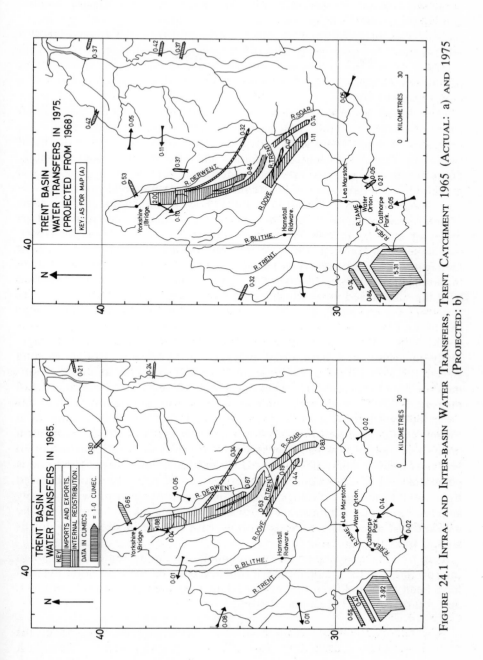

FIGURE 24.1 INTRA- AND INTER-BASIN WATER TRANSFERS, TRENT CATCHMENT 1965 (ACTUAL: a) AND 1975 (PROJECTED: b)

reduced by designing reservoirs with spare storage capacity (Linsley, Kohler, and Paulhus, 1949). Hence the hydrological regime downstream experiences reduced flow magnitudes. This effect is exemplified by the River Blythe at the Hamstall Ridware gauge (SK109191), downstream from the Blithfield reservoir. After the reservoir filled (1952–1954) the annual amplitude of pentad stream-flow averages decreased from about 1 m^3/s to about 0.3 m^3/s (Anderson, 1975).

The altered magnitude-frequency characteristics of floods downstream from reservoirs has the most significant effect on the channel and its processes. Analysis requires at least daily data, and may proceed in one of two ways. Given sufficient data before and after reservoir development, flood frequency analysis may be undertaken for the two periods and magnitudes compared at similar frequencies. Data obtained prior to reservoir filling are rarely sufficient in quantity, however. Furthermore, problems of interpretation arise if land use changes accompany reservoir development, or changes in rainfall intensity and storm frequency occur (Howe, Slaymaker, and Harding, 1967). Water Authorities responsible for reservoir management monitor reservoir storage changes to allow estimation of daily 'naturalized' discharges downstream. Actual gauged flows can be compared with the naturalized flows which would have occurred in the absence of the reservoir. For example, the 1971 measured and naturalized hydrographs of the River Blythe, indicate that storage reduced peak flow by about 35%.

The upper Derwent is intensely developed for surface water storage. Daily measured and naturalized flows at the Yorkshire Bridge gauging station (SK198851) are available from 1913 to 1933. This follows development of the Howden and Derwent dams in 1912 and 1916. Naturalized and measured flows show similar temporal variations during this period (as indicated by the moving averages of Figure 24.2). Nevertheless, while little overall change occurred in naturalized monthly mean discharge, regression analysis of gauged flow shows a decline of 0.5 m^3/s (a 20% reduction in 20 years) as water demand has increased abstraction from storage. The trend is generalized by a regression, but is more probably a step function obscured by variability in the hydrologic system: increased demand on water supply sources motivates administrative processes leading to an increase in the amount abstracted. The effect is a progressive decline in streamflow downstream from the reservoir sites.

This trend, averaging *c.* 1% per annum, should be remembered when interpreting results of magnitude–frequency analysis. Annual flood maxima have been extracted from the series of daily measured and naturalized flows, and return periods (*r*) estimated from:

$$r = \frac{n+1}{m} \tag{1}$$

where *n* is the number of years and *m* the rank of an individual event.

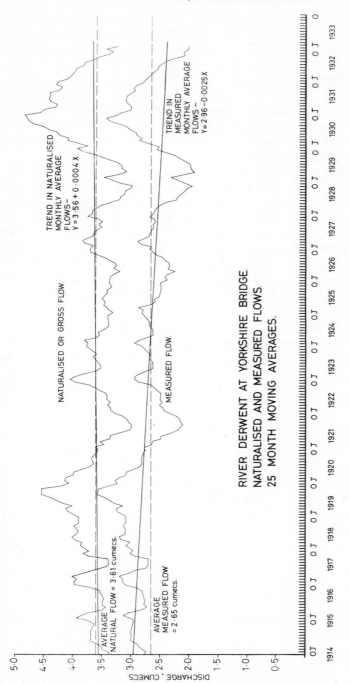

FIGURE 24.2 TRENDS IN GAUGED AND NATURALIZED FLOWS

River Derwent, Yorkshire Bridge, 1913–1933

Figure 24.3 illustrates the reduction in flood magnitude caused by the upstream reservoirs during this period. Mean annual flood was reduced by 15% and the return period of natural mean annual flood (31 m^3/s) was lengthened from 2.33 years to 4.1 years in the gauged flow record. Storage of floods was most efficient when the reservoirs were drawn down, in summer. Separate analysis of summer floods between July and September illustrates this. From Figure 24.3, a gauged flow of 5-year return period is 33 m^3/s (*c.* 14% below the naturalized flow). A similar measured discharge in summer is of 10.5 years recurrence interval, and represents a reduction of 22% from the gross, naturalized flow.

The upper Derwent now experiences even greater disruption of the natural regime. In 1943, the Ladybower reservoir was completed, with a storage capacity of 28.5 million cubic metres. The 1971 hydrographs (Figure 24.4) show a maximum daily naturalized flow of 28.1 m^3/s, which had a return period of 1.75 years in the 1913–1933 data. However, this summer flood was completely absorbed by upstream storage. The maximum gauged flow (6.6 m^3/s) occurred in winter, indicating a 75% reduction in peak flow in 1971. Recent analysis of inflows to Ladybower reservoir shows that by using 20% of total storage for flood protection, Derwent floods can be

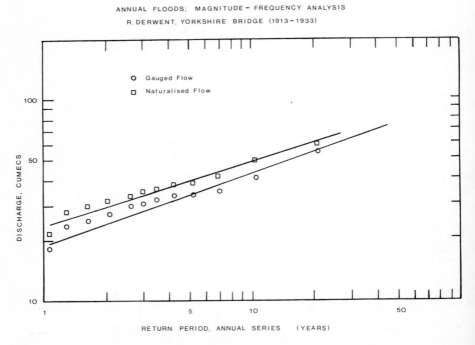

FIGURE 24.3 MAGNITUDE–FREQUENCY ANALYSIS

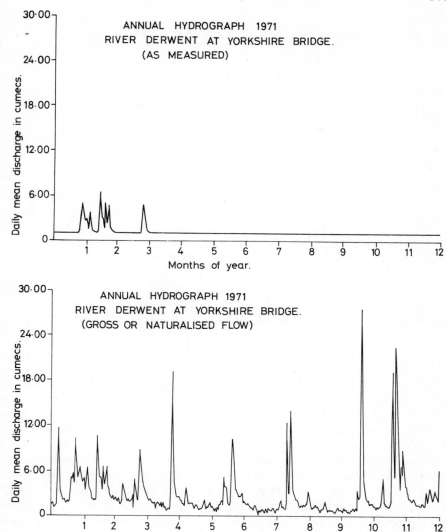

FIGURE 24.4 MEASURED AND NATURALIZED HYDROGRAPHS

River Derwent, Yorkshire Bridge, 1971

virtually eliminated for even the most severe events. Such protection decreases downstream, but the permeable nature of the tributary catchments enables almost complete protection for settlements downstream to Matlock. Figure 24.5 illustrates the effect of this protection at Matlock (where only 18% of the catchment is reservoired), using as examples the floods of 1965

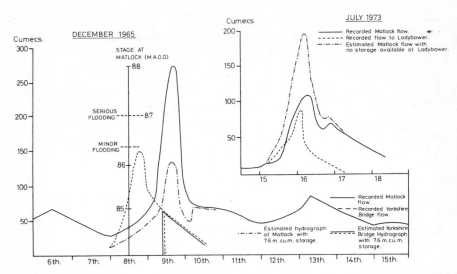

FIGURE 24.5 FLOOD STORAGE CAPACITY OF LADYBOWER RESERVOIR, RIVER DER-
WENT

The diagrams illustrate the fortuitous protection afforded in July, 1973, and the way
in which regulatory drawdown could provide protection at Matlock from an
event equivalent to that of December 1965

and 1973. The former caused flooding 1.8 m deep in the town, but had the
proposed flood drawdown curve for Ladybower been operated, this would
have been prevented. The 1973 storm precipitated 130–150 mm in 36 h, at
a time when the reservoir was naturally drawn down. The level of
Ladybower rose 5 m overnight, and the Derwent downstream recorded
normal winter levels while nearby streams experienced 400-year rainfall
events. This illustrates the effect of reservoir operating policy on downstream flow regime. Most large reservoirs can perform secondary flood-
protection roles, and are increasingly being so used as water engineers
utilize the joint value of the regulatory and supply functions. The attendant
risks of supply failure are small, and can often be offset at far smaller cost
than conventional flood alleviation works. Thus increasingly, streams with
reservoirs will experience a reduced frequency of bankfull flows, with
consequent effects on downstream channel processes (see Chapter 9).

Import Areas

Flood magnitudes in urban catchments are higher than in comparable
non-urbanized basins, because of the rapid time of concentration of urban
runoff. However, imported water contributes to this increase in flood peaks,

after entering the natural channel system through water treatment works. Under dry weather flow conditions, stream-flow in urban catchments in the Midlands and southeast England is mainly imported water. The River Tame at Lea Marston (SP 206935) contains about 90% effluent in its 95% duration flow (6.6 m³/s in 7.4), and the River Avon at its confluence with the Severn about 65% (Severn-Trent Water Authority, 1974; Ledger, 1972). This problem increases as more water is imported, reduced in quality, and added to low natural base flows reduced by lack of recharge. Figure 24.6 illustrates the trend in monthly runoff as a proportion of rainfall at Lea Marston on the Tame. At Trent Bridge, Nottingham, the Trent contained, in 1962–1964, 15.8 m³/s of effluent, and low summer flows may increase from 21 to 27.5 m³/s by the year 2000.

Increased flood peaks in towns are primarily caused by paving and sewering of the catchment; Leopold (1968) estimates two- to threefold increases in mean annual flood where 50% of a small catchment (2.59 km²) is paved and sewered. Hollis (1974) argues that smaller, frequent floods experience the greatest increase (220% for the maximum monthly flood). High magnitude, low frequency, events are proportionately affected least since these floods occur mainly in saturated catchments whose runoff characteristics are similar to the impervious urban area. Harlow New Town has not noticeably increased flood magnitude in the Canon's Brook at recurrence intervals of 20 years or more (Hollis, 1974).

Imported water contributes to this varying proportional increase in magnitude. Addition of a roughly constant amount of imported water increases the smaller, more frequent, floods by a greater percentage. Although urbanization reduces natural base flow, because recharge and interflow are

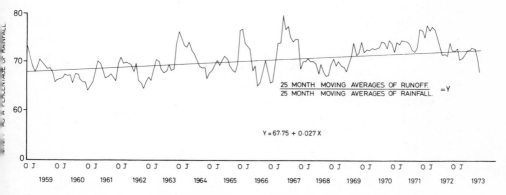

FIGURE 24.6 TRENDS IN 25-MONTH MOVING AVERAGES OF RUNOFF AS A PERCENTAGE OF 25-MONTH MOVING AVERAGES OF RAINFALL

River Tame, Lea Marston, 1957–1974. The trend reflects increasing amounts of imported 'runoff'

K. S. Richards and R. Wood

reduced, imported water provides continuous 'base flow' on which the sharp urban flood peaks are superimposed. The duration curve for Lea Marston on the River Tame illustrates this continual base flow (Figure 24.7). Magnitude–frequency analysis of flows in urban streams is complicated, since stream-flow will reflect the recent 3% per annum increase in water supply (Trent River Authority, 1968). Furthermore, sampling of annual maximum floods over several years includes the effects of increasing paved area. Hence a partial duration flood series for a short period is preferable, and justifiable because independence of events sampled is maintained as urban floods recede almost as rapidly as they rise. Extracting 22 independent flood peaks above 25 m^3/s from the 1971–1973 Lea Marston (River Tame) record permits the magnitude–frequency analysis of Figure 24.8. Lea Marston is downstream from Minworth (SP 165920) water reclamation works, where the imported water is added to stream-flow. The addition of an average of 5.2 m^3/s of imported water in this period increases mean annual flood by 8.5%, and the proportional increase is greater for high frequency floods. Those with 0.5-year return periods are increased by 15%. The annual series suggests similar percentage increases.

While these increases are smaller than those caused by high urban runoff rates, they contribute to the increased frequency of hydrological events capable of transporting sediment and initiating bed and bank erosion. The effect of imported water on flood frequency is complementary to the altered

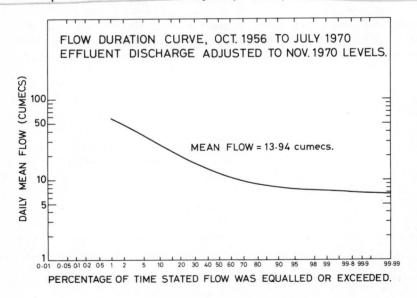

FIGURE 24.7 FLOW DURATION CURVE

River Tame, Lea Marston, October 1956–July 1970

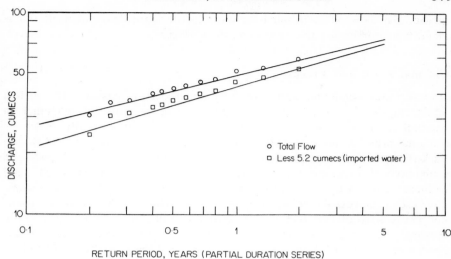

FIGURE 24.8 MAGNITUDE–FREQUENCY ANALYSIS

River Tame, Lea Marston; measured flows, and measured flows less imported water
(1971–3, partial duration series)

runoff characteristics of urban catchments. At Lea Marston, a 'natural' mean annual flood of $51\,\text{m}^3/\text{s}$ recurs on average every 1.25 years because of imported 'base flow' rather than every 1.83 years.

Imported water only achieves geomorphological significance when added to the natural network through sewers and water treatment facilities. Apart from natural channels, which are altered in detail by culverting and realignment, there are two man-made urban networks whose interconnection varies between and within cities. Older combined sewer systems carry both foul and storm waters and, after rain, increased flow demands storm sewage overflows (Lester, 1967). This inevitably causes serious pollution of water courses. In newer urban areas separate storm and foul systems are installed, so that a lesser increase in flow in the foul system occurs after rain. However, the front roof of a house often drains to a storm sewer and the back to a foul sewer. Thus some increase in flow in the foul water system after rain is a common occurrence, and an increasing tendency is to provide temporary storage to avoid overloading facilities at treatment works. Storm water excesses are increasingly subjected to similar storage. For example, plans for the urban drainage of Milton Keynes include off-stream balancing lakes in the Ouzel valley and on-stream lakes on Loughton Brook (which drains to the Ouse). These link with upstream gauges to provide a sophisticated flood control network (Davies, 1974). Thus control of qualitative and

quantitative aspects of urban hydrology increasingly demands further engineering intervention in the hydrological process.

Channel Form and Process

In supply catchments with reservoirs and receiving basins within urban
areas, the main observed effect of altered flood magnitude is a change in
channel capacity. In urbanized basins, increased channel width or cross-
sectional area has been identified by Hammer (1972), Leopold (1973), and
Park (Chapter 8), although the relative effects of urbanization on floods and
sediment yields confuse the issue somewhat. Engineers recognize the need
to increase downstream channel capacity in urban drainage schemes (Shaw,
1974). In mid-west American reservoired streams, degradation of the channel bed is often observed (Leopold, Wolman, and Miller, 1964, pp. 454–6).
However, 'highland' streams impounded in Britain generally have lower
concentrations of load to deposit, and coarse, armoured beds. Therefore
reduction in channel capacity may be observed rather than degradation
(Gregory and Park, 1974).

Simple changes in width or cross-sectional area are unlikely in either case.
In urban areas channel widening may be a rapid process, but a complex,
variable cross-section may be necessary to accommodate low flows without
silting problems, and floods without bank erosion. The required section is
often achieved artificially, but may evolve naturally as depositional bars
develop to be frequently inundated, removed, and reformed. Adjustment to
reduced flood peaks below reservoirs is a slower process, involving depositional berms to narrow the channel. These are stabilized by vegetation and
confine the flow to approximately the new mean annual flood level. Under
certain circumstances reduced competence prevents the necessary sediment
transport (Petts, personal communication, 1976 and Chapter 9), so that
adjustment occurs in the water cross-section alone, and the only evidence of
altered hydraulic regime is a reduced frequency of bankfull and overbank
flow, and a reduced rate of bank migration.

Where relatively complex sections evolve, single-valued parameters such
as width or area vary in significance: this may explain some contradictions in
studies by Hammer (1972) and Leopold (1973). Furthermore, channel shape
(as opposed to size) responds to alterations in sediment and sediment load.
Although channel changes may be identified by extrapolating relationships
between upstream channel capacity and basin area (Gregory and Park,
1974), the conventional hydraulic geometry approach (Leopold and Maddock, 1953) has the advantage of correlating a morphometric parameter to
the process variable, a discharge measure. It may be extended to include the
effects of bank sediment, in multivariate relationships of the form,

$$W = 33.1 Q_m^{0.58} B^{-0.66} \qquad (2)$$

(Ferguson, 1973) where Q_m is mean annual flood and B is bank per cent silt-clay. Downstream from a reservoir, the removal of bed load and reduction of competence may cause depositional berms with a high silt-clay content which would encourage a narrower section than the reduced peak discharge predicts. Urbanization, however, need not affect pre-existing flood plain sediment, which is a semi-independent inherited control of the cross-section. If bank sediment is unchanged during a period of channel widening after urbanization upstream, the width change (ΔW) can be predicted from

$$\Delta W = (\Delta Q_m)^b \tag{3}$$

where ΔQ_m is the change in mean annual flood, and b is the exponent in the downstream hydraulic geometry width relation. Similarly, changes in depth and velocity are predicted from:

$$\Delta d = (\Delta Q_m)^f \tag{4}$$

and

$$\Delta v = (\Delta Q_m)^m \tag{5}$$

so that

$$\Delta w \cdot \Delta d \cdot \Delta v = \Delta Q \tag{6}$$

Thus a discharge change is accommodated by the three degrees of freedom of the flow cross-section. Width may change less than expected, because depth and velocity also change. Figure 24.9 illustrates the magnitude of change in width for various changes in discharge: if mean annual flood doubles, width increases 1.2 to 1.5 times, depending on the pre-existing equilibrium channel geometry (the normal range of observed values of b being between 0.3 and 0.6). Reduction of mean annual flood, by a factor of 0.85, as experienced on the lower Derwent between 1913–1933, would cause a practically imperceptible reduction of width of from 0.90 to 0.96 of the original value, again depending on the original equilibrium state.

The effect of reservoir storage on flood magnitude depends on the sequence of flows; winter floods are affected less than summer ones. Conventional correlations of mean annual flood to 'bankfull' channel dimensions ignore the time of occurrence of peak discharge although this may bear on bank erosion. For example, frost action on bank material in winter (Hill, 1973) may weaken the banks and make them more susceptible to erosion, so that a peak annual flood in summer has less morphological erosional significance than an identical winter peak. Since the effect of the reservoir on flood magnitude is less apparent in winter, downstream channel changes may be less than anticipated. Furthermore, the flashy floods experienced in urban catchments do not thoroughly saturate the banks, which remain relatively stable (Harvey, 1974). Channel widening in this circumstance may be less apparent. Frequently, therefore, the increased (urban) widths or reduced (reservoir) widths fail to lie outside confidence bands established for

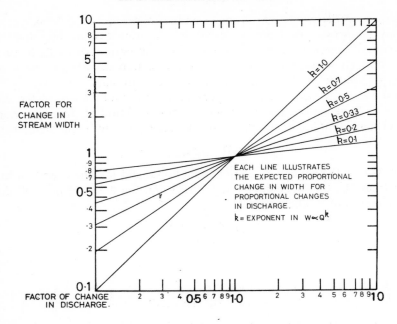

FIGURE 24.9 CHANNEL WIDTH CHANGES

Expected proportional changes of width for given proportional changes in discharge,
dependent on prior equilibrium hydraulic geometry

width-discharge or width-channel length relationships developed for sections
upstream from town or reservoir. Prediction of channel changes awaits
improved data on sedimentological changes as well as discharge changes,
and may be best achieved using multivariate process-based relations such as
equation (2).

The regime of urban catchments creates many channel design problems.
Discharge of effluent into the channel demands a low-flow section sustaining
sufficiently high velocities to transport sediment without silting. Flood peaks,
however, must be accommodated in larger sections without excessive velocities and rapid bank erosion. Lack of information about the required
channel geometry has caused problems. Nixon (1959) quotes the widening
of the Tame from 12.2 m to 21.4 m, and subsequent shoaling, in-channel
vegetation growth, and narrowing to 15.3 m. Drastic enforced changes in
channel capacity were obviated by progressive natural channel changes
during urban development. The range of flows experienced demands complex sections, and this is reflected in channel design (Figure 24.10). The
section adopted at Water Orton (SP 169915) contains low flows to maintain
sediment transport; high flows spread over berms and dissipate their energy
in overcoming the roughness of the section. The alternative (Medrington,

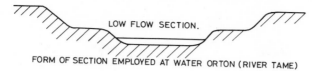

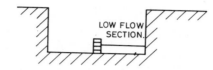

See text for explanation.

FIGURE 24.10 TYPICAL CHANNEL DESIGNS FOR URBAN STREAMS

1959), confines low flows on one side of the central divide, while higher discharges spill over to fill the section. These designs mimic natural channels with marginal berms or central bars, which stabilize if vegetation growth occurs at low flow. Nevertheless, the information available to rationalize such design is limited, and most practical regime theory (Ackers, 1972) is based on simple sections.

In flood, rivers draining urban areas carry polluted water if partially separate sewer systems overflow. Rating curves for an individual flood in the Tame are illustrated in Figure 24.11 (data from Lester, 1967). The passage of the flood is characterized by rapidly increasing suspended sediment as sewers are flushed and overflows occur, and stream bed sludge is thrown into suspension. In the early stage, the exponent of the suspended solids curve is 2.75, a rapid increase of concentration by any standards. The other parameters demonstrate the poor water quality in this stage of the flood. Once initial scouring of sewers has occurred, overflows cease, cleaner natural runoff increases in proportion, and the water quality on the falling limb improves. These are therefore well defined hysteresis loops in all rating curves. These variations in water quality reflect the nature of channel sediment in highly urbanized rivers, where conventional parameters such as dissolved load and suspended mineral sediment are virtually irrelevant. The range of solute concentrations in the Exe (Walling and Webb, 1975) is *c.* 25 to *c.* 650 mg/l which is less than that in Coventry's sources of treated water supply (31 to 1181 mg/l; Severn-Trent Water Authority, 1973). This parameter is necessary when analysing chemical denudation in 'natural' catchments, but less relevant in urban catchments. Suspended sediment load reflects erosion during the construction phase (Wolman and Schick, 1967; Walling and Gregory, 1970), but rivers like the Tame carry substantial loads of organic sediment, which may encourage unstable channel geometry.

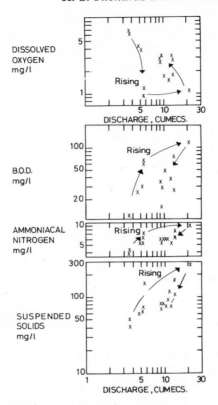

FIGURE 24.11 WATER QUALITY VARIATIONS DURING A FLOOD

River Tame, Water Orton

The organic and mineral sediment suspended by flood discharge is deposited when the flood recedes, but frequent flooding and poor water quality preclude stabilization by vegetation. Nevertheless, between floods deposition of marginal or central bars helps confine the flow and maintain some sediment transport. This instability is attested by the hydraulic geometry of the Water Orton gauging station on the River Tame (Figure 24.12). Considerable low flow scatter occurs as flood scour causes frequent detailed changes in the channel bed. A band containing observed depths and velocities up to about 8 m^3/s implies about 25% variation, while at higher discharges less than half this variability occurs. This reflects a basically rectangular section between near-vertical banks, but low flow deposition of bars within that section. These deposits are suspended during floods (Figure 24.11), so that frequent changes in the low flow stage–discharge relation are common (Ledger, 1972).

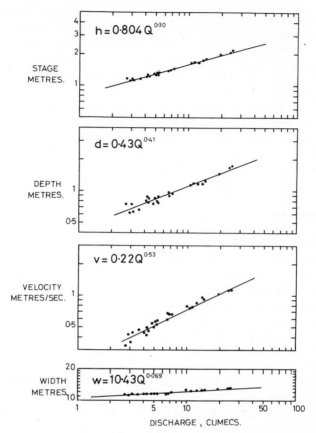

FIGURE 24.12 HYDRAULIC GEOMETRY AND STAGE–DISCHARGE RELATIONS
River Tame, Water Orton

Conclusion

A tentative conclusion is that man-induced changes in channel geometry are frequently less apparent than might be expected, and remain unpredictable. Reservoir storage has always provided some flood regulation, but its effect often decreases rapidly downstream. The Derwent at Yorkshire Bridge was substantially reservoired before construction of Ladybower dam (25% of upstream channel length), but mean annual flood was only reduced by 15%. The greater effect noted by Gregory and Park (1974) suggests that the impact of a reservoir depends on the initial natural storage provided by the unreservoired catchment. Redistributed water may increase floods by similar proportions in urban catchments, contributing to the higher frequency of channel forming events caused by urban runoff. Nevertheless, it is impossible to predict changes in channel sections without considering the pre-existing equilibrium, and alterations in sediment load as well as discharge. At present, there are few general rules about the effect of intervention by man on sedimentological phenomena (Leopold, 1968).

In urban streams, sedimentological aspects are closely related to qualitative considerations, and future environmental engineering of channel processes is necessary to improve water quality. The Severn-Trent Water Authority is considering as a long term policy the development of the Trent as a source of supply. However, increasing degradation of the Tame jeopardizes this policy. One scheme to improve the Tame envisages a series of purification lakes in disused gravel pits, which would serve to store floodwater and remove sediment (Trent River Authority, 1968). The water would be returned to the river after the flood has subsided, with a much reduced load. This addition of relatively clean water, with lower suspended sediment concentrations, would encourage bed and bank erosion. Future development of regulating reservoirs and changes in the operating rules of direct supply reservoirs will increase the impact of these structures on hydrologic regime. Controlled releases to provide drawdown will protect downstream reaches from bankfull or overbank flows, but may increase the frequency of flows above sediment transport threshold. Furthermore, these releases may transmit more sediment through the reservoir.

These continuing examples of environmental engineering suggest that few significant streams in Britain will remain without some form of conscious or unconscious disruption by human activity, either geomorphologically or hydrologically. Traditional areas of enquiry may become less relevant. Solutes and sediments added artificially to streams already confuse estimates of denudation rates (Edwards, 1973). Water transfers cause rivers to carry water whose dissolved load reflects the geology, land use, and precipitation of remote catchments, as well as degradation by industrial and domestic pollutants. The products of chemical denudation in supply catchments are transferred elsewhere. Some traditional geomorphic concepts may be questioned in view of this developing situation. Nevertheless, other models may be of considerable value. Techniques for analysing sediment hydrographs (Walling, 1974, 1975) may be applied to unstable urban channel sections and developed for quality control purposes. The flexibility of hydraulic geometry investigations may be employed to analyse channel forms with complex cross-sections (Richards, 1976), to aid design of channels in urbanized rivers. In particular, application of geomorphological models to problems of prediction, design, and control may become increasingly desirable.

Acknowledgements

The authors acknowledge the assistance of the Severn-Trent Water Authority, particularly in making available substantial amounts of data. The opinions expressed in the paper are those of the authors, and do not represent a statement of the official views of the Authority. The authors would also like to express their thanks to Shirley Addleton for drawing the diagrams.

References

Ackers, P. 1972. River regime: research and application. *J. Inst. Water Eng.*, **26,** pp. 257–81.

Anderson, M. G. 1975. Demodulation of streamflow series. *J. Hydrol.*, **26,** pp. 115–21.

Davies, L. H. 1974. Problems posed by new town development with particular reference to Milton Keynes. Paper 2, In *Rainfall, runoff and surface water drainage of urban catchments*, Proc. Res. Colloquium, Dept. Civ. Eng., Bristol Univ. April 1973. C.I.R.I.A.

Dury, G. H. 1965. Theoretical implications of underfit streams. *U.S. Geol. Survey. Prof. Paper 452-C.*

Edwards, A. M. C. 1973. Dissolved load and tentative solute budgets of some Norfolk catchments. *J. Hydrol.*, **18,** pp. 201–17.

Ferguson, R. I. 1973. Channel pattern and sediment type. *Area*, **5,** pp. 38–41.

Gregory, K. J. and Park, C. C. 1974. Adjustment of river channel capacity down- stream from a reservoir. *Water Resour. Res.*, **10,** pp. 870–3.

Hammer, T. R. 1972. Stream channel enlargement due to urbanization. *Water Resour. Res.*, **8,** pp. 1530–40.

Harvey, A. M. 1974. The influence of flow regime on the relationships between sediment type and channel shape: unpublished paper presented at Br. Sedimen- tol. Res. Group meeting: *The Geometry of River Deposits*. October 1974.

Hill, A. R. 1973. Erosion of river banks composed of glacial till near Belfast, Northern Ireland. *Z. Geomorph.*, **17,** pp. 428–442.

Hollis, G. E. 1974. The effect of urbanization on floods in the Canon's Brook. Harlow, Essex In Gregory, K. J. and Walling, D. E. (Eds), *Fluvial Processes in Instrumented Watersheds*. Inst. Br. Georg. Spec. Publ. No. 6, pp. 123-39.

Howe, G. M., Slaymaker, H. O., and Harding D. M. 1967. Some aspects of the flood hydrology of the upper catchments of the Severn and Wye. *Trans. Inst. Br. Geogr.*, **41,** pp. 33–58.

Knighton, A. D. 1975. Variations in at-a-station hydraulic geometry. *Amer. J. Sci.*, **275,** pp. 186–218.

Ledger, D. C. 1972. The Warwickshire Avon: a case study of water demands and water availability in an intensely used river system. *Trans. Inst. Br. Geogr.*, **55,** pp. 83–110.

Leopold, L. B. 1968. Hydrology for urban land planning—a guidebook on the hydrologic effects of urban land use. *U.S. Geol. Survey*, Circular 554.

Leopold, L. B. 1973. River channel change with time: an example. *Geol. Soc. Amer. Bull.*, **84,** pp. 1845–60.

Leopold, L. B. and Maddock, T. 1953. The hydraulic geometry of stream channels and some physiographic implications. *U.S. Geol. Survey. Prof. Paper 252.*

Leopold, L. B., Wolman, M. G., and Miller, J. P. 1964. *Fluvial processes in Geomorphology*: Freeman, San Francisco.

Lester, W. F. 1967. Effect of storm overflows on river quality. In *Storm Sewage Overflows*. Inst. Civ. Eng., symposium, pp. 13–21.

Linsley, R. K., Kohler, M. A., and Paulhus, J. L. H. 1949. *Applied Hydrology* McGraw-Hill, New York.

Medrington, M. N. 1959. Discussion of a study of the bankfull discharges of rivers in England and Wales, by M. Nixon. *Inst. Civ. Eng. Proc.*, **14**, pp. 397–9.

Nixon, M. 1959. A study of the bankfull discharges of rivers in England and Wales. *Inst. Civ. Eng. Proc.*, Paper 6322, pp. 157–74.

O'Riordan, T. and More, R. J. 1969, Choice in water use. In Chorley, R. J. (Ed.) *Water, Earth and Man*. Methuen, London. pp. 547–573.

Rango, A. 1970. Possible effects of precipitation modification on stream channel geometry and sediment yield. *Wat. Resour. Res.*, **6**, pp. 1765–70.

Rees, J. 1969. Thirsty land with a temperate climate. *Geogrl Mag. Lond.*, **42**, pp. 136–45.

Richards, K. S. 1976. Intra-section variation in the width–discharge relation. *Geol. Soc. Amer. Bull.*, **87**, pp. 199–206.

Schumm, S. A. 1968. River adjustment to altered hydrologic regimen— Murrumbidgee River and palaeochannels, Australia. *U.S. Geol. Surv. Prof. Paper 598*.

Schumm, S. A. and Lichty, R. W. 1965. Time, space and causality in geomorphology. *Amer. J. Sci.*, **263**, pp. 110–9.

Shaw, T. L. 1974. Requirements of drainage design for large urban areas. Paper 13. In *Rainfall runoff and surface water drainage of urban catchments*, Proc. Res Colloquium Bristol Univ. April 1973, C.I.R.I.A.

Simons, M. 1969. Long term trends in water use. In Chorley, R. J. (Ed.), *Water, Earth and Man*. Methuen, London, pp. 535–44.

Severn-Trent Water Authority 1974. *Water Quality 1973*.

Trent River Authority 1968. *Water resources—a preliminary study*.

Walling, D. E. 1974. Suspended sediment and solute yields from a small catchment prior to urbanization. In Gregory, K. J. and Walling, D. E. (Eds.) *Fluvial processes in instrumental watersheds*. Inst. Br. Geogr. Spec. Publ. no. 6, pp. 169–92.

Walling, D. E. 1975. Solute variation in small catchment streams; some comments. *Trans. Inst. Br. Geogr.*, **64**, pp. 141–7.

Walling, D. E. and Gregory, K. J. 1970. The measurement of the effects of building construction on drainage basin dynamics. *J. Hydrol.*, **11**, pp. 129–44.

Walling, D. E. and Webb, B. W. 1975. Spatial variation of river water quality: a survey of the River Exe. *Trans. Inst. Br. Geogr.*, **65**, pp. 155–71.

Wolman, M. G. 1967. A cycle of sedimentation and erosion in urban river channels. *Geogr. Annlr.*, **49A**, pp. 385–95.

Wolman, M. G. and Schick, A. P. 1967. Effects of construction on fluvial sediment: urban and suburban areas of Maryland. *Wat. Resour. Res.*, **3**, pp. 451–62.

K. J. GREGORY

Professor of Geography
University of Southampton

25

Channel and Network Metamorphosis in Northern New South Wales

Geomorphological studies of channel changes during the last decade have tended to concentrate upon either channel geometry, or channel pattern, or drainage network. Channel adjustments may result from changes of climate or from man's impact, and the adjustments of channel form which can follow changes of river channel process can be visualized as adjustments of size, shape, and composition of basin network, channel pattern and channel geometry (Table 25.1). Future research should be able to demonstrate more clearly the interrelations between these adjustments (Table 25.1) as a basis for an integrated understanding of the metamorphosis of channels and network. Progress towards this objective is particularly desirable in Australia where hydrologic regimes are characteristically flashy and where great progress in the investigation of compound river channel cross-sections has been made (Woodyer, 1974). This problem is approached by outlining general approaches offered to express the relations between the three aspects of the fluvial network, by reviewing the progress made against the Australian background, and by illustrating metamorphosis of the network and channels in an area of northern New South Wales.

Progress Towards a General Approach

A model of induced aggradation was proposed by Strahler in 1956 and this represented an important attempt to integrate several spatial aspects of the fluvial morphology of the basin. The consequence of accelerated erosion producing gullies was seen as more extensive flood plains downstream. Strahler employed the Pi theorem as a method of approach whereby drainage density (D_d), maximum basin relief (H), and ground surface or channel gradient (θ) were combined in a geometry number (HD/θ), and

TABLE 25.1 POTENTIAL ADJUSTMENTS OF CHANNEL MORPHOMETRY
(After Gregory, 1976d)

Change of	Landform aspect of drainage basin		
	River channel cross-section	River channel pattern	Channel network
Size	Increase or decrease of channel capacity	Increase or decrease in meander wavelength; change of spectra	Increase or decrease in network extent and density
Shape	Adjustment of shape such as width/depth ratio	Adjustment of sinuosity	Alteration of drainage pattern
Composition	Material transported through channel (and material of bed and banks) may change	Alteration from single thread to multi-thread channel or vice versa	Change in composition of network

runoff intensity (Q) and an erosion proportionality factor (Ke = mass rate of removal of debris per unit area) gave a Horton number $Q(HD/\theta, QKe) = 0$, but there was no inclusion of channel capacity or channel planform in this model. More recently, in a study of the Driftless area of Wisconsin, Trimble (1976) has provided a specific example of network change and has suggested that since 1938 overall sedimentation rates have decreased by almost half, that channel enlargement has been dominant in tributaries, and that the stream bed in the main valley has aggraded as much as the overbank flood plain. A similar approach figures in Wolman's model of the consequences of urbanization for river channels (Wolman, 1967), but although stable and unstable channel phases were identified there was no inclusion of network and channel capacity, although capacity has been used in more recent studies of specific cross-sections in other areas (e.g. Hammer, 1972; Gregory and Park, 1976).

The general problem has been identified by Chorley and Kennedy (1971 Figure 6.26). They argued that the most probable network was one in which stream channels exist at the foot of each infinitesimally small slope so that a high density of small channels exists. The most efficient situation would be where the drainage area is drained by a single, very large capacity channel. However most natural networks occur with an intermediate density reflecting the amalgam of climatic, topographic, pedologic, lithological, and biotic controls (Gregory and Gardiner, 1975). This approach to the channel network underlines an important task for fluvial geomorphology, and one which is analogous to the notion of the limiting least-energy expenditure and equal-energy expenditure models advocated by Langbein and Leopold (1964) for application to channel geometry and the river long profile. A necessary extension of this general model (Chorley and Kennedy 1971, Figure 6.26) is the need to relate channel density and channel size quantitatively, and also to include the facts that the drainage network is composed of several components and shows temporal fluctuations. Not only do short-term network fluctuations have to be included but allowance should also be made for long-term changes. In a study of Lake Ontario watersheds, Ongley (1975) proposed that 'the complexity of the network hierarchy represents a geometry formed in response to the most intense network-forming climatic regime experienced during its existence.' He further argued that if contemporary networks are not in equilibrium with contemporary climatic regimes and land management practises, then they may be either expanding or in a state of static disequilibrium. The current imbalance between network form and process could be manifested by agricultural disturbance which promoted destruction of natural channels and the creation of artificial elements, by accelerated erosion leading to extending networks and headward growth, and by networks which are over-developed for present conditions, for example where the original network was either formed or modified in early deglacial times. Ongley (1975) contended that the networks in southern Ontario are unlikely to have achieved a new homeostasis in the short period since colonization despite the extent of deforestation. However, whether any change in the network can be detected depends upon the way in which the network is delimited and analysed, and also upon whether consideration is accorded to the size of the channels as well as to their density. It is imperative that we know the magnitude of events necessary to occasion adjustments and Schumm (1973) has demonstrated the need to identify thresholds as a basis for interpreting change.

Further approaches to the achievement of integration between indices have been afforded by studies of particular aspects of the network, by the development of complex indices of basin character, and by relations between channel capacity and indices of basin size.

Studies of particular aspects of the network have been the basis for a number of developments, including relationships between channel pattern and geometry as a basis for the interpretation of underfit streams (Chapter 18), of hydraulic geometry, and of channel metamorphosis. A comparison of the streams flowing north and south from the Blue Mountains of Jamaica was effected by Gupta (1975) using analyses based upon at-a-station hydraulic geometry relationships. This approach could be the basis for the investigation of temporal change but inclusion of network extent is not easily accomplished. Valuable studies by Schumm (1969, 1971) have been based upon relations established between river sinuosity, weighted mean percent silt clay (M) in the channel bed and banks, channel width, width/depth ratio (F), stream discharge, and sediment discharge, and these have been the basis for interpretations of channel metamorphosis. The basic relationships have not yet been evaluated for a wide variety of areas, and when tested some of the relationships require modification. Riley (1975) has shown that the relation $F = 255\,M^{-1.08}$ does not apply to clay-bedded streams in the Namoi–Gwydir distributary system of eastern Australia.

Furthermore the approach does not embrace the extent of the drainage network and the density of channels.

Complex indices of basin character include the geometry number HD/θ (Strahler, 1956) referred to above, and other indices proposed to express the dimensions of a drainage network including not only the extent but also the size of channels. Ruggedness number is the product of drainage density and relief (HD_d) but does not explicitly incorporate stream channel size. Similarly the stream gradient index proposed by Hack (1973) as the product of channel slope at a point (S) and channel length of the main stream above that point (L) in the form SL, does not incorporate channel size. A development of indices by Flint (1974) included stream order but not capacity, and indices proposed by Lustig (1965) for the purpose of analysing sediment yields also did not include channel size. Many such complex indices therefore do not provide a completely adequate index of fluvial geometry, they suffer from the disadvantage that such hybrid indices give values to which it is difficult to ascribe a precise physical meaning, but they have been useful in statistical analyses of water and sediment yield—the purpose for which many such complex indices were designed.

Alternative attempts to express the character of the drainage network have been elaborated in relation to hydrograph generation based upon stream ordering and network topology. Thus Surkan (1974) has simulated the effects of large storms travelling over drainage networks by specifying an average flow velocity for each segment of the network. The flow velocity for a segment reflects channel size and channel character but channel capacity is not incorporated in the technique which depends essentially upon the length and arrangement of channels. A similar approach had been developed by Rastogi and Jones (1969) who modelled hydrographs according to the composition of the network and also included consideration of channel roughness. Further developments of this kind are necessary for the purposes of improved physical models (e.g. Singh, 1975) and such developments should be useful for study of channel change.

A comprehensive expression for fluvial channel landform has often been sought by resort to the relation between stream channel capacity and drainage area (e.g. Park, Chapter 8, this volume) or to that between channel width and drainage area (Day, 1972). This technique has been employed for applications of the ergodic hypothesis. However drainage area is not necessarily a good surrogate for discharge or for channel length, and an alternative index of channel size can be provided by the relation between the total channel length above particular sites and the channel capacities at those sites. Because this combines *channel* geometry and network *morphometry* it is proposed as an index of channel morphometry. The basic task is illustrated in Figure 25.1 by a comparison of two basins of identical area but with different network densities. If channel capacity is measured at a number of sites throughout each basin and related to the drainage area

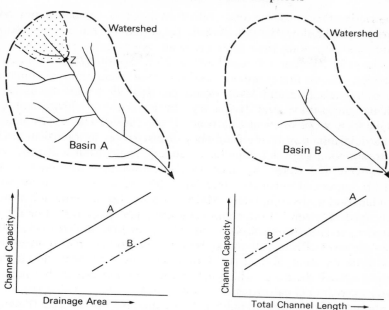

FIGURE 25.1 RELATIONS OF CHANNEL CAPACITY WITH DRAINAGE AREA AND WITH TOTAL CHANNEL LENGTH

Two relationships are contrasted for data obtained at a number of sites from basin A with a high drainage density and from basin B with a low drainage density. Whereas the relation between channel capacity and drainage area does not allow for channel size in relation to total stream length, this is incorporated into the relation between channel capacity and total channel length

above each site, the relation between channel capacity and drainage area may be very different from that between total stream length ($\sum L$) and channel capacity (C) (Figure 25.1). To express the volume of the channel between network limits n_1 and n_2 we can integrate the equation $C = a \sum L^b$ so that $\int_{n_1}^{n_2} C = a \int_{n_1}^{n_2} \sum L . \, \mathrm{d} \sum L$ and this will provide a value for the volume of the channel reach. Whereas this improves upon the drainage area–channel capacity relation, it does not include channel slope and a future necessary refinement will be to incorporate a value of slope into the expression. Values of stream power (Bagnold, 1966) have hitherto been calculated for particular reaches and Chih Ted Yang (1972) has expressed unit stream power, $\mathrm{d} Y/\mathrm{d} t$, in terms of the longitudinal distance (X), time (t) and elevation above a datum (Y) as $\mathrm{d} Y/\mathrm{d} t = \mathrm{d} X/\mathrm{d} t \cdot \mathrm{d} Y/\mathrm{d} X = VS$ where $V =$ average water velocity and $S =$ energy slope. A similar approach applied to the network could be the basis for an index of network power above a particular site.

The need for a closer integration of analyses of channel geometry and network morphometry in Australia is suggested by the interpretations of channel geometry that have been achieved. Analyses of river channel cross sections at a number of locations in New South Wales demonstrated (Woodyer, 1968) the frequent occurrence of compound channel cross-sections in which certain levels could be associated with discharges of a particular magnitude and frequency. In a series of subsequent papers (McGilchrist, Woodyer, and Chapman, 1968, 1969, 1970, 1972) techniques for the determination of the frequency of discharges to a particular level in the cross section were developed and refined. The existence of these several bench levels underlines the need to know how they are associated with different degrees of network extension so that network metamorphosis may be understood more completely. More recent studies have demonstrated the significance of several flows comprising dominant discharge. For the Cumberland river basin three flows were shown (Pickup and Warner, 1976) to occur: the most effective discharge with respect to bed load transport which has a return period of 1.15 to 1.40 years; the 1.58-year flood; and the natural bankfull discharge which has a return period usually between 4 and 10 years on the annual series.

Although many Cumberland basin streams are not incised (Pickup and Warner, 1976), streams elsewhere in New South Wales have been considered as being incised (Dury, 1968; Woodyer, 1968). Incision has been ascribed to a modified runoff regime following nineteenth century settlement (Dury, 1968). In the Georges river basin south of Sydney, urban and rural

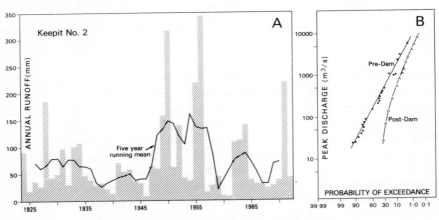

FIGURE 25.2 DISCHARGE BELOW KEEPIT DAM

Annual runoff downstream of the dam (catchment area 5676 km² completed 1960) is shown in A and the flood frequency before and after dam construction (B) is based upon Laurenson *An Economic Study of Keepit Dam.* Prepared by International Engineering Consortium. Issued by WCIC, December, 1969

activity between 1880 and 1920 caused so much silt loss that sediment deposited by floods smothered the oyster beds of the estuary (Munro and coworkers, 1967). With increasing modification of river regimes consequent upon dam construction, the effects upon bed degradation and upon the channel capacities may extend many kilometres downstream from the dam (Australian Water Resources Council, 1969). The flows along many rivers have been altered and below Keepit dam the flood frequency curves (Figure 25.2) before and after dam construction (Laurenson, 1969) indicate the hydrological changes which may be expected to have an influence upon channel capacity. Many of the causes of channel metamorphosis since 1800 in New South Wales may be associated with the land use changes which affected runoff and sediment production. Holmes (1946) summarized the types of soil erosion as including early stages manifested by raindrop impact, sheet erosion, rill erosion. gully erosion, river bank erosion, and road water erosion together with coast erosion, wind erosion, and mass movements of the soil. Specific examples of the transformation of creeks, which were formerly a chain of ponds, into deeply incised water courses due to trampling by stock were instanced for Monaro (Hancock, 1972), and in a study of high mountain catchments accelerated erosion was attributed (Australian Academy of Science, 1957) to roadmaking, logging, building of dams and aqueducts, grazing and trampling, and to fire and insects.

Northern New South Wales: New England

Ample evidence of land use change over a century and a half exists in New England and in the basin of Commissioner's Waters (383 km^2) chosen for study. This basin, which extends from the basalt plateau over the new England granite and also over other rocks, principally greywackes and argillites, has a mean annual precipitation of 850 mm and it includes areas of substantial land-use change. Two dams were constructed for water supply, numerous stock ponds or farm dams exist with an average density between 39 and 116 per 1000 km^2 (WCIC, 1971), and the urban area of Armidale is also included (Figure 25.3). In the basin of Dumaresq Creek the consequences of land management in the catchment have been expressed in upstream erosion, siltation, and flood problems as shown by McConnell, Monteith, and Berman (1966). They produced a map showing the distribution of erosion classes in the basin and suggested that the rate of growth of gullies in basalt material had slowed down, and a comparison of 1943 and 1956 aerial photographs (Hopley, 1965) also indicated that gully erosion is now slow. The effects of the urban area of Armidale can also affect runoff and, more particularly, the amount of sediment delivered to Dumaresq Creek (Douglas, 1975). The consequences of urban effects and of rural land management have created changes on Dumaresq Creek within and downstream from Armidale (Burghardt, Loughran, and Warner, 1968; Douglas,

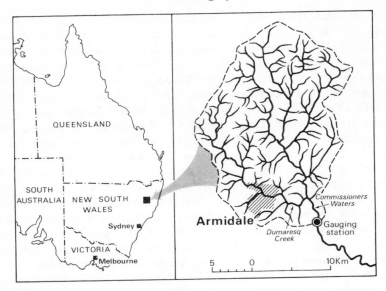

FIGURE 25.3 LOCATION OF COMMISSIONER'S WATERS DRAINAGE BASIN

1976), but no study has hitherto investigated the network of Dumaresq Creek in the wider context of the basin of Commissioner's Waters. An intuitive summary of the hydrological and geomorphological changes is offered in Table 25.2

Present Channel Morphometry

To coordinate investigation of the drainage network and channel capacity a field survey was made of channels throughout the basin of Dumaresq Creek and extended to the whole of the Commissioner's Waters basin by reference to black and white aerial photographs. The channel types recognized were classified as shown in Figure 25.4. In addition the channel-cross section was surveyed at 48 sites throughout the Commissioner's Waters drainage basin to provide data on channel size which could be related to the extent of the network above each site. The definition of channel cross-section and of a 'bankfull level' is complicated by the fact that some sections include up to seven elements in the compound cross section (Gregory, 1976a). A number of methods have been proposed for the consistent determination of river channel capacities (Gregory, 1976b). A method found to be useful throughout this area, where bedrock channels occur in the headwaters and large boulders occur in downstream channel sections, is to employ lichen limits to indicate a consistent channel capacity from one area to another. It has been shown that lichen limits appear to be broadly consistent hydrologically from one area to another and that they represent

TABLE 25.2 POTENTIAL CAUSES OF CHANNEL AND NETWORK METAMORPHOSIS IN NEW ENGLAND, NEW SOUTH WALES

Type of change	Hydrological consequence	Possible consequences for fluvial landforms		
		Channel geometry	Channel pattern	Drainage network
Clearing of eucalypt woodland over 140 years	Increased runoff	Channels enlarged and incised	Enlarged in alluvial sections	Definite channels replace grass-covered depressions
Overgrazing	Increased surface runoff	Channels eroded and enlarged	Enlarged?	Gullying and extension of network
Stock pond construction	Reduced runoff if pond not full			Definite channels may be instituted down valley of stock pond
Dam construction		Erosion immediately downstream, reduced channel capacity	Decreased?	
Road construction	Increased runoff from road surface	Locally enlarged but may be affected by increased sediment supply	Local effects only	Extended locally by roadside gullies
Urbanization	Increased runoff from impervious surfaces and stormwater drainage. Some storage by tank storage pre-1950	Small channels enlarged and increased sediment supply may reduce capacities of larger channels	Increased frequency of flood plain inundation	Parts of network may be regulated,

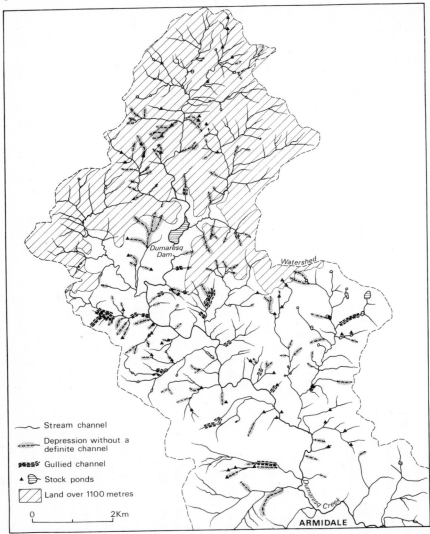

FIGURE 25.4 CHANNEL TYPES IN PART OF DUMARESQ CREEK DRAINAGE BASIN
The classification was based upon field survey and black and white and colour air
photographs at 1 : 10,000 scale

levels equalled or exceeded at least once each year (Gregory, 1976c).
Although the lichen limit capacity is lower than the morphological bankfull
level, it is helpful to indicate the latter level. The lowest lichen limit may
correspond approximately to the most effective discharge as interpreted
elsewhere by Pickup and Warner (1976). The relationship between channel
capacity to the lichen limit and drainage area is shown in Figure 25.5. For

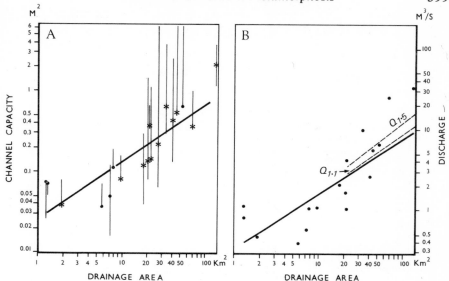

FIGURE 25.5 CHANNEL LICHEN LIMIT CAPACITY AND ESTIMATED LICHEN LIMIT
DISCHARGES RELATED TO DRAINAGE AREA

Lichen limit sites are shown as asterisks in A. Estimated discharges plotted in B were
calculated using the Manning equation and for comparison the 1.5– and 1.1–year
recurrence interval discharges are based upon data from nine gauging stations

each site the channel capacity to the limit was surveyed in the field, the slope
was surveyed in the field or estimated from 1 : 25,000 maps, and the
roughness was estimated for each section. It was then possible to employ the
Manning equation to estimate the lichen limit discharge at each site, and
these estimated discharges were related to drainage area in Figure 25.5. For
purposes of comparison the discharges with recurrence intervals of 1.1 and
1.5 years based upon regional flood frequency analysis using the annual
series are also shown (Figure 25.5).

The relation between lichen limit capacity and drainage area provides a
representation of the present channel morphometry throughout the basin
but it requires improvement in two ways. Because the lichen limit is lower
than the 'bankfull level', the relationships based upon the morphological
bankfull level need to be developed. Furthermore drainage area is not
necessarily the most sensitive surrogate for discharge so that total channel
length may be preferable for the reasons elaborated above. In Figure 25.6
channel capacity is plotted against drainage area for Dumaresq Creek and
the major subdivisions of the basin are also demonstrated. Whereas a single
regression line can be fitted to the scatter of points (Figure 25.7), it is
possible that two regression lines are more appropriate, and that these
reflect the way in which the channel capacity and drainage area relationship

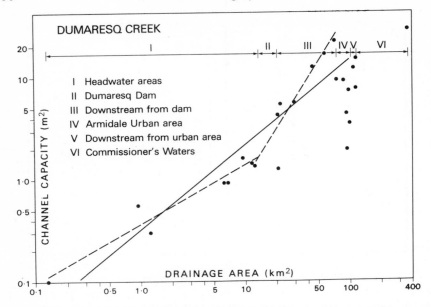

FIGURE 25.6 CHANNEL CAPACITY RELATED TO DRAINAGE AREA FOR DUMARESQ
CREEK

The solid line is the regression line fitted to all points whereas the dashed lines
distinguish the headwater and main channel relationships

differs in the headwater areas where vegetation exercises a more significant
influence than elsewhere in the basin. A similar pattern of points occurs
when total channel length is plotted against channel capacity (Figure 25.7)
for Dumaresq Creek. This relationship was calculated for the entire basin of
Commissioner's Waters and it is then possible to demonstrate the differences in total volume of the network using the technique described above.
The two equations for the entire basin are:

$$C_{BF} = 0.1561 \sum L^{0.7640}$$
$$C_{lichen} = 0.1610 \sum L^{0.6581}$$

The total channel length to the lowest site surveyed is 1107.38 km so that
the two volumes of the channel network may be calculated by integrating
each equation between the limits $\sum L = 0$ and $\sum L = 1107$ km. Thus channel
volume to the bankfull level = 20,738 m²/km = 20,738,000 m³ and channel
volume to the lichen limit = 10,832 m²/km = 10,832,000 m³. The foregoing
relationships and equations are all significant at the 99.9% level and
describe the present channel morphometry in the basin analysed. Multiple
regression equations were also calculated to include channel slope (S) in the
form $C_{BF} = aL^b S^c$.

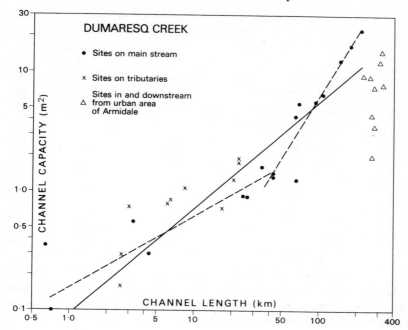

FIGURE 25.7 CHANNEL CAPACITY RELATED TO TOTAL CHANNEL LENGTH FOR DUMARESQ CREEK

The solid line is the regression line fitted to all points but the dashed lines distinguish the headwater and main channel relationships

Adjustment of Channel Morphometry

Three types of change have occurred in the study basin. First there has been gullying and channel incision, although the volume of this cannot easily be determined from the relationships above because gully, or incised, channel cross-section area cannot be related significantly to drainage area or to channel length. Second, there may have been change downstream of dams and stock ponds, and third, some form of change would be expected within and downstream from the urban area of Armidale. The second and third types can be examined in relation to the equations described above.

Substantial morphological adjustment of the river channel downstream of Dumaresq dam (constructed 1898) and Puddledock dam (1943) would not be expected because of the incidence of rock in the channel cross-sections. However, if the capacity of the lichen limit is compared with that expected from the regional relationship (Table 25.3) the lichen capacity is substantially lower than expected. This is ascribed to the lower frequency of peak flows that has characterized the period since dam construction. The headwater sites surveyed were also classified according to their situation in relation

TABLE 25.3 RESIDUALS FROM REGRESSION EQUATION FOR CHANNEL CAPACITY TO LICHEN LIMIT

Site number		Channel capacity to lichen limit		
		1. From equation	2. Actual	Ratio 2 : 1
18	Below Dumaresq Dam	2.50	1.89	0.72
19		2.52	1.26	0.50
49	Below Puddledock Dam	3.68	2.26	0.61
	Density of stock ponds (km^{-2})			
4	3.39	1.81	0.92	0.51
1	3.13	0.94	0.55	0.59
24	2.99	1.16	0.74	1.51
32	2.81	0.92	0.81	1.09
33	2.13	0.75	0.83	1.11
47	2.02	2.56	1.41	0.55
46	1.94	2.63	3.89	1.48
45	1.80	1.59	1.30	0.82
49	0.79	2.89	2.26	0.79
37	0.66	1.42	1.19	0.84
34	0.42	1.26	0.50	0.40
35	0	0.49	0.74	1.51

to stock ponds so that the actual lichen limit capacity could be compared with the regional relationship (Table 25.3). This demonstrates that there is tendency for channels to be smaller than expected from the basins with a high density of stock ponds and to be positive residuals from the multiple regression in areas with few or no stock ponds. This is certainly a conservative estimate because the multiple regression was computed for all rural data. This method cannot be completely conclusive, however, because although stock-pond construction may have led to a reduction of peak flows and their frequency, thus giving a lower capacity depending upon the efficiency of the bywash channel (Kelson 1968), the channel network may also have changed. In the headwater areas the presence of a broad vegetated depression lacking a definite channel may have been transformed after stock pond construction into a depression with a channel. In this case the development of the bywash channel may have encouraged the development of a definite channel downstream.

Armidale provides an urban influence upon Dumaresq Creek which has had more obvious implications for the channel morphometry. Channel cross-sections were surveyed at five sites on Dumaresq Creek as it flows through the urban area of Armidale and also at four sites downstream. Many of these channel cross-sections are multiple in form and it is usually possible to identify a smaller channel (2 in Table 25.4) within the overall channel (1 in Table 25.4) below the flood plain level. Earlier work (Burghardt, Loughran, and Warner 1968) had concluded that the channel capacity of Dumaresq Creek through Armidale was smaller than expected but the actual capacities can now be evaluated against the regional pattern. In Figure 25.6 the capacities within and downstream from the urban area are significantly smaller than expected, and this conclusion is affirmed by the

TABLE 25.4 CHANNEL CAPACITY OF DUMARESQ CREEK THROUGH ARMIDALE, NEW SOUTH WALES

Site number	Comments	1. Capacity before flood plain level (m²)	2. Capacity of channel (m²)	3. Channel capacity according to regional equation (m²)[a]	Ratio 2:3	Ratio 1:3
17	Incised section above urban area	22.9	16.9	14.30	1.18	1.63
16	Martin's Gully tributary	20.2	9.4	14.68	0.64	1.38
15	Drainage from four roads ⎫ Urban	12.8	9.2	15.59	0.58	0.82
14	Drainage from three roads ⎬ area	15.9	4.4	15.65	0.28	1.02
13	Drainage from five roads ⎭	8.7	2.0	15.66	0.13	0.55
12	Two small tributaries	7.4	3.3	16.06	0.21	0.46
11	Two tributaries, one ⎫ Downstream	13.8	3.5	20.21	0.17	0.68
26	from urban area ⎬ from Armidale	12.2	12.2	21.15	0.58	0.58
28	⎭	15.3	3.0	21.49	0.14	0.71
27		7.8	7.8	23.82	0.33	0.33

[a] $\log X_{BF} = 0.5138 \log \sum L - 0.5508 \log S + 0.2164$, where X_{BF} = channel capacity (m²); $\sum L$ = total contributing stream length (km); S = channel slope (m/km) $r = 0.95$, $N = 40$.

relation between channel capacity and total channel length (Figure 25.7). Whereas the urban sites are compared with sites upstream on Dumaresq Creek in Figures 25.6 and 25.7, they have been compared with the regional pattern from the entire basin in Table 25.4. This data indicates that the outer channel has a capacity which decreases downstream and which is often less than the expected value, and at one site is as little as 0.13 of the capacity expected (Table 25.4). It is therefore concluded that the capacity of the channel passing through the urban area is significantly less than indicated by the regional relationship (Figures 25.5 and 25.6, and equation of Table 25.4) and as little as one-fifth of the capacity expected. Calculating the volume of the channel passing (p. 393) through and downstream of the urban area according to the regional equation gives an expected channel volume of 190 m²/km or 190,000 m³. However, integrating the equation fitted to most of the actual urban capacities (Figure 25.7) gives a volume of 110 m²/km or 110,000 m³. The actual channel volume therefore appears to be only 58% of the volume expected from the regional relationship.

This apparent reduction in capacity could however be a result of the ergodic hypothesis, in which space is substituted for time, because the rural relationship above the urban area may not reasonably be extrapolated below the urban area. This limitation has been overcome to some extent by selecting sites throughout a large basin containing the urban area. To investigate further the reduction in capacity, it was possible to estimate discharges in order to compare them with the regional relationship and to use records at one site to demonstrate changes over time. The Manning equation was employed to estimate discharges for the inner channel within the urban area and this gave discharges for the channel at five sites in the range 2.85–13.89 m³/s. According to the regional flood frequency curve, a discharge with a recurrence interval of 1.1 years would be 39.2 m³/s, and for 1.5 years would be 57.95 m³/s. The estimated channel discharges are therefore below a third of the expected regional values. At the site of Stephens Bridge, Armidale, surveys were undertaken in 1927 and 1967–8 (Armidale City Council Surveyors Department), a further one in 1964 (McConnell, Monteith, and Berman, 1966) and these could be compared with a survey by the author in 1975. Comparison of these four surveys (Figure 25.8) allows estimation of the channel capacities at different dates and it appears that a progressive reduction has occurred from 65 m² in 1927, to 33 m² in 1964, 18.5 m² in 1967–8, and 8.7 m² in 1975. Up to 1964 it had been noted (McConnell and coworkers 1966) that up to 1 m of sediment had accumulated in the channel cross section at this location.

The Causes of Channel Morphometry Adjustment in Urban Areas

The reduction in channel capacity within and downstream from the Armidale urban area seems to be confirmed by comparison with the spatial

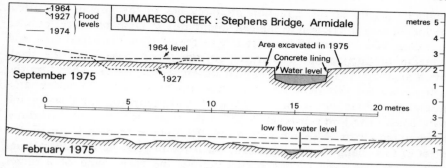

FIGURE 25.8 SECTIONS SURVEYED ACROSS DUMARESQ CREEK AT STEPHENS BRIDGE, ARMIDALE

The 1975 survey is compared with earlier surveys kindly made available by Armidale City Council

pattern, by estimated discharges compared with the regional pattern, and by comparison of successive surveyed cross sections. Studies elsewhere have indicated that there may be an increase in channel capacity downstream of the urban area (Hammer, 1972; Gregory, 1976d) but the amount of change has sometimes been difficult to establish (Hollis and Luckett, 1976), and in some studies empirical measurements have indicated that a reduction in channel capacity may occur with urbanization (Leopold, 1973; Emmett, 1974) although this decrease may later be succeeded by an increase in capacity. It has been suggested that siltation and flood problems in Armidale have been increasing since 1927 (Figure 25.8) and were the result of upstream erosion following many years of exploitive land management practises in the basin (McConnell, Monteith, and Berman, 1966). Additional influences may be found within the urban area. The direct influences upon the channel have included the incidence of fences, bridges, culverts, and also the growth of herbaceous vegetation along the channel margins, all of which have encouraged siltation. The urban area has also affected stream channel processes significantly but less directly. Runoff has certainly been increased by the urban area (Douglas, 1975), and in addition to the drainage at the sides of roads entering between the survey sites (Table 25.4) it has been estimated that the street pattern created about 100 new points of entry for both water and sediment to the channel (Burghardt, Loughran, and Warner 1968). In addition to the extended network provided within the city, the greater runoff percentage has been achieved as a result of the increase in the impervious ground area and the runoff from the roofs of buildings. The earlier practise of collecting roof water in rain water tanks has now been largely superceded since the advent of a piped water supply network (Burghardt Loughran, and Warner 1968). The reason for a reduction in channel capacity in this situation probably lies in the size of the urban catchment area in relation to the catchment area of Dumaresq Creek above Armidale. Whereas runoff has certainly increased from the urban area, and sediment production has also increased, the runoff peak from the small urban area considerably precedes that on Dumaresq Creek. Therefore the high suspended sediment concentrations draining from the urban area prompt deposition during the lower velocities of prehydrograph flows on the main creek. The lower channel slope on the main creek is probably also a contributing factor. Although urban runoff is greater than rural runoff, the effect of this change is not significant when the increase of a small hydrograph is superimposed on, or occurs before, the rising limb of a larger hydrograph. A further factor which may be significant is the calibre of the material into which the channel is cut. The increased provision of fine material from the urban area giving the recent siltation (Figure 25.8) means that the shear strength of the accreting sediment is less than formerly, so that a smaller channel capacity may figure despite the higher discharges. This situation can arise because a lower shear strength is associated with a lower competent velocity and channels will not be developed above a particular size because of bank collapse.

Conclusion

It is argued that future studies of channel change need to stress the relation between channel geometry, pattern, and network. This can be achieved by relating channel cross-sectional geometry at a number of sites in a basin to the total stream length above each site. Although further refinement of the method is necessary to include slope and to allow for different components of the drainage network, it provides a basis for interpretation of changes in the Commissioner's Waters drainage basin of New South Wales. The channel capacities below stock ponds and below dams are lower than expected, and the river channel within and downstream from the urban area of Armidale is substantially lower than expected, and as little as 0.13 of the expected size and 58% of the expected volume. The reduced channel capacity within and below the urban area is confirmed by bankfull discharges, calculated using the Manning equation, which are as little as one-third of the discharges expected from the regional flood-frequency pattern. Further support is provided by analysis of a channel cross-section within the urban area where 1 metre of sediment has accumulated in the channel cross section since 1927.

The accumulation of sediment alone is not sufficient to account for a reduced channel capacity. Sediment could accumulate whilst the channel carved into the flood plain maintained the same capacity. Because channel size is thought to be adjusted to flow regime, then if flow regime is maintained, channel size should remain the same despite sedimentation.

However, the stable relation between channel size and flow regime depends upon the existence of sediment which is constant in character. If finer and/or more friable sediment accrues, as is likely from an urban area, then because the shear strength is reduced substantially a smaller channel may exist with an unchanged flow regime. The presence of reservoirs and dams has reduced peak flows and has been thought to lead to a reduction in channel capacity. Conversely, the increased runoff from an urban area can prompt the development of a greater channel capacity. These two generalizations obtain only if sediment supply and sediment calibre is substantially unchanged. If a new sediment source is revealed by an urban area, it is quite likely that if the material has a lower shear strength than the original bank material then the channel size can become smaller despite the increased peak flows. This may

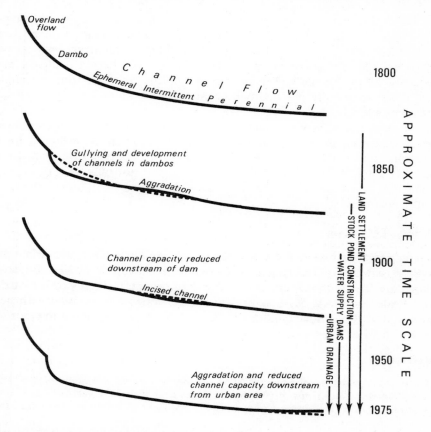

FIGURE 25.9 NETWORK METAMORPHOSIS IN COMMISSIONER'S WATERS BASIN: A
PROVISIONAL MODEL

The long profiles are positioned with reference to the time scale explain why channels have first been observed to reduce but then perhaps to increase in size.

The sequence of changes in Commissioner's Waters drainage basin is tentatively proposed in Figure 25.9. Before land settlement the sediment sources were restricted and channel capacity was adjusted to the prevailing flow regime throughout the basin. Soon after land settlement, stock density and land pressure instigated some gullying which released the sediment for fills downstream and which were later to become incised sections. Channel capacities have effectively been reduced downstream of dams and stock-ponds, but the composition of the channel network may have been adjusted at the same time so that the influence is not so easily detected. As greater runoff and direct modification of the channel has taken place during urbanization, the channels have decreased in size due to the availability of sediment and the type of sediment, which, with a reduced shear strength, can support a lower channel capacity for a particular regime. A channel cut in material with a very low shear strength will have a relatively low capacity value whereas a channel cut in resistant bedrock will have a relatively high capacity value. Events which are capable of changing the former channel will occur frequently but bedrock channels can be modified by much less frequent events.

For many years the consequence of man's influence upon runoff has been found to appear in the occurrence of increased flooding. A further factor to keep in mind is that increased flooding can arise because of a decrease in channel capacity, and such a decrease can occur not only as a result of a change in flow regime but also as a result of a change in sediment character and availability.

Acknowledgement

The research reported in this paper was conducted in 1975 when the author visited the Department of Geography, University of New England, Armidale. The author is grateful to that department, to Professor I. Douglas and Professor D. A. M. Lea, and to Mr W. Johns who assisted with field survey. The Royal Society 20th I.G.C. fund kindly made a grant towards the cost of field expenses.

References

Australian Academy of Science. 1957. *A report on the condition of the High Mountain Catchments of New South Wales and Victoria*, Canberra.
Australian Water Resources Council, 1969. *Sediment sampling in Australia Hydrological Series No. 3*, Canberra.
Bagnold, R. A. 1966. *An* approach to the sediment transport Problem from General physics. *U.S. Geol. Surv. Prof. Paper 422-I.*
Burghardt, J., Loughran, R. J., and Warner, R. F. 1968. *Some preliminary observations on streamflow and wash load in Dumaresq Creek at Armidale, New South Wales.* University of New England Research Series in Applied Geography, No. 18.

Chorley, R. J. and Kennedy, B. A. 1971. *Physical Geography. A systems approach.* Prentice-Hall International Inc., London.

Day, T. J. 1972. The channel geometry of mountain streams. In Slaymaker, H. O. and McPherson, H. J. (Eds), *Mountain Geomorphology*, British Columbia Geographical Series No. 14, pp. 141–150.

Douglas, I. 1975. Flood waves and suspended sediment pulses in urbanized Catchments. *The Institution of Engineers, Australia, National Conference Publication*, pp. 61–64

Douglas, I. 1976. Urban hydrology. *Geogr. J.* **142**, pp. 65–72.

Dury, G. H. 1968. Footnote in Woodyer (1968), p. 117.

Emmett, W. W. 1974. Channel changes. *Geology*, **2**, pp. 271–272.

Flint, J. J. 1974. Stream gradient as a function of order, magnitude and discharge. *Water Resour. Res.*, **10**, pp. 969–973.

Gregory, K. J. 1976a. Bankfull determination and lichenometry. *Search*, **7**, pp. 99–100.

Gregory, K. J. 1976b. *The determination of river channel capacity.* New England Research Series in Applied Geography, No. 42.

Gregory, K. J. 1976c. Lichens and the determination of river channel capacity. *Earth Surface Processes*, **1**, pp. 273–285.

Gregory, K. J. 1976d. Changing drainage basins. *Geogr. J.*, **142**, pp. 237–247.

Gregory, K. J. and Gardiner, V. 1975. Drainage density and climate. *Z. Geomorph.*, **19**, pp. 287–298.

Gregory, K. J. and Park, C. C. 1976. Stream channel morphology in north west Yorkshire. *Revue Geomorph. Dyn.*, **25**, pp. 63–72.

Gupta, A. 1975. Streamflow characteristics in eastern Jamaica, an environment of seasonal flow and large floods. *Amer. J. Sci.*, **275**, pp. 828–847.

Hack, J. T. 1973. Stream-profile analysis and stream-gradient index. *Jour. Research U.S. Geol. Surv.*, **1**, pp. 421–429.

Hammer, T. R. 1972. Stream channel enlargement due to urbanization. *Water Resour. Res.*, **8**, pp. 1530–1540.

Hancock, W. K. 1972. *Discovering Monaro: A study of man's impact on his environment* Cambridge University Press.

Hollis, G. E. and Luckett, J. K. 1976. The response of nature river channels to urbanization: two case studies from southeast England. *J. Hydrol.*, **30**, pp. 351–363.

Holmes, J. M. 1946. *Soil erosion in Australia and New Zealand*, Angus and Robertson Ltd., Sydney.

Hopley, D. 1965. In R. F. Warner (Ed.), *A preliminary report on the geography of Dumaresqshire* Council of Dumaresqshire, Armidale, 1965, Chapter 6.

Kelson, K. D. 1968. The design and construction of small dams. *Water on the Farm*, Water Research Foundation of Australia Report No. 25, pp. 10–21.

Langbein, W. B. and Leopold, L. B. 1964. Quasi-equilibrium states in channel morphology. *Amer. J. Sci.*, **262**, pp. 782–94.

Laurenson, E. M. 1969. Flood mitigation benefits. In. *International Engineering Service Consortium, an economic study of Keepit dam, NSW*, WCIC, Appendix E.

Leopold, L. B. 1973. River channel change with time. *Geol. Soc. Amer. Bull*, **84**, pp. 1845–1860.

Lustig, L. K. 1965. Sediment yield of the Castaic Watershed, western Los Angeles County, California—a quantitative geomorphic approach. *U.S. Geol. Surv. Prof. Paper 422-F*, pp. F1–F23.

McConnell, D. J., Monteith, N. H., and Berman, R. W. 1966. Soil conservation and changing land use in a small New England catchment. *N.S.W. J. Soil Conservation Service*, **22**, pp. 29–41.

McGilchrist, C. A., Woodyer, K. D., and Chapman, T. G. 1968. Recurrence intervals between exceedances of selected river levels. 1. Introduction and a markov model. *Water Resour. Res.*, **4**, pp. 183–189.

McGilchrist, C. A., Woodyer, K. D., and Chapman, T. G. 1969. Recurrence intervals between exceedances of selected river levels. 2. Alternatives to a Markov model. *Water Resour. Res.*, **5**, pp. 268–275.

McGilchrist, C. A., Woodyer, K. D., and Chapman, T. G. 1970. Recurrence intervals between exceedances of selected river levels. 3. Estimation and use of a prior distribution. *Water Resour. Res.*, **6**, pp. 499–504.

McGilchrist, C. A., Woodyer, K. D., and Chapman, T. G. 1972. Recurrence intervals between exceedances of selected river levels. 4. Seasonal streams. *Water. Resour. Res.*, **8**, pp. 435–443.

Munro, C. H., Foster, D. N., Nelson, R. C., and Bell, F. C. 1967. The Georges River Hydraulic, Hydrologic and Reclamation studies. *Univ. New South Wales Water Research Laboratory Report 101.*

Ongley, E. D. 1975. Disequilibrium characteristics in fluvial networks: Denudation rates for Lake Ontario Watersheds. In Johnson, P. G. (Ed.) *Fluvial Processes* Occasional Papers Dept. Geog. and Regional Planning, University of Ottawa, pp. 39–54.

Pickup, G. and Warner, R. F. 1976. Effects of hydrologic regime on magnitude and frequency of dominant discharge. *J. Hydrol.*, **29**, pp. 51–75.

Rastogi, R. A. and Jones, B. A. 1969. Simulation and hydrologic response of a drainage net of a small agricultural drainage basin. *Trans. Amer. Soc. Agr. Engineers*, **12**, pp. 899–908.

Riley, S. J. 1975. The channel shape—grain size relation in eastern Australia and some palaeohydrologic implications. *Sedimentary Geology*, **14**, pp. 253–258.

Schumm, S. A. 1969. River metamorphosis *J. Hydraul. Div. Amer. Soc. Civ. Engrs.*, **95**, pp. 255–273.

Schumm S. A. 1971. Channel adjustment and river metamorphosis. In Shen. H. W. (Ed.) *River Mechanics*, Vol. I. H. W. Shen, Fort Collins, Colorado, pp. 5–1 to 5–22.

Schumm, S. A. 1973. Geomorphic thresholds and complex response of drainage systems. In Morisawa, M. E. (Ed.) *Fluvial Geomorphology* State University of New York, Binghamton, pp. 299–310.

Singh, V. P. 1975. Hybrid formulation of kinematic wave models of watershed runoff. *J. Hydrol.*, **27**, pp. 33–50.

Strahler, A. N. 1956. The nature of induced erosion and aggradation. In Thomas, W. L. (Ed.), *Man's role in changing the face of the earth.* Chicago, pp. 621–638.

Surkan, A. J. 1974. Simulation of storm velocity effects on flow from distributed channel networks. *Water Resour. Res.*, **10**, pp. 1149–1160.

Trimble, S. W. 1976. Modern stream and valley sedimentation in the Driftless Area, Wisconsin, U.S.A. *International Geography '76*, **1**, pp. 228–231.

WCIC (Water Conservation and Irrigation Commission). 1971. *Water resources of New South Wales*, Sydney, Government Printer.

Wolman, M. G. 1967. A cycle of sedimentation and erosion in urban river channels. *Geogr. Annaler*, **49A**, pp. 385–395.

Woodyer, K. D. 1968. Bankfull frequency in rivers. *J. Hydrol.*, **6**, pp. 114–142.

Woodyer, K. D. 1974. River morphology. In *Progress in Australian Hydrology 1965–1974*, Australian National Committee for UNESCO, Canberra, pp. 64–70.

Yang, C. T. 1972. Unit stream power and sediment transport. *J. Hydraul. Div. Amer. Soc. Civ. Engrs.*, **98**, pp. 1805–1826.

W. FROEHLICH

Institute of Geography
Polish Academy of Sciences, Kraków

L. KASZOWSKI

Institute of Geography
Jagiellonian University, Kraków

and

L. STARKEL

Institute of Geography
Polish Academy of Sciences, Kraków

26

Studies of Present-day and Past River Activity in the Polish Carpathians

The entire Polish Carpathians, except for the crystalline-limestone high-mountainous Tatra, are characterized by a relief of middle mountains and extensive foothills. They are made up of sandstone–shaly flysch, predominantly of low resistance. About 50% of the mountain area is under forest whilst as little as 10–20% is farmed land. The main type of sediment transport is as suspended load. The western part of the area including the river Dunajec is marked by high relief energy with relief between 300 and 800 m, and a considerable share of resistant sandstones, so that the bed-load transport in the channels is apparent.

Rivers of the middle-mountain Beskidy are characterized by a rainfall–snow–ground water regime and by runoff concentration during meltwater and summer rainfall floods. Bankfull stages occur in summer and are the result of downpours lasting a few days or, more rarely, of torrential rains. Catastrophic floods appear more than once every 10 years when the maximum unit flows attain between 1000 and 6000 $I\,s^{-1}\,km^{-2}$ (Figuła, 1966; Froehlich, 1975). Low water stages usually occur in autumn and winter when the unit flow decreased to approximately $10\,I\,s^{-1}\,km^{-2}$.

Studies of fluvial processes and river deposits in the Polish Carpathians have been in progress for several decades. Since 1960 the Institute of

W. Froehlich, L. Kaszowski, and L. Starkel

Geography, Jagiellonian University, Kraków, have been systematically un-
dertaking studies of channels and of fluvial processes in the basins of small
Carpathian streams (Kaszowski, 1970, 1973; Niemirowski, 1974). Since
1968 the Department of Physical Geography, Institute of Geography of the
Polish Academy of Sciences, has followed earlier investigations on the
evolution of valley-floors in the Last Glacial and Holocene, by their studies
of erosion, transport, and sedimentation in larger flysch catchment basins
(INQUA, 1972; Froehlich, 1975). This review of the studies is based mainly
on the results of these two research teams. These studies include three aims:
classification and typology of river channels from the view of the dynamics

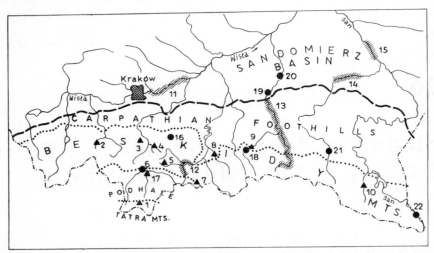

FIGURE 26.1 AREAS OF DETAILED INVESTIGATIONS OF PRESENT-DAY FLUVIAL PRO-
CESSES AND YOUNG QUATERNARY ALLUVIAL FORMS AND SEDIMENTS

Areas of detailed measurements of present-day channel forms and processes: 1,
Biały stream (Kaszowski, 1973); 2, Stryszawka stream (Figure 26.3); 3, Tenczyński
stream (Figure 26.3); 4, Mszanka, Porębianka and Konina streams (Figure 26.3); 5,
Jaszcze and Jamne streams (Niemirowski, 1974); 6, Lepietnica stream (Figure 26.3);
7, Biała and Czarma Woda streams (Figuła, 1966); 8, Kamienica Nawojowska
stream (Froehlich, 1975); 9, Ropa near Szymbark (Gil and Słupik, 1972; Soja,
1977); 10, Hoczewka stream (Figure 26.3).
Areas of measurement of downcutting and changes of river pattern in historical
times: 11, Wisła near Kraków (Trafas, 1975); 12, Dunajec in Baskidy Mts.
(Froehlich, Klimek, and Starkel, 1972; Klimek, Trafas, 1972); 13, Wisłoka river
(Klimek, 1974); 14, Wisłok river (Strzelecka, 1958); 15, Lower San river
(Szumański, 1972).
Localities of dated Last Glacial and Holocene alluvial sediments: 16, Dobra (Klimas-
zewski, 1971); 17, Na Grelu (Klimaszewski, 1961); 18, Szymbark (Changes, 1972);
19, Podgrodzie (Niedziałkowska, 1977); 20, Brzeźnica (Mamakowa and Starkel,
1974); 21, Besko (Changes, 1972); 22, Tarnawa (Ralska-Jasiewiczowa and Starkel,
1975)

and trends of their development; the determination of the course of fluvial transport and its role in the shaping of channels and floodplains; and reconstruction of fluvial processes in the Last Glacial and Holocene, using analysis of fluvial landforms and deposits (Figure 26.1).

Typology and Moulding of River Channels

The river network in the Polish Carpathians is composed of streams and rivers of first to ninth order according to Strahler's classification. Their long profiles are characterized by varying degrees of maturity achieved in different lithological–structural as well as climatic and tectonic conditions. Valley-heads are marked by gradients up to 100‰ or even more, and the width of channels is between 4 and 10 m. Streams of fourth to seventh order have gradients of between 4 and 30‰, with the width of their channels up to 50 m and the width of floodplain up to 200 m. The lower reaches of the Carpathian valleys (eighth–ninth order) are up to 150 m wide, and their flood plains up to 1 km wide.

The main Tatra streams have inherited their channels from proglacial rivers which mould the Pleistocene valley-floors. Those channels are not now subject to considerable modification because the regime of present-day streams is essentially different from their regime in the Pleistocene. Even during a flood brought about by an unprecedented diurnal precipitation of 300 mm (Hala Gąsienicowa, 30 June 1973), the Tatra channels were not transformed to a substantial degree. Most of the channels in the flysch portion of the Carpathians, except for their heads, have been shaped comparatively recently—in the past centuries, or even in the past few decades. Several erosional and accumulational phases in the Holocene and during the last centuries, have been produced by natural causes (climatic changes, neotectonic movements) as well as by anthropogenic influences (deforestation of catchment basins, introduction of root plants, channel improvement works, exploitation of gravel from the channels), and these have produced a segmented structure of the channel long profiles (INQUA, 1972; Kaszowski and Krzemień, 1977). Three fundamental kinds of reaches may be distinguished: *erosional*, with a trend towards deepening and horizontal stabilization; *transportational*, with a trend towards local bank erosion and point-bar accumulation, in addition to upbuilding of floodplains during major floods, and *depositional*, with the least horizontal stability and a tendency towards intense lateral erosion, deposition, and redeposition of sediments within the braided channels, partly also on the floodplain (Kaszowski, Niemirowski, and Trafas, 1976).

The distribution of channel reaches fulfilling definite morphodynamic functions in a river system, according to their location within the system is shown in Figure 26.2. Within the Carpathian channels the most often

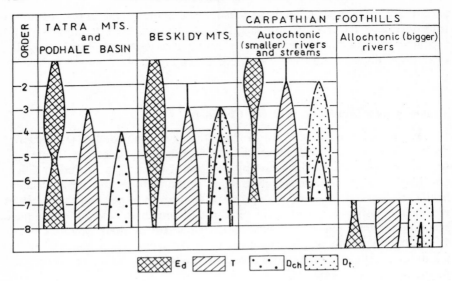

FIGURE 26.2 MORPHODYNAMIC FUNCTIONS OF CARPATHIAN STREAMS AND RIVERS IN RELATION TO THEIR ORDER

Ed, of deep erosion; T, of sediment transport; D_{ch}, of deposition within the channel D_t, of deposition on the floodplains

repeated sequences of the reaches are: Ed-T-D and Ed-D (Figure 26.3). Equally frequently, however, the channel systems are marked by a more composite structure (Figure 26.3, streams Hoczewka or Mszanka) which corresponds to a differential resistance of rock strata (e.g. Hoczewka), or which can be a result of successive phases of aggradation and rejuvenation within a channel brought about, among other things, by man's alterations in the environment of a basin and of the channels themselves (Figure 26.4; Kaszowski and Krzemień, 1977).

A tendency towards channel deepening is revealed by streams, of first to fifth orders, within afforested areas of middle-mountain relief (the Beskidy, the Tatra; Figure 26.2). A frequent occurrence is where two reaches clearly moulded by deep erosion are separated by a transport reach (Figure 26.3; Mszanka, Konina, Stryszawka). The erosional character of the lower reaches is often of anthropogenic origin (Kaszowski and Krzemień, 1977). Over the areas less-intensively uplifted in the Quaternary, a trend towards deep erosion is exhibited by all the rivers (Starkel, 1972). In the foothill areas relief deepening is nearly exclusive to the channels of first and second orders. The intensity of moulding and the development of rocky or debris channels depends mainly on the frequency of inundation, and the floods with a recurrence interval of 3–5 years are erosionally effective.

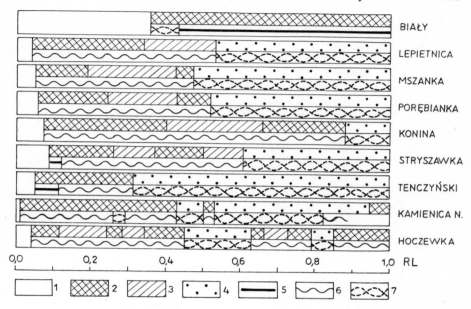

FIGURE 26.3 STRUCTURE OF CHANNELS OF THE CHOSEN CARPATHIAN STREAMS

1, Denudational reach; Channel reaches related to their morphodynamic function: 2, of deep erosion; 3, of transport; 4, of deposition; Channel reaches related to their pattern: 5, straight; 6, crooked; 7, braided; RL: scale of relative length

Channels moderately stable in plan are characteristic of the streams and rivers of third to eighth order in middle-mountain areas, and in foothill areas. Their pattern is sinuous or meandering and they fulfill a transport function. The floodplain which accompanies them, especially in the foothill areas, is raised at floodtime on the side of a channel by levees and on the side of the slopes by proluvial plains at the foot of slopes. A floodplain eventually becomes a suprainundation plain while within the ever-deeper and ever-capacious channel a new floodplain is being formed (Klimek and Starkel, 1974).

Most unstable are the channel reaches moulded by accumulation, as they are accompanied by lateral erosion and redeposition of sediments. These are characteristic of deforested basins (Niemirowski, 1974). The highest portions of the river systems were least deforested and therefore head-streams have preserved their natural deep-erosional moulding regime. They often merge directly into the reaches overloaded by debris and anastomosing within deposits (Figure 26.3). In the Beskid valleys, depending on the extent of basin deforestation, even reaches of third to fifth order are unstable and they grow upstream especially after great floods. Depositional

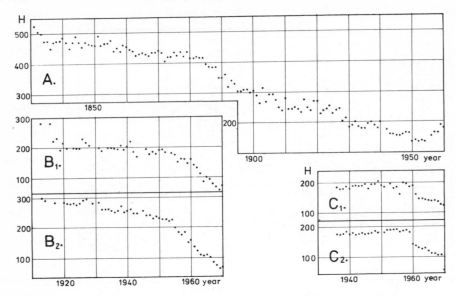

FIGURE 26.4 Changes of the Minimum Annual Water Stages, H(cm), of the Chosen Carpathian Rivers

A, Wisła in Kraków (after Trafas); B_2, Wisłoka in Łabuzie, and B_2, Brzeźnica (after Klimek); C_1, Kamienica Nawojowska at Łabowa, and C_2, Nowy Sącz (after Froehlich)

reaches of the braided channels display a disturbance in the natural sequence of morphodynamic reaches of a channel system in the forest belt of mountains of the moderate climatic zone. Their development has taken place most frequently at the cost of transport reaches. The consequence of the deforestation of foothill regions has been the more rapid upbuilding of the floodplains by fine sediments brought down by soil erosion in the basins; the transportational functions of the channels themselves being preserved. This process is still operative.

A separate category includes those reaches of channels subject to control works or lined technically, in which conditions of the course of fluvial processes have been altered. River control works have been carried out in south Poland in various periods, beginning in the second half of the nineteenth century. Incorrectly lined channels undergo especially strong transformation during great floods, and the improvement structures themselves may be destroyed (Zietara, 1968). Flood prevention works have also been the reason for an intensification of the process of channel deepening. For example, the Vistula in the region of Cracow had deepened its channel by about 2.5 m at a rate of some 4 cm year (Trafas, 1975). The intensification of deep erosion during the past 25 years has been brought about by new control works as well as by intense exploitation of debris from river beds. The intensity of deep erosion is of similar magnitude in the Wisloka, the Ropa, the Kamienica and in the other rivers, and amounts to some 6–8 cm/year (Figure 26.4; Froehlich, 1975; Klimek and Starkel, 1974; Soja, 1977).

The spatial picture of river channels and of their morphodynamic functions in the territory of the Carpathians is a fairly composite one. This results from the natural process of channel adaptation to varying conditions in the catchment basin environment, and reflects alterations of varying scale caused by man's economic activity. A feature of that adaptation is a tendency towards lengthening erosional reaches in mountain parts of river systems. Evidence for this is provided by exposures of bedrock in numerous braided channels, or even by distinct dissection of depositional reaches. This is a return to a tendency similar to that prevailing before deforestation.

In the lower sections of rivers, deep erosion is being intensified, although this is not the natural trend but rather is the result of anthropogenic alterations in the channels due to the exploitation of debris from their beds and to flood control works.

The Course of Fluvial Transport

Carpathian rivers are characterized by varied mechanisms of load transportation. Of the three kinds of transport, that of chemical character is of little importance in moulding the channels but plays a crucial role in the denudation balance of flysch catchment basins (50–90 tonnes/year, after Figuła, 1966). In the years without major inundations, the solute load accounts for more than 50% of the total mass of removed material (Froehlich, 1975).

Most of the physical weathering products are carried away in suspension. An exception here is found in the Tatra mountains which are deprived of fine-fraction covers, and where a small amount of material is removed in suspension (Kaszowski, 1973). In the flysch Carpathians the sources of suspended matter are thick clayey-silty waste-mantles, with an admixture of debris, which cover the slopes. Part of the transported material is also derived from crumbed alluvia and from eroded rock-cut benches. Agricultural use of thick silty waste-mantles of the Carpathian foothills is the cause of the greatest supply of suspended matter. A smaller supply is observed within the Beskid farming belt and the lowest values are recorded in the forest belt of those mountains (Figuła, 1966).

Transport in suspension is characterized by great variability throughout a year, this being an expression of periodical changes in the conditions of debris supply as well as of river capacity. During low water stages, turbidity is small and not more than 10 g/m^3 but it rises to a few tens of thousands of g/m^3 during catastrophic floods such as those of 1934, 1958, 1970, and

1973. Specially high values of turbidity are observed in small catchment basins under farming during torrential rains and this is due to the processes of soil liquefaction on slopes (Gil and Słupik, 1972).

The size and pattern of spates affects the grain-size composition of material transported in suspension. According to Pasternak and Cyberski (1973), the Carpathian rivers transport substantial quantities of clayey particles during the summer floods rather than during winter–spring ones.

In the periods between inundations, part of the material transported is deposited away from the channel current. This material, in addition to that brought down from field roads, contributes to an outpacing of the peak water-discharge by the peak of the suspended sediment hydrograph. With similar conditions of spates, the intensity of transport in suspension is

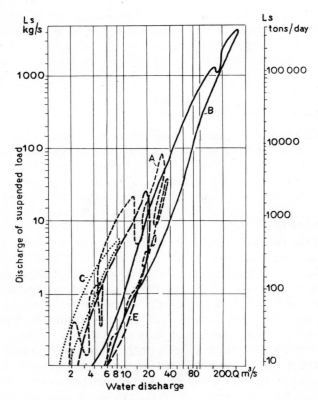

FIGURE 26.5 RELATION BETWEEN THE WATER DISCHARGE (Q) AND THE LOAD OF SUSPENDED MATERIAL (L_S)

During inundations due to thawing A, 17–31 March, 1971, and during those due to rainfalls: B, 18–23 July, 1970; C, 24–26 August, 1971; D, 6–8 October, 1970, in the Kamienica Nawojowska (Nowy Sącz)

considerably varied. This is reflected in the shape and pattern of the loops which relate discharge and suspended load (Figure 26.5). In the type of climate characteristic of this country, a great diversity of spates is observed. This is to be seen in the shapes of flood waves and in the conditions of material supply (Froehlich, 1975).

A considerable amount of the annual load is transported by springtime meltwaters and summertime rainfalls (Brański, 1968; Cyberski, 1969; Froehlich, 1975). During the spates more than 90% of the yearly amount of suspended load is removed. In the years marked by a continental winter, i.e. frosty and snowy, and a continental summer, i.e. dry, the springtime runoff peak exceeds that of summer (Figure 26.6). Rapid spring snowmelt usually causes great inundations in these years which carry away the prevailing part

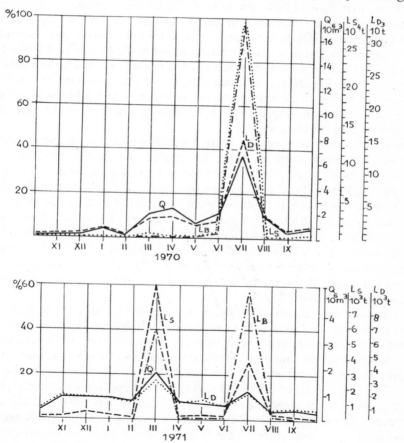

FIGURE 26.6 MONTHLY VALUES OF THE RUNOFF OF WATER (Q) AND OF LOADS OF DISSOLVED MATERIAL (L_D), OF SUSPENDED MATERIAL (L_S), OF BED MATERIAL (L_B) IN THE KAMIENICA NAWOJOWSKA (NOWY SACZ) IN THE YEARS 1970 AND 1971

of the yearly load of material. However in years with an oceanic or transitional type of winter and with an increased cyclone circulation in summer (oceanic summer), frequent and rapid rainfall spates or even catastrophic floods occur, and their peak discharge in summer is often greater than that of spring. Most of the yearly load of material is then carried away in summer (Figure 26.6). According to calculations based upon direct measurements of turbidity, the average amount of material removed from 1 km^2 varies from 5 to 250 tonnes (Brański, 1975; Figuła, 1966; Jarocki, 1957). The proportion of the total load removed in suspension can exceed 85% (Froehlich, 1975).

The supply of material is usually derived from side tributaries, channels, roads, and field paths which from time to time act as periodical streams, in addition to that derived from arable land. Under the influence of slopewash, the displacement of remarkable masses of material down the slopes of the flysch Carpathians (Gerlach, 1966) is not fully reflected in the suspended load of the rivers. Most of the material washed downslope is redeposited at its foot as well as in the valley floors, hence it does not reach the channels (Starkel, 1972). Therefore, calculation of the denudation rate in the Carpathians, according to measurements of the suspended load of seventh and eighth order rivers and expressed in tonnes/km^2 or m^3/km^3 is not exact and can lead to false conclusions.

Suspended material is usually carried away outside the mountains or is deposited on floodplains, in water reservoirs, and in the channels. An important role in the rate and mechanism of floodplain growth is played by riverside vegetation (Froehlich, Klimek, and Starkel, 1972). The thickness of floodplain covers increases from a few cm in the interior of the mountains to a few metres in the foreland (Starkel, 1960; Klimek, 1974). The bottoms of normal and flood channels in accumulation reaches are strewn with boulder–gravelly–sandy alluvia. The transport of coarse bed load takes place exclusively at a time of major floods. During smaller floods sandy–silty material is transported along the bottom, leading in consequence to the infilling of debris-catching reservoirs.

The present-day supply of material for bed-load transport is from the active undercuts of valley sides and terraces. It is also derived from the deepening of channels cut across bedrock. Material supplied is therefore either sharp-edged or rounded. Debris in the valleys of the Carpathian rivers achieves mature shapes after a few initial kilometres of transport (Figure 26.7). The trend is disturbed by the supply of fresh, unrounded debris which comes from the eroded rock-cut benches, or is supplied by short tributaries.

Extensive gravel-bars of the Beskid valleys may indicate that the transport of bed load is playing a great role, but it appears from measurements of transport using enamel-painted pebbles that the transport of dragged material takes place over short distances (Froehlich, 1975; Kaszowski, 1973; Niemirowski, 1974). Along the whole length of a river the amount of

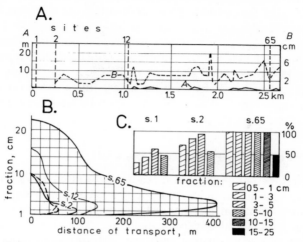

FIGURE 26.7 DOWNSTREAM DIFFERENTIATION OF THE BED MATERIAL TRANSPORT IN THE CHANNEL OF BIAŁY STREAM (TATRA MTS.) INVESTIGATED BY MEANS OF THE PAINTED DEBRIS METHOD

A. Transport during the flood on 27 August, 1966, with discharge of 1.7 m³/s and peak water velocity of 2.1 m/s (*A*, maximum distance of the transport of 3–5 cm fraction; *B*, maximum fraction transported during the flood). B. Curves of the maximum distance of the transport during the flood on 18 July, 1968, with discharge 7.6 m³/s and peak velocity of 3.5 m/s (each curve is one site; s. = site). C. Amount of painted debris in separate fractions displaced during the flood on 18 July, 1968

bottom material which is put in motion is frequently very large, but the amount of material removed through a specific cross-section is small.

In particular parts of the Carpathians in the channel reaches the transport of bed load varies both qualitatively and quantitatively. Nearly continuous bottom transport of the sandy fraction takes place in river channels of the Carpathian foothills. Bottom transport in the Beskid rivers is confined to a few days yearly. The least frequent bottom transport of boulder fractions characteristically occurs in the Tatra streams (Kaszowski, 1973).

Nowadays, the greatest displacement of bed load occurs in channels of the Beskid rivers that are involved in an intense exploitation of their alluvia. In the depositional reaches of the rivers of sixth to eighth order, and locally of a braided pattern, bottom transport leads to the reworking of alluvia derived by lateral migration of a channel or from the backward dissection of debris banks slowly progressing downstream.

The bed material carried away from the Beskid basins is small and constitutes some 2–12% of the total load (Froehlich, 1975). Based upon 10 years' observations, Figuła stated (1966) that the mean annual bed load derived from a small afforested Beskid basin (the Czarna Woda) is

6.8 tonnes/km^2 and that from a more deforested basin of the Biala Woda amounts to 18.1 tonnes/km^2.

At present, the amount of alluvia removed from the channels of the Carpathian rivers is much greater than the possible rate of natural regeneration. This results in a general shortage of bed load the supply of which is impeded from valley heads by hydrotechnical lining. This occasions some excesses of energy due to the river underloading and eventually leads to deepening of the channels.

Reconstruction of Fluvial Processes in the Past

The relationships between terrace landforms and cover deposits make it possible to distinguish the older and younger elements in the valley floors. Their facial differentiation, mechanical composition, and their relationship with the deposits of lateral transport, including slope deposits or those from side tributaries, gives an insight into the conditions of supply and river discharge. The microrelief of terrace plains including oxbow lakes, bars, and the character of edges of the higher terraces, makes it possible to determine some parameters of the former channels. Radiocarbon dating at several sites, in addition to paleobotanical records, culture horizons, and historical

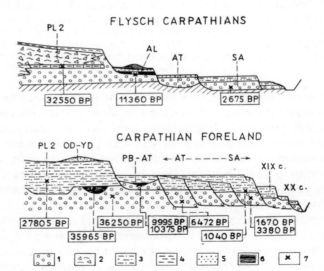

FIGURE 26.8 SCHEMATIC CROSS-SECTIONS OF THE UPPER PLENIGLACIAL, LATE GLACIAL AND HOLOCENE TERRACES IN THE UPPER AND LOWER COURSES OF THE CARPATHIAN VALLEYS

1, Gravels of channel facies; 2, solifluctional deposits; 3, muds of floodplain facies; 4; deluvial deposits; 5, aeolian sands; 6, peat and oxbow-lake sediments; 7, position of ^{14}C datings; PL 2, upper Pleniglacial; OD, Older Dryas; YD, Younger Dryas; AL, Allerod; PB, Preboreal; BO, Boreal; AT, Atlantic; SA, Subatlantic

sources, all give information about the absolute age of deposits and land-forms, and make it possible to reconstruct the processes of erosion and accumulation from the pleniglacial until the present time.

The plain of river accumulation derived from the last glacial in some Carpathian valleys averages 10 to 15 m in height. In the interior of the mountains it rises towards the slopes, and the interfingering of its coarse-grained alluvia with solifluction and deluvial deposits is evidence for a simultaneous supply of material from the slopes into the channels (Klimas-zewski, 1971; Starkel, 1969; Figure 26.8). The presence of distinct gravelly horizons from an interpleniglacial period (Dobra 32,550±450 B.P.; Klimas-zewski, 1971) points to washing away and existence of broad channels at that time, whilst the commonly occurring thick solifluction deposits, with clayey-debris character at the top, indicate an intensification of the supply from the slopes and vertical growth of the cover at a later time.

In the foreland of the mountains, the plain of the pleniglacial terrace that was built up by floods during conditions of tundra and permafrost (Brzeznica 27,805±330 B.P.; Mamakowa and Starkel, 1974), with dune sands at the top, indicates a different river regime and this is also evidenced by the systems of braided channels, as has been stated by Szumański (1972) in the lower San valley (Figure 26.9). These deposits and forms are also certainly

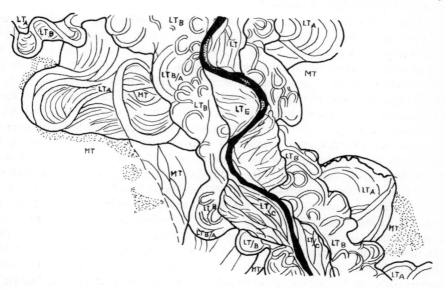

FIGURE 26.9 GEOMORPHOLOGICAL OUTLINE OF THE SAN VALLEY NEAR LEŻAJSK
(AFTER SZUMAŃSKI, 1972)

MT, Middle terrace with traces of a braided river (Pleniglacial), LT_A, low terrace with great meanders (Late Glacial); LT_B, low terrace B with three generations of small meanders: I, II, III (Holocene); $LT_{B/A}$, low terrace A with muds of low terrace B; LT_C, low terrace C with traces of braided river (eighteenth–nineteenth century)

an indication of major oscillations of discharge thus pointing to a dryer climate. Under such conditions with the decreased supply from the mountains and probably with icings, the deepening and broadening of channels was taking place in braided rivers; these changed into meandering ones at the close of the glacial period. Evidence for this is afforded by erosional plains cut both in the foreland of the mountains and in their interior, their most frequent heights reaching between 4 and 6 m above the level of present-day river beds. The plains are overlain by late glacial alluvia and organic remnants (na Grelu—Klimaszewski, 1961; Ralska-Jasiewiczowa and Starkel, 1975). Meanders are sometimes cut into bedrock (Klimek and Starkel, 1974), and their radius of curvature at the close of their formation is 3–4 times greater than that of the meanders encountered within Holocene floodplains. Greater discharges, and the migration of bends together with deposition of bars had finished fairly rapidly on the border between the Younger Dryas and Holocene and is indicated by numerous datings derived from a layer in contact with the overlying peats and muds (10,300–9900 B.P.).

The Preboreal and Boreal periods are characterized by the stabilization of processes, the formation of peats and clayey muds, both at the front of and within the mountains, this being accompanied by the development of sinuous and shallow channels.

About 8400 B.P. a phase of inundation, simultaneous erosion, and sedimentation began that is recorded in the entire Carpathians (Ralska-Jasiewiczowa and Starkel, 1976) as can be best observed in the profile at Podgrodzie (Niedziałkowska and coworkers, 1977). With the complete afforestation of the Carpathians, the vertical growth of alluvial plains by thick fans from the tributaries indicates a considerable downcutting of the channels in their upper courses together with material supplied by landslide processes (Changes, 1972). At the same time deep erosion continued because, for example, the peatbog at Tarnawa on the upper San has not displayed muddy insertions since as early as 7840 ± 100 B.P., and the incisions descending below the present-day bed are infilled by deposits dated 6472 ± 100 B.P.(Klimek and Starkel, 1974). Major floods, which had probably been more sizeable than the present ones because they were capable of overdeepening river channels by 4–6 m, probably created comparatively broad braided channels. In the following few thousand years the channels of the lower courses were meandering in pattern as is shown both by alluvia of the channel and floodplain facies; these are found close to one another, and were produced by lateral accretion of the floodplain (Klimek and Starkel, 1974), as well as by many oxbow lakes of small radius within the 6–10 m terrace plain.

Man's economy in the Neolithic period involved the burning of forests and ploughing, since at least 4500 B.P. (Ralska-Jasiewiczowa and Starkel, 1976), which disturbed water circulation and increased sediment supply to the channels. In an inner-mountain basin at Besko, an increase in the share of sandy fraction corresponds well with the appearance of, and increase of, the pollen of cereals and weeds (Changes, 1972). Organic remnants at the bottom of gravels found on the margin of the Ropa valley floor, their age put at 2675 ± 60 B.P. (period of the Lusatian culture) and the position of the channels some 2–3 m higher in the nineteenth century provide proof that the phases of increased soil cultivation probably coincided with the downcutting of broad channels of a braided type in the mountains (Starkel, 1977). At the same time, although it is difficult to be very precise, climatic changes must have been expressed in the frequency of flooding. Proof of this is provided by the remnants of settlements in the Vistula valley of late Roman times and of the Middle Ages, which were sited as low as the present channel level (Starkel, 1972).

River discharges from the fifteenth to seventeenth centuries must have been inconsiderable and the frequency of inundations low because in the lower courses of Carpathian rivers, meandering channels can be observed with radii 3–5 times smaller than those of the nineteenth century (Strzelecka, 1958; Trafas, 1975). Trafas has calculated that in the Vistula valley the mean radius of those older bends must have been 228 m with a reconstructed average flood discharge $Q = 268 \, \text{m}^3/\text{s}$, while that of the Vistula in the nineteenth century amounted to 961 m with $Q = 762 \, \text{m}^3/\text{s}$. According to an analysis of older maps, the change in question had also occurred in the interior of the mountains and in the Dunajec valley by the end of the eighteenth century and this is attributed to the increase in potato cultivation in the Carpathians (Klimek and Trafas, 1972). Both in the mountains and in the foreland, the braided rivers came into being, their channels overloaded not only with suspended load but with bed load too, and this was certainly assisted by a greater density of roads. Aggradation of valley floors had then started.

The second half of the nineteenth century brought in flood-control works by straightening and shortening of rivers as well as the embankment of the majority of the lower courses. The increased gradients caused rivers to cut down. In this way, in the former reaches of braided rivers a new floodplain came into being, its height amounting to between 3 and 5 m at present (Szumanski, 1972; Klimek and Starkel, 1974). Numerous river reaches in the mountains underwent horizontal stabilization and deepening of their channels in solid rock (Changes, 1972).

The past 30 years have been characterized by a general trend towards downcutting of the channels both in mountain reaches as well as in the foreland (Figures 26.2 and 26.4). Downcutting has been accelerated in many channels, such as the Wisloka, by the exploitation of gravels from the channel immediately after the last war (Klimek and Starkel, 1974), and in others by the construction of dams which decreased the downvalley supply of material. Similar effects can be also due to a restriction of farming in the upper portions of the basins which is favoured by a major frequency of inundations (Soja, 1977). With a considerable amount of suspended load, accumulation proceeds to raise floodplain levels. In the meandering channels subject to downcutting, lateral migration favours the lateral growth of the floodplain. The speed of aggradation reaches 1.5 m in 15 years (Klimek, 1974).

Studies of the channel changes in the Holocene, in historical times and at present indicate that the channels of the major Carpathian rivers are of composite origin. Direct interpretation of the tectonic trends is not possible. The evolution of channels can be learned by complementary studies of processes, the age of deposits, and landforms both within the floodplain and the higher terrace benches (Figure 26.10). Most of the Carpathian valley reaches experienced at least two evolution cycles in the Holocene, namely aggradation followed by erosion. The latter 'cycle' was associated with various forms of man's interference. The Carpathian rivers constantly attempt to adjust their channels either to maximum discharges during periods of increased frequency of floods and intensified erosional activity, or to mean discharges during lesser oscillations in discharges accompanied by rarer inundation.

A feature of the present time is the decrease of debris supply into the rivers contemporaneously with more rapid floods as the result of lack of the general change in land use. An effect of this is that the process of channel deepening with a comparative increase of suspended material in the balance of sediment transport. The foreseen intensification of forest economy accompanied by a rise of forest exploitation may involve, with inconsiderable lowering of the rapidity of inundations, a return to aggradation in the river channels similar to that of the nineteenth century.

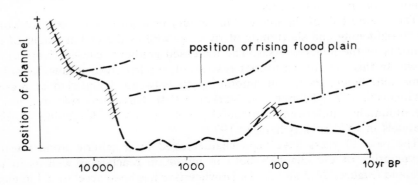

FIGURE 26.10 SCHEMATIC DRAWING SHOWING THE VERTICAL SHIFTING OF CHANNELS AND DEVELOPMENT OF FLOODPLAIN SEQUENCE IN LOWER AND MIDDLE COURSES OF CARPATHIAN VALLEYS

Shading indicates the reconstructed periods of braided river systems

References

Brański, J. 1968. Water silt charge and transport of suspended load in the Polish rivers. *Prace PIHM*, **95**, pp. 49–67.

Brański, J. 1975. Assessment of the Vistula catchment area denudation on the basis of bed load transport measurements. *Prace IMiGW*, **6**, pp. 5–58.

Cyberski, J. 1969. Sedimentation of bed load in the Rożnów basin. *Prace PIHM*, **96**, pp. 21–42.

Figuľa, K. 1966. Investigations on transport of solids in mountain and submontane watercourses with different structure and utilization. *Wiadomości IMUZ*, **6**, pp. 131–145.

Froehlich, W. 1975. The dynamics of fluvial transport in the Kamienica Nawojowska. *Prace Geograficzne IG PAN*, **114**.

Froehlich, W., Klimek, K., and Starkel, L. 1972. The Holocene formation of the Dunajec valley floor within the Beskid Sadecki in the light of flood transport and sedimentation. *Studia Geomorph. Carpatho-Balcanïca*, **6**, pp. 63–83.

Gerlach, T. 1966. Development actuel des versants dans le bassin du haut Grajcarek (in Polish). *Prace Geograficzne IG PAN*, **52**.

Gil, E. and Słupik, J. 1972. The influence of the plant cover and land use on the surface runoff and wash down during heavy rain. *Studia Geomorph. Carpatho-Balcanica*, **6**, pp. 181–190.

INQUA, 1972. Changes in the paleography of valley floors of the Vistula Drainage basin during the Holocene. *Excursion guide-book, Symposium of the INQUA Commission on Studies of the Holocene*, Poland, Sept. 12–20.

Jarocki, W. 1957. *Ruch rumowiska w ciekach*, Wyd. Morskie, Gdynia.

Kaszowski, L. 1970. Methods of investigation of contemporary fluvial processes applied in the Kraków centre. *Studia Geomorph. Carpatho-Balcanica*, **4**, pp. 65–81.

Kaszowski, L. 1973. Morphological activity of the mountain streams (with Biały Potok in the Tatra Mts. as example). *Zesz. Nauk. U. J., Prace Geogr.*, **31**.

Kaszowski, L. and Krzemień, K. 1977. Structure of mountain channel systems as exemplified by chosen Carpathian streams. *Studia Geomorph. Carpatho-Balcanica*, **11**.

Kaszowski, L., Niemirowski, M., and Trafas, K. 1976. Problems of the dynamics of river channels in the Carpathian part of the Vistula basin. *Zesz. Nauk. U.J., Prace geograficzne*, **43**, pp. 7–37.

Klimaszewski, M. 1961. Southern Poland. *Guide-book of INQUA Excursion*, Vol. III, Warsaw.

Klimaszewski, M. 1971. The effect of solifluction processes on the development of mountain slopes in the Beskidy (Flysch Carpathians). *Folia Quaternaria*, **38**.

Klimek, K. 1974. The structure and mode of sedimentation of the floodplain deposits in the Wisłoka Valley. *Studia Geomorph. Caarpatho-Balcanica*, **8**, pp. 135–151.

Klimek, K. and Starkel, L. 1974. History and actual tendency of floodplain development at the border of the Polish Carpathians. *Abh. d. Akademie d. Wiss. in Gottingen, Math.-Phys. Klasse, III*, **29**, pp. 185–196.

Klimek, K. and Trafas, K. 1972. Young-Holocene changes in the course of the Dunajec river in the Beskid Sądecki Mts. *Studia Geomorph. Carpatho-Balcanica*, **6**, pp. 85–92.

Mamakowa, K. and Starkel, L. 1974. New data about the profile of Young-Quaternary deposits at Brzeźnica in Wisłoka Valley, Sandomierz Basin. *Studia Geomorph. Carp.-Balc.*, **8**, pp. 47–59.

Niedziałkowska, E., Skubisz, A., and Starkel, L. 1977. Lithology of the Eo- and Meso-holocene alluvia in Podgrodzie upon Wisłoka river. *Studia Geomorph. Carp. Balc.*, **11**.

Niemirowski, M. 1974. The dynamics of contemporary river beds in the mountain streams. *Zeszyty Nauk. U. J., Prace Geogr.*, **34.**

Pasternak, K. and Cyberski, J. 1973. Granulometric composition of the carried load in waters of Carpathian rivers as compared to the quality of the substrata in their catchment basins. *Probl. Zagosp. Ziem Górskich*, **12,** pp. 131–152.

Ralska-Jasiewiczowa, M. and Starkel, L. 1975. The basic problems of paleogeography of the Holocene in the Polish Carpathians. *Biul. Geol. U. W.*, **19,** pp. 27–44.

Soja, R. 1977. Deep erosion of the Carpathian river in the light of measurements of the transversal profile. *Studia Geomorph. Carpatho-Balcanica*, **11.**

Starkel, L. 1960. The development of the flysch Carpathians relief during the Holocene. *Prace Geograficzne IG PAN*, **22.**

Starkel, L. 1969. L'evolution des versants des Carpates a'flysch au Quaternaire. *Biul. Peryglacjalny*, **18,** pp. 349–379.

Starkel, L. 1972. Trends of development of valley floors of mountain areas and submontane depressions in the Holocene. *Studia Geomorph. Carpatho-Balcanica*, **6,** pp. 121–133.

Starkel, L. 1977. Last Glacial and Holocene fluvial chronology in the Carpathian valleys. *Studia Geomorph. Carpatho-Balcanica*, **11.**

Strzelecka, B. 1958. Historyczna dokumentacja niektórych młodszych zmian hydrograficznych na brzegu Karpat. *Czasop. Geogr.*, **29,** p. 4.

Szumański, A. 1972. The valley of lower San river in the Sandomierz Basin. *Excursion Guide-book of Sympoisim of the INQUA Commission on Studies of the Holocene*, pp. 58–68.

Trafas, K. 1975. Changes of the Vistula river bed east of Cracow in the light of archival maps and photointerpretation. *Zesz. Naukowe U. J., Prace Geograficzne*, **40.**

Ziętara, T. 1968. Part played by torrential rains and floods on the relief of Beskid Mts. *Prace Geograficzne IG PAN*, **60.**

M. J. KIRKBY

Professor of Physical Geography
University of Leeds

27

Maximum Sediment Efficiency as a Criterion for Alluvial Channels

The General Hypothesis

Many systems show an indeterminacy in their physical equations, and yet are observed to settle into one or more modal states. These modes can often be analysed as representing minima of total system energy, which may be thought of as depressions in an energy surface. In general it may be argued that as a system changes over time, it will be 'captured' by these minima, and that only large events can have enough energy to lift the system out of a depression in the energy surface. The system will therefore spend most of its time at or near the bottom of some depression. If, therefore, a series of systems is observed at one time, most will lie in one or other of these depressions, which will be described as modal states of the system.

This approach has been applied to river systems by a number of authors (e.g. Bagnold, 1960) and to engineering problems as standard practice, but has recently been dominated in the geomorphological literature by attempts to apply the concept of *landscape entropy* (Leopold and Langbein, 1962), and the subsequent concept of *minimum variance* (Langbein and Leopold, 1964, 1966; Scheidegger and Langbein, 1966) which is closely related to a maximum likelihood criterion. It is argued here that the original concept of landscape entropy is based on a false premise, and that the minimum variance concept, if applicable, has not been presented with sufficient attention to the relevant constraints. In principle, however, a maximum likelihood approach should provide a *quantum* analogue to the *classical* methods which are considered here.

River systems share with other natural environmental systems a great complexity, although it is less than for many other examples. In complex systems within which many processes take place, the criterion of maximum efficiency normally applies only to a single process, i.e. the limiting process. A river system is involved in transporting its water and a range of sediment sizes, in eroding its bed, and in providing a series of important ecological niches, among other increasingly irrelevant processes. It is argued that, for an alluvial channel, transportation of sediment is the limiting process. The general hypothesis of this paper is, therefore, that in the medium term the channel is adjusted, over a period, to carry the sediment load provided to it as efficiently as possible. The selection of a suitable, graded, time-span is critical to the application of this hypothesis. In the very short term the river is able to transport its load selectively, and it is only over a moderate period of, say, several decades, that it must carry the sediment provided to it from upstream and from the sideslopes, so that this sediment does not accumulate indefinitely. In the very long term, the river can no longer be thought of as a system which is near equilibrium. It must, instead, be considered as an evolving system in which perhaps the limiting process becomes the rate of bed erosion. In the moderate term, however, this activity can largely be ignored, but it is necessary to remember that by ignoring downcutting, it is impossible to predict the river long profile, as its evolution is no longer within the system studied.

Bed Forms

An analogy with the theory of road traffic flow has been applied (Langbein and Leopold, 1968) to pools and riffle sequences and is briefly repeated here. Figure 27.1 schematically represents an empirical relationship between mean grain velocity, $\bar{v}$, and the concentration of sediment on the stream bed, c_s. It is argued that this will be a steadily decreasing function as concentration rises, due to increasing interference effects between the grains, and will become zero at some limiting upper concentration of 1. It may be seen that the total sediment transport S is given by $c\bar{v}$, and that it will be zero at $c = 0$ and $c = 1$, with a maximum at some point in between. It may also be shown that the spread of individual grain velocities will lead to a grouping of grains, analogous to queues of moving vehicles on a road, and that the velocity of the group is given by the slope of the tangent to the curve in Figure 27.1 (b). Hence at point P corresponding to concentration $c = ON = ON'$, the average vehicle velocity is given by P'N' in Figure 27.1(a) and the group or kinematic wave velocity by the slope of the line RPQ in Figure 27.1(b).

It may be seen that the peak of the curve in Figure 27.1(b) is associated with a horizontal tangent, i.e. a zero kinematic wave velocity. The argument of Langbein and Leopold (1968) is that the observed constancy of riffle position during appreciable movement of stones in the riffles shows that each riffle is a group of stones with zero kinematic wave velocity. It therefore follows that they correspond to the peak of the curve in Figure 27.1(b), that is to maximal transport rate for the riffle gravels. This illustrates the general hypothesis about efficiency for sediment transport, and

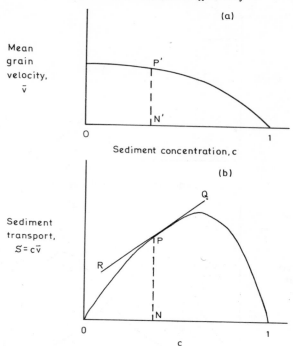

FIGURE 27.1 HYPOTHETICAL VARIATION OF GRAIN VELOCITY AND TOTAL SEDI-
MENT TRANSPORT FOR A GIVEN MATERIAL, AS SEDIMENT CONCENTRATION ON THE
BED VARIES

also extends it. Since river sediment usually spans a considerable range of
grain sizes, the efficiency criterion may be held to apply mainly to the grain
sizes which are carried with the greatest difficulty. In general this critical
grain size is intermediate between the modal size carried (which must match
that of the source material) and the maximum size present. Figure 27.2
illustrates schematically the influence of increasing grain size in reducing
grain velocity summed over the full distribution of flows at a site (Figure
27.2(a) ignoring concentration effects). Figure 27.2(b) shows how this effect
modifies the source material and channel-bed material. The distributions of
grain sizes are related by: Sediment transport of size $d \propto f_s(d) \propto f_B(d) \cdot \bar{v}(d)$.
Hence the bed material mode is coarser than that for the source material,
and it is this grain size which tends to be carried with greatest difficulty.

It is the coarsest elements of this bed material which provide the roughness elements for the river flow. The peak of the channel bed material curve
may be considered the dominant grain size with respect to the need for
efficiency of sediment transport. The roughness diameter and this dominant
diameter may differ little, or by a ratio of 100:1 or more, according to the
amount of coarse debris which the river receives from its catchment area.

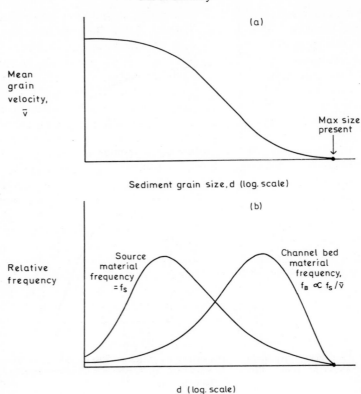

FIGURE 27.2 HYPOTHETICAL VARIATION OF GRAIN VELOCITY (IGNORING CONCENTRATION EFFECTS) WITH GRAIN SIZE AND ITS EFFECT ON SORTING THE MATERIAL CARRIED

The ratio of these sizes is considered further (below) in the context of hydraulic geometry.

The argument concerning the need for efficient transport of coarse material may perhaps be carried further in considering meandering rivers. Comparison with meandering flows in the Gulf Stream, on glaciers, and elsewhere suggests that the tendency to meander, and the geometry of meanders, is related to hydraulic rather than sediment characteristics. Nevertheless, it is argued that the way in which meanders in gravel-bed rivers migrate is a response to efficient transport of the gravel, at least in qualitative terms. It may be observed that because the gravel is concentrated at crossings between meander bends (which have the average spacing of five widths characteristic of pool and riffle sequences), it plays little part in the process of bank erosion or deposition. It is therefore held within the channel and not normally abandoned within the floodplain deposits. It is consequently permanently available for transportation, in strong contrast to the finer grained materials stored for long periods in the flood plain alluvium. It is argued, therefore, that within a given flood plain system, the gravel sediment is dispersed for most efficient transportation at all times.

Where channels are in sand, the river bed is commonly formed into ripples, dunes, or antidunes. These bed forms may also be considered as groups of particles, exactly as in the argument for riffles described above. If sand sediment concentration is steadily increased within a river of constant discharge then a sequence of forms, from none to ripples to dunes to a plane bed to antidunes, is observable, the plane bed occurring at a Froude Number of approximately 1.0, as the water flow changes from tranquil to rapid. In Figure 27.1(b), the increasing sediment concentration is seen to lead to kinematic wave velocities which are at first positive (ripples and dunes) and then become zero (plane bed) and reverse (antidunes). The association of the plane bed with a Froude Number of 1 is thought, once more, to indicate maximum efficiency at this point. This is certainly true in that the total flow energy of the water is a minimum at this point, though the evidence is less clear for the rate of sediment transport. Once more it is argued, though rather more speculatively, that the channel-bed forms are related to the efficiency of the sediment transport, with a maximum hypothesized for a plane bed with a Froude Number of about 1.0.

Analysis of a Channel Reach in Graded Time-spans

The implications of assuming maximum sediment efficiency are laid out more fully in this section with respect to the hydraulic geometry of a channel reach. It will be shown which of the standard empirical relations rely critically on this relationship, and which rely on other and external information. In making these deductions, it is important to recognize the implications of choosing a graded time-span, some of which have already been mentioned. As a simplifying assumption, the flow is considered to proceed at a constant dominant rate, for a proportion of the time which may vary downstream. It is also assumed that the dominant size of bed material carried remains more or less constant, even though the roughness diameter may change appreciably downstream. The graded time-span is thought to impose values of water and sediment discharge, sediment dominant grain size, and valley slope (equal to maximum channel slope). The channel regime will be analysed in terms of equations for discharge, Darcy–Weisbach roughness, and bedload transport, and the most efficient channel cross-sections and roughness diameters derived.

The governing equations will be taken in the forms:

$$Q = wrv = \left(\frac{2g}{ff}\right)^{\frac{1}{2}} wr^{\frac{3}{2}}s^{\frac{1}{2}} \tag{1}$$

where Q = total dominant water discharge, w = channel width, r = channel hydraulic radius (depth), v = mean water velocity, s = channel slope, and ff = Darcy–Weisbach roughness.

$$\frac{1}{\sqrt{ff}} = 1.77 \ln(r/d_{84}) + 2.0 \tag{2}$$

where d_{84} = roughness diameter for bed material

$$S = 8(g\Delta d^3)^{\frac{1}{2}} w \left(\frac{rs}{\Delta d} - 0.047\right)^{\frac{3}{2}} \tag{3}$$

This is the Meyer-Peter and Müller (1948) equation for bedload transport where S = total sediment transport rate, Δ = sediment submerged density/water density ≈ 1.65, and d = sediment dominant diameter.

Writing the dimensionless ratios $r/d_{84} = u$, $d_{84}d = \lambda$, and $S/Q = C$, The following expression for the volumetric sediment concentration, C, is obtained:

$$\lambda C = 3.43 \frac{(u\lambda s - 0.077)^{\frac{3}{2}}}{u^{\frac{3}{2}}(\lambda s)^{\frac{1}{2}}(1.77 \ln u + 2.0)} \tag{4}$$

giving the family of curves shown in Figure 27.3. In this figure it may be seen that, except at very high concentrations, there is a minimum slope at which the sediment may be carried, and that these minima lie on the maximum efficiency curve:

$$\lambda S = \frac{0.115 \ln u + 0.208}{u} \tag{5}$$

This curve is also shown in Figure 27.3. It should be noted that the area of the curves for $u < 1$ is somewhat suspect, as in this region the roughness equation (2) is being applied to stones which stick out of the water ($r < d_{84}$)! Figure 27.3 and equation (5) will be used in several ways below, and are crucial to what follows. Although the sediment transport equation (3) used above is one of many in the literature, it is believed that the results shown in Figure 27.3 and in what follows are broadly valid for any reasonable choice of equation.

From equation (1), channel cross-section (for a given discharge) may be seen to decrease steadily as depth increases along a line of constant slope in Figure 27.3. Hence, on the right of the maximum efficiency curve a small local increase of depth leads both to a reduction of sediment concentration (keeping slope constant) and to a reduction of channel cross-section. These two effects reinforce one another, the loss of sediment from the flow filling the channel, and therefore the channel cross-section is unstable, becoming deeper and narrower until the banks collapse. Similarly a local reduction of channel depth leads to shallowing and enlargement of the channel in an unstable fashion. Thus this whole region to the right of the maximum

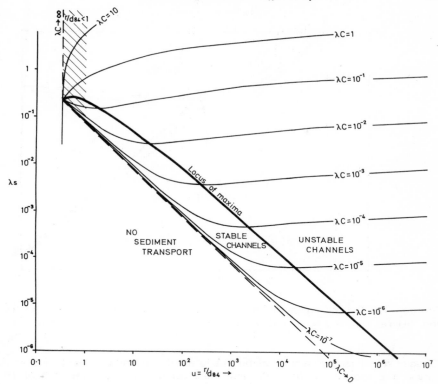

FIGURE 27.3 THE RELATIONSHIP BETWEEN SEDIMENT CONCENTRATION AND CHANNEL SLOPE, DEPTH AND GRAIN SIZE

Heavy line is locus of maximum concentration. Shaded area is where roughness elements project from the water

efficiency curve is one of unstable channels, of which braids are the most obvious morphological example. In a similar way the region to the left of the maximum efficiency curve is stable, as local increases of depth lead to an increase in concentration and a reduction in cross-section, which counterbalance one another. This region is typified by meandering channels.

To convert this diagram into a more practically useful one, allowance must be made for possible variations in the ratio $\lambda = d_{84}/d$. Figure 27.4 has been derived from Figure 27.3 by transposing curves across and downwards equally, so that Figure 27.4 shows slope, s, plotted against $u\lambda = r/d$, which is a direct measure of depth (as d is assumed constant). Scales have been labelled for a notional $d = 10$ mm. It is clear from Figure 27.4 that the stable/unstable channel transition depends only to a slight extent on roughness diameter of the bed material. Comparison with Leopold and Wolman's (1957) paper on braided and meandering streams shows excellent agreement

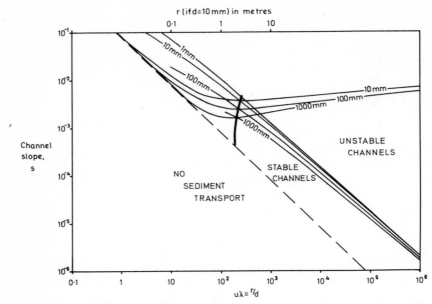

FIGURE 27.4 THE RELATIONSHIP BETWEEN SEDIMENT CONCENTRATION AND CHANNEL SLOPE, DEPTH AND ROUGHNESS GRAIN SIZE. ASSUMING DOMINANT GRAIN SIZE, $d = 10$ MM

Oblique set of curves shows variation of locus of maximum concentration as d_{84} varies. The second set shows the individual curves for a concentration of 10^{-3} (1000 p.p.m.) as d_{84} varies. Heavy line shows locus of maxima where they intersect

with their empirical curve, and supports the conclusion of their work and that of more recent authors (e.g. Ferguson, 1971) that the transition is minimally dependent on bed material grain size.

Figure 27.4 also traces the effect of changing λ on the individual concentration curves for the example of $C = 10^{-3}$ (1000 p.p.m.). It may be seen that the same amount, C, and size, d, of sediment may be carried on lower and lower slopes as λ is increased. These curves therefore indicate that the maintenance of the coarsest available material on the stream bed is the most efficient way of carrying the whole of its load, even though its dominant size is very much smaller.

Hydraulic Geometry

A great deal of work has been done in observing and deducing power relationships between discharge and measures of channel cross-sectional geometry. Although the exponents obtained vary within a somewhat limited range for several reasons, it is important that the hypothesis of maximum efficiency should be tested for relevance as an explanation of them.

At-a-Station Exponents

For all practical purposes, the behaviour of a river within its cross-section is specified by equations (1) and (2) above. For most purposes d_{84} and s may be taken as fixed as stage varies. Expressing equation (2) in approximate power-function form,

$$\frac{1}{\sqrt{ff}} \propto r\beta \tag{6}$$

where β takes an average value of 1/6 for large rivers $(10 < r/d_{84} < 10^4)$ but may be larger for small rivers $(0.48$ for $1 < r/d_{84} < 10)$. Substituting in equation (1) and using the conventional exponents $(w \propto Q^b; r \propto Q^f; v \propto Q^m)$:

$$b + (1.5 + \beta)f = 1$$
$$m = (0.5 + \beta)f \tag{7}$$

This relationship is shown in Figure 27.5 (stippled area), and compared with measured data (from Leopold and Maddock, 1953, Appendix B). It is seen that measured values of f cover a similar range, but that b values vary less

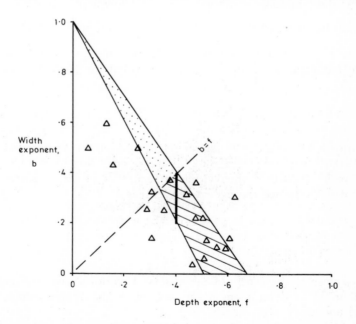

FIGURE 27.5 AT-A-STATION VARIATION IN HYDRAULIC GEOMETRY EXPONENTS

Shaded areas show increasing probable areas. Predicted points are from Leopold and Maddock (1953, Appendix B)

than the corresponding amount predicted by equation (7). Most of these differences, and most of the variation along the trend line, are thought to be causally related to the shape of the channel cross-section. If discharge, calculated for each vertical from equation (1), is summed across cross-sections of irregular shape, then marked divergences may be found. Points with $b = 0.4$–0.6 and $f = 0$–0.2 for example arise from channel-within-channel cross-sections such as those for the rivers to which those data points refer (e.g. Leopold and Maddock, 1953, Appendix B; Cheyenne River near Hot Springs, and Rio Grande at San Acacia). It is argued that the channel cross-section is primarily related to two largely external factors, the frequency distribution of flows and bank stability. Channel-within-channel cross-sections are an example of the first factor. The second factor produces, for example, characteristic rectangular cross-sections in some lithologies (i.e. $b \simeq 0$) and uniform side-slopes (i.e. $b = f$) in others. Most river cross-sections tend to plot in this lower, shaded portion of the area in Figure 27.5.

Perhaps the most typical cross-section is trapezoidal. The response of this cross-section may be analysed most clearly if other variables are expressed as approximate power functions in terms of depth, or hydraulic radius, r. Thus β in equation (6) averages about $\frac{1}{6}$ and tends towards zero for very large r. At the same time width varies as r^0 for low depths, and rises towards r^1 as depth becomes very large. Thus $\dfrac{w}{\sqrt{ff}}$ in equation (1) behaves approximately as $r^{1.5}$ over a large range of depths, the increase in the width exponent balancing the decrease in the exponent for $1/\sqrt{ff}$. From equation (1) then:

$$Q \propto \frac{w}{\sqrt{ff}} \cdot r^3/2 \propto r^{2.5} \qquad (8)$$

This in turn leads to a depth exponent averaging 0.40. From equation (6), $m = 0.20$–0.40, with an average of 0.27 corresponding to $\beta = 1/6$, and by subtraction $b = 0.20$–0.40 (heavy line in Figure 27.5) with a corresponding average of 0.33. Hence the expected averages are:

$$b = 0.33$$
$$f = 0.40 \qquad (9)$$
$$m = 0.27$$

which lies close to the centre of the observed scatter shown in Figure 27.5, and is thought to differ from it on account of the channel-shape factors discussed above.

It should be noted that the argument here is based not on arguments about sediment transport, efficient or otherwise, but solely on the assumption that channel cross-section is partially constrained by factors which are external to the channel hydraulics and sediment transport.

Downstream Exponents

In the downstream sense the locus of expected exponents may be calculated on the assumption of maximum sediment efficiency and constant dominant grain size, d. The expression for maximum efficiency (equation 5) may be approximated by the power function:

$$\lambda S \propto u^{-(1-\beta)} \tag{10}$$

where β has the same values as above. Suppose, without sacrificing generality, that:

$$r \propto Q^f, \tag{11}$$

where Q is understood to refer to some dominant discharge. Then equation (10) may be expanded to:

$$\left(\frac{r}{d_{84}}\right)^{-(1-\beta)} \propto \left(\frac{d_{84}}{d}\right)_{\cdot\ \underline{s}} \tag{12}$$

or

$$s \propto Q^{-(1-\beta)f} d_{84}^{-\beta} \cdot d \tag{13}$$

Remembering that d is assumed constant, and substituting for width in equation (1):

$$w \propto Q f f^{\frac{1}{2}} r^{-\frac{3}{2}} s^{-\frac{1}{2}}$$
$$\propto Q(r/d_{84})^{-\beta} r^{-\frac{3}{2}} s^{-\frac{1}{2}}$$
$$\propto Q^{[1-(1.5+\beta)f+\frac{1}{2}(1-\beta)f]} d_{84}^{1.5\beta}$$
$$\propto Q^{[1-f(1+1.5\beta)]} d_{84}^{1.5\beta} \tag{14}$$

Writing $w \propto Q^b$ as before, and writing $d_{84} \propto Q^{-\gamma}$ for some γ which is usually positive:

$$b = 1 - (1+1.5)f - 1.5\beta\gamma \tag{15}$$

This describes the locus of probable values for the downstream exponents as β and γ vary; and is shown in Figure 27.6. It shows the area covered at $\beta = 1/6$ for $0 < \gamma < \frac{1}{2}$, and compares it with the data for rivers used by Leopold and Maddock (1953, Figures 5–9). In large sand-bed rivers, the value of d_{84} should refer to dune rather than grain dimensions, which may increase downstream, leading to negative values for γ. The fair agreement between observed and expected spreads allows the low exponent for velocity to be deduced as a result of the tendency to high sediment efficiency.

In small rivers, with larger values of β up to 0.5 or more, the possible spread of values becomes wider, and much more sensitive to the downstream variation in d_{84}. In these cases another useful limit is given, where

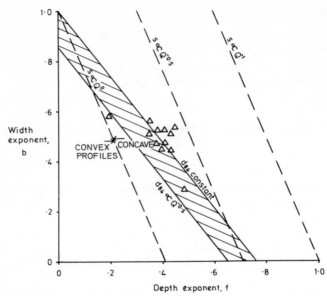

FIGURE 27.6 DOWNSTREAM VARIATION IN HYDRAULIC GEOMETRY EXPONENTS

Shaded area shows prediction. Points are from Leopold and Maddock (1953, Figures 5–9)

appropriate, by the condition that the long-profile is convex. This is given from equation (13) as:

$$(1-\beta)f - \beta\gamma > 0 \tag{16}$$

Substituting this condition in equation (15) gives concavity (provided discharge increases downstream) if:

$$b > 1 - 2.5f \tag{17}$$

This limiting line is also shown in Figure 27.6. Lines of greater concavity lie parallel to this line as also indicated in Figure 27.6. For the low exponents of slope with respect to discharge, which are normally found in large rivers, the main grouping of points seen in Figure 27.6 is readily explained, in the neighbourhood of:

$$S \propto Q^{-0.3}$$

The efficiency criterion may be rewritten to give an expression for sediment concentration, which is as effective a measure of sediment yield per unit area (even if the duration of dominant flow is unknown and variable) as bedload. It is:

$$\lambda C \propto \frac{1}{u}\left(\frac{1.77 \ln u + 2.00}{1.77 \ln u + 3.18}\right)^{\frac{1}{2}}$$

This expression is proportional to u, raised to a power of -0.93 to -0.997 for $1 < u < 10^4$, and differs little from -1.0. Substitution for u and gives approximately

$$C \propto Q^{-f}$$

Data for bedload yield are too few to be reliable, but it is likely that the negative exponent has a larger value than the exponent for total sediment yield, which is often taken (Brune, 1948) as:

$$\propto A^{-0.15} \propto Q^{-0.2}$$

In terms of causation, it is thought that in the graded time interval considered here, the values for bedload yield and slope are externally determined, and it is their downstream variation which determines the other hydraulic geometry exponents. Hence for a channel of constant slope and bedload concentration, it is predicted that $b = 1$, $f = 0$, implying a widening channel of constant depth, and constant roughness diameter, d_{84}. Similarly for a channel of constant slope and a normal variation in bedload concentration $(C \propto Q^{-0.4})$, it is predicted that $b = 0$, $f = 0.4$, implying a deepening channel of constant width, with a marked decline in roughness diameter downstream.

Conclusion

In the short term, a river channel cannot control its slope or hydraulic geometry to a marked degree, but is able to modify its sediment load at will. There is therefore no question of applying maximum efficiency criteria. In the longer term, however, the channel must adapt to the load provided by hillslopes and tributaries within its catchment. Analogy with other partially indeterminate systems argues that the channel will tend towards carrying its bedload as efficiently as possible, at least in its downstream sections where this function is the one it can do least easily. As an analytical device, the frequency distribution of flows has been replaced by a single dominant flow acting for part of the time. Although this device may introduce some errors due to the non-linearity of the processes, it is thought to provide a useful way of predicting overall channel behaviour at both meso- and macro-scales. This paper has shown the feasibility of this approach, particularly in its application to downstream variations in hydraulic geometry. The principle of maximum efficiency, or minimum work, is a well established principle in engineering practice, and it seems sensible for fluvial geomorphologists to test its validity more fully before abandoning it in favour of principles which apply only to the landscape.

References

Bagnold, R. A. 1960. Some aspects of the shape of river meanders. *U.S. Geol. Surv. Prof. Paper* 282–E.

Brune, G. 1948. Rates of sediment production in midwestern United States. *Soil Conservation Service Technical Publication* **65.**

Ferguson, R. I. 1971. Theoretical models of river channel patterns. *Unpublished Ph.D. thesis,* University of Cambridge.

Langbein, W. B. and Leopold, L. B. 1964. Quasi-equilibrium states in channel morphology. *Amer. J. Sci.,* **262,** pp. 782–794.

Langbein, W. B. and Leopold, L. B. 1966. River meanders—theory of minimum variance. *U.S. Geol. Surv. Prof. Paper, 422–G,* pp. H1–15.

Langbein, W. B. and Leopold, L. B. 1968. River channel bars and dunes—theory of kinematic waves. *U.S. Geol. Surv. Prof. Paper 422-L,*

Leopold, L. B. and Langbein, W. B. 1962. The concept of entropy in landscape evaluation. *U.S. Geol. Surv. Prof. Paper 500–A.*

Leopold, L. B. and Maddock, T. Jr. 1953. The hydraulic geometry of stream channels and some physiographic implications. *U.S. Geol. Surv. Prof. Paper 252.*

Leopold, L. B. and Wolman, M. G. 1957. River channel patterns: Braided, meandering and straight. *U.S. Geol. Surv. Prof. Paper 282–B.*

Meyer-Peter, E. and Müller, R. 1948. Formulas for bed-load transport. *Proc. 3rd Meeting of Int. Assoc. for Hydraulics Research,* Stockholm.

Scheidegger, A. E. and Langbein, W. B. 1966. Probability concepts in geomorphology. *U.S. Geol. Prof. Paper 500–C.*

Index

Subjects indexed below are restricted to major references. Material in illustrations and tables are indicated by page references in italics. Locations are not included unless they are elaborated in detail in the text and authors are not included.

Aeolian processes 80
Allogenic channel changes 178–179
Alluvial balance positive 78
  channels 429–441
  deposits 85
  fans 80, *81*, 82, 307, *311*, *312*, 318
  sediments 174
Anabranching 168, 207, 218
Anaglaciation 291
Anastomosing 168, 178, 215, 217, 415
Annual series 65, 67, 72
Antidunes 34, 35, *37*, 433
Armouring 34, 52, 146, 160, 380
Autogenic channel change 175–178

Bank erodibility 20
  erosion 194, 310
  highly cohesive 51
  manipulation 1
  sediment 197
  undermining 200
Bankfull discharge 152, 157, 286, 287, 290, 301
  flow 26, 376
  level 396, 398
  stage 69, 411
Beatton river, Canada 250–259

Bed armouring 52
  forms 31, 32, 34, *35*, 36, 328, 430–433
  elevation changes *55*, 58
  material calibre 36
  material sampling 231
  sediments 34
Bedload equations 31
Bollin-Dean, Cheshire 102
Boundary conditions 238
  layer streaks 37
  layer theory 297
  shear stress 24, 36
Braided river channels 78, *80*, *82*, 83, 168, 177, 178, 201–220, 361, 413, *415*, 421, 423, 424, 425
Braiding measure 172
Burrator reservoir, Somerset 128
Bywash channel 402

Carpathians 411–426
Cataglaciation 291
Catastrophic events 61
  floods 411
Channel bend evolution 26
  bywash *209*
  change classes *209*, 213, 305, 369

compound cross sections 157, 160, 389, 394, 396, 402, 438
conditions 58
degradation 122, 146
erosion and deposition rate 51
forming discharge 61, 70, 179
metamorphosis 389–408
migration 175
morphometry 392
multiple 361
sedimentation 363
shape 58, 123
sinuosity 27, 171, 240, 391
stabilization schemes 122, 416
states *116*
structure 325
widening 52,
width changes *382*
width-depth ratio 20, 325
Channel capacity 390, 396
and drainage area *126, 127, 393, 400*
and total channel length *393, 401*
Channel morphology 301–304
Channel volume 393, 400
Channel cross-section changes *51*, 58, *106, 107, 110*, 128–141, 147–160, 328–333, 380–385, 401–406
types of 92
depth increases 137
pattern changes 84, 175–179, 207–220, 265–280, 355–363
planform changes 167–184
width increases 137
Channelization 1
Climatic geomorphology 91
Competence 348
Complex response 5, 16
Computer Program ERODE 53
Confined channel patterns 221–233
Continuity equation 329, 330
for sediment transport 47, 49, 50
Cound catchment, Shropshire 194
Crawfords Creek, eastern New South Wales 52
Crevassing 178, 361
Critical boundary shear stress 20
Cross spectral analysis 268
Current ripples 34
Cut-offs 175, 177, 185, 186, *187, 190*, 228, 235, 271,
chute 177

Dam construction 1, 145, 395, 401, 425
degradation downstream of 47
effect on sediment load 5, 146
scour downstream of 58
Darcy–Weisbach friction coefficient 19, 24, 37, 433, 434
Debris cones 307, *311*
Deflation depressions *81*
Deforestation 85
of flood plain 241
Degradation below dams 47, 52
channel 122
Dendrochronology 29, 158, 174
Derwent River, Derbyshire 147, *148, 150*, 372–375
Discharge channel-forming 61, 123, 394
events of modest magnitude 61
ratios 65, 70
and mean suspended-sediment concentration *66*
Disequilibrium relationships 97
Distributary channel networks 337–352
Disturbed periodic model 244–247
Dominant discharge 95, 151, 249, 394
Drainage density 153, 298 389
net retraction 290
network 297
Dry valleys 290, 298
Du Boys equation 32
Dune behaviour *39*
movement downstream 38
wavelength 38
Dunes 34, 36, *37*, 433
aeolian *81*

East Devon 265–280
Electron microscope 82
Enlargement ratio index 133
Entrainment threshold of bed sediment 24, 32
Entropy 429
Environmental engineering 386
geomorphology 121
Ephemeral channels 96
streams 317–335
Equifinality 94
Equilibrium channel 21
Ergodic hypothesis 4, 392, 404

Erodibility bank   20
  floodplain   361
Erosion balance   417
  classes   202
  dynamics model   329
  nails   304
  patterns of   *57*
  pins   193, 231, 265
  prevention 279
  response   203
Euclidean space   112
Event frequency   301–314
Exhaustion effects   34

Feedback mechanisms   161
Flood   319, 341, 424
  control works   425
  deposits   4
  discharges   359, 373
  frequency   203, 372, *374*, 395, 399, 425
  increase   377
  magnitudes   343
  mean annual   374, 379, 381
  most probable annual   61
  recurrence interval   62
  routing   298
  thousand year   62
Floodplain   61, 420
  morphology   357
  scrolls   236
  sediments   361
Flow direction distortion   217
  duration curve   *378*
  frequency distribution   441
  phases   318
  profile   48, 58
  regulation   151
  types hierarchy of   96
  unifrom steady   17, *18*, 22, 348
  unsteady   318
  variability   58
Flume experiments   57, 93, 172, 227, 238
Fluvial accumulation processes   85
  deposits, dating of   158
  regime   17, 20
  transport   417
Fossil valleys   75
Friction   coefficient,   Darcy-Weisbach   19

Frost structures   76
Froude number   37, 177, 433

Grain size distribution   75
  velocity   *432*
Grains Gill   301–314
Gully   137, 303, 310, 389
  systems   318
Gullying   298, 408
Gumbel distribution   342
  graph   65

Helically   spiralling   secondary motion   22
Hernád river   185–190
Holocene accumulation   *81*, 413
Human activity   85, 121, 175, 413
Hurricane Agnes   62
  floods   62
  sediment discharge   64
Hydraulic autogeometry   94
  controls   36
  exponents   *106*, 437, 439, *440*
  geometry   21, 91–100, 101–119, 330, 381, 384, *385*, 386, 391, 433, 436–441
  regime altered   380
  variables   36, 97
Hysteresis effect   97, 327
  loops   104

Incised streams   394
Indeterminacy 92, 95, 104, 429
  of response   97
Inter-basin water transfer   370, *371*
International Hydrological Decade   121
Irrigation diversions   1

Jacobian   332

Kinematic   analysis   of   meander evolution   29
  waves   328, 430
Knickpoint   47, 54
  passage of   56

Lag between wavelength maximum and peak discharge   38
Levee   81,
  construction   1
Lichen limits   396
Lichenometry   158

Land use alterations 2
changes 3
Lead-lag 318
Long profile changes *54*

Manning equation 157, 160, 399, 404, 406
Mass Movement 71
Meander curvature, radius of 83
growth 31
initiation 176
irregularity 247
laws 249, 251
loop development *30*
migration 109, 235–248, 426
parameters 268
planform analysis 251, *253*
processes 168
scrolls 175
translation 271
wavelength 70, 95, 170
Meandering causes of 249
channels 79, 85, 168, 425, 426, 435
streams 61, 432
valleys 3, 171
Meanders *80*, 83
bedrock 283, *284*, 424
confined 222–233
entrenched 283
evolution of 23, 29, 186–192, 193–201, 235–247
generations of *80*, 83, 423
kinematic analysis 29
ingrown 283
stacked pattern 240
Meyer-Peter and Müller equation 32, 49, 58, 434
Minimum variance 95, 115, 117, 118, 429
Mining activity 353–367
wastes model for dispersal of *354*
Misfit channels 93
Mississippi river 38, 173, 239
Most probable annual flood 61

Namoi–Gwydir distributary system 337–352, 391
Neocatastrophic thought 62, 72
Neotectonic movements 85, 413
Network adjustments 297
metamorphosis 389–408
power 393
processes 297

New England 395–408
Non-uniform, unsteady behaviour 317

Ordnance Survey maps 267
Osage type underfitness 95
Overbank deposition, rate of 4
delivery of sediment 70
flow 67
Ox bows 83, 422, 424
lake *188*, 235

Palaeochannels 71, 93, 281, 286, 369
Palaeohydrology 3, 289, 297, 369
Palaeohydrological regime 75
Partial duration series 62, 67, 68, 72, 203, *212*, 377, *379*
Peak discharges 341
changes of 96
lower 2
Permafrost 76, 78, 79, 423
Perennial channels 96
Phase difference 38
Plane beds with parting lineations 34, 35, 36, *37*, 325
Pleistocene terrace *81*
valley floors 413
Pleniglacial 78, 79, *82*, 83, 422, 423
Pluviation 288
Point bars 24, 174, 176
Pool and riffle 21, 34, 176, 194, 281, 327
Precipitation excess, changes in 71

Quartz grains texture 82
Quasi-equilibrium 113, 114, 118

Rate of bank erosion 231, 235
bend development 27
change of meander loop central angle 29
erosion of deposition in channel 51
fluvial denudation 61, 70, 321, 420
meander arc erosion 193–205
overbank deposition 4
Rating curves 383, *384*
Rational formula 286
Recurrence intervals of floods 62, 301, 399, 414
annual series 65
Reduction ratio index 131
Regime fluvial 17
hydrologic 123
river 20
theory 3, 125

Relaxation paths 96, 179
Reservoir construction 1, 3, 123
  effect on discharge 372, *373*, *394*
  effect on sediment load 5, 96
  reduced channel size below 128,
    145–161, 395–404
  seepage 147
  storage 370, 376
Resistance channel 19
Reticulate 168
Riffle 21, 34, 327
Ripple height 36
Ripples 34, 36, *37*, 433
Rising stages 34
River channels *80*
  arid and semi arid 98
  braided 78, 80, 83; *see also* Braided
    river channels
  degraded 3
  development 84
  metamorphosis 147, 169, 179, 240,
    297
  migration plains 222
  pattern changes 76, 96, 161–184,
    265–280
  planforms 161
  regulation 5, *81*
  transitory pattern 83
  types of 96, 413–417
River construction works 1, 185, 416
  equilibrium form 236
  metamorphosis 3, 5, 7, 236
  protection measures 224
  regime 20
  regulation 2
  system 16
  training design 185
  underloading 422
Roadway runoff 140
Roughness features 32
Rounding of quartz grains 75
Routing 145, 298
Ruggedness number 392
Runoff mean annual, increase of 70
  trends *377*

Sand bed channels 71
  river 34, 37
Sandomierz Basin 75
Sandurs 218
Scour and fill 34, 323, 326–328
  downstream of dams 58

Secondary flow 172
  motion 22
Sediment balance of reach 49
  delivery per flood wave 65
  discharge Agnes floods *64*
  efficiency 429–441
  mobility 33
  organic 383
  percentage delivered by overbank
    flows 67
  production 304
  polluted 359
  restricted supply 52
  size 322
  yield classes 306
Sediment transport 34, 36, 430
  and bedforms 31
  discontinuities 47, 51, 52, 54, 57
  formulae 32, 33, 49
  local transport rate 33
Shear stress boundary 24
  critical boundary 20
Silt-clay, percent 152, 345, 381, 391
Similarity patterns 348–351
Simulation model 238
  of river channel erosion 47
Sine-generated curve 238, *239*
Sinuosity 27, 171, 240, 391
  change 175, 274
Slope wash 79
Spectral analysis 171, 240, 243–247,
    249, 252, 254, 259, 268
  cross 268
Solute load 61, 417
Stable channel 21, 24, 31
Stage changes 38
  discharge relation *385*
Stockponds 401
Stream ordering 392, 413
Stream power 32, *35*, 36, 97, 172,
    176, 323, 329
Stream shrinkage 287
Strickler equation 48
  bed roughness coefficient 49
  particle roundness coefficient 49
Survival length 331
Suspended load 32, 61
  sediment concentration, mean 65,
    66, 304
    at-a-station 69
  sediment hydrograph 418

Tame river 377–379

Terrace   422
  deposits   158
  Pleistocene   *81*
Thalweg   21
Thermal erosion   78
Thousand year flood   62
Thresholds   5, 96, 161, 179, 201, 203,
  212, 235, 236, 307, *309*, 313, 319,
  391
  of bank resistance   94
  of sediment entrainment   32
Tithe maps   267
Tracers   365
Transient behaviour   97
Transmission losses   96, 317, 324–326,
  327, 330, 333

Underfit   streams   62,   70,   172,   222,
  261, 281–293, 369, 391
  meanders   236
Underfitness   71, 171, 174
  Osage type   95, 179, 225, 226, 282
Uniform steady flow 17, *18*, 31
    in a uniformly curved channel   22
  models   40

Universal soil loss equation   329
Unsteady open channel flow   318
  equations   328
Urban catchments   376, 379, 386
  effects 395, 402, 404–406
Urbanization   92, 123, 369–387, 390,
  408
Urban drainage schemes   380

Valley meanders   95, 171
Valleys fossil   75
Variance analysis   259
Vistula valley   75

Water quality variations   383
Water redistribution   369–387
  supply areas   370–376
  surface slope   21
Weser river   38, *39*
White River, Indiana   241–244
Width–depth ratio   20, 325

Ystwyth river   355–363

BRITISH GEOMORPHOLOGICAL RESEARCH GROUP

The British Geomorphological Research Group was founded in 1961 to encourage research in geomorphology, to undertake large-scale projects of research or compilation in which the co-operation of many geomorphologists is involved, and to hold field meetings and symposia. Membership is open to all persons interested in geomorphological research and details can be obtained from the Honorary Treasurer. The Group is a Study Group of the Institute of British Geographers.

Publications of the group

TECHNICAL BULLETINS

Available from Geo Abstracts Ltd., University of East Anglia, Norwich, NR4 7TJ.

1. Field methods of water hardness determination. Ian Douglas, 1969
2. Techniques for the tracing of subterranean drainage David P. Drew and David I. Smith, 1969
3. The determination of the infiltration capacity of field soils using the cylinder infiltrometer. Rodney C. Hills, 1970
4. The use of the Woodhead sea bed drifter. Ada Phillips, 1970
5. A method for the direct measurement of erosion on rock surfaces. C. High and F. K. Hanna, 1970 (out of print)
6. Techniques of till fabric analysis J. T. Andrews, 1970
7. Field method for hillslope description. Luna B. Leopold and Thomas Dunne, 1971
8. The measurement of soil frost-heave in the field Peter A. James, 1971
9. A system for the field measurement of soil water movement. Brain J. Knapp, 1973
10. An instrument system for shore process studies. Robert M. Kirk, 1973
11. Slope profile survey. Anthony Young with Denys Brunsden and John B. Thornes, 1974
12. Electrochemical and fluorometer tracer techniques for streamflow measurements. Michael Church, 1974
13. The measurement of soil moisture. L. F. Curtis and S. Trudgill, 1975
14. Drainage basin morphometry. V. Gardiner, 1975
15. The use of electrode instrumentation for water analysis. A. M. C. Edwards, A. T. McDonald and J. R. Petch, 1975
16. Instruments for measuring soil creep. E. W. Anderson and B. L. Finlayson, 1975
17. Shorter technical methods (I). M. P. Mosley, E. K. Isaac, S. T. Trudgill and B. L. Finlayson, 1975

Current Register of Research, 1970–71
Current Register of Research, 1972–73
Current Register of Research, 1974–75 (£1.00)
A Bibliography of British Geomorphology. Ed. K. M. Clayton. Philip, 1964

SPECIAL PUBLICATIONS

Available from the Institute of British Geographers, 1 Kensington Gore, London, S. W. 7.

Slopes: Form and Process. Compiled by D. Brunsden, 1971 (£6.60)
Polar Geomorphology. Ed. R. J. Price and D. E. Sugden, 1972 (£6.60)

Fluvial Processes in Instrumented Watersheds. Eds. K. J. Gregory and D. E. Walling, 1974 (£6.60)

Progress in Geomorphology: Papers in honour of David L. Linton. Eds E. H. Brown and R. S. Waters, 1974 (£6.60)

OTHER PUBLICATIONS

Spatial Analysis in Geomorphology. Ed. R. J. Chorley. Methuen, 1972 (£6.00)

The Unquiet Landscape. Eds D. Brunsden and J. Doornkamp. David and Charles, 1974 (£6.50)

Nearshore Sediment Dynamics and Sedimentation. Eds J. R. Hails and A. P. Carr. Wiley, 1975 (£11.75)

Geomorphology and Climate. Ed. E. Derbyshire. Wiley, 1976 (£14.00)

Earth Surface Processes. A Journal of Geomorphology. The Journal of the British Geomorphological Research Group. Wiley, 1976